Organic Waste Recycling

Second Edition

Chongrak Polprasert
Environmental Engineering Program
Asian Institute of Technology
Bangkok
Thailand

JOHN WILEY & SONS
Chichester · New York · Brisbane · Toronto · Singapore

National 01243 779777
International (+44) 1243 779777

Other Wiley Editorial Offices

John Wiley & Sons, Inc., 605 Third Avenue,
New York, NY 10158-0012, USA

Jacaranda Wiley Ltd, 33 Park Road, Milton,
Queensland 4064, Australia

John Wiley & Sons (Canada) Ltd, 22 Worcester Road,
Rexdale, Ontario M9W 1L1, Canada

John Wiley & Sons (Asia) Pte Ltd, 2 Clementi Loop #02-01,
Jin Xing Distripark, Singapore 0512

Library of Congress Cataloging-in-Publication Data

Polprasert, Chongrak.
Organic waste recycling : technology and management / Chongrak Polprasert. — 2nd ed.
p. cm.
Includes bibliographical references and index.
ISBN 0 471 96434 4
1. Organic wastes — Recycling. I. Title.
TD804.P65 1996
628.4′458 — dc20 95-49527
CIP

British Library Cataloguing in Publication Data

A catalogue record for this book is available from the British Library

ISBN 0 471 96434 4 (Ppc) ISBN 0 471 96482 4 (Pbk)

Typeset in 10/12pt Times by Techset, Salisbury, Wiltshire
Printed and bound in Great Britain by Bookcraft (Bath) Ltd
This book is printed on acid-free paper responsibly manufactured from sustainable forestation, for which at least two trees are planted for each one used for paper production.

Organic Waste Recycling

Second Edition

Contents

Preface (to the second edition)

Organic wastes are those originated from humans, animals, agricultural, and agro-industrial activities, which, if not properly treated, can cause serious environmental damages to water bodies and soil. This book is intended to be an introductory text on organic waste recycling technologies, their problems and potentials, and the related management aspects. Chapter 1 deals with the general problems and needs for waste recycling, while Chapter 2 describes the characteristics of organic wastes including the pollution and diseases associated with these waste materials. Chapters 3–9 address the various waste recycling technologies which are low-cost and employ natural/biological processes in waste stabilization and resource recovery. Chapter 10 covers the planning, institutional development and regulatory aspects of waste recycling. Metric units are used throughout the book, and where another unit is mentioned a conversion into metric unit is made.

In the second edition of this book, up-to-date information on organic waste recycling is included. A section on waste minimization and clean technology is added in Chapter 2; the application of constructed wetlands in Chapter 7; and regulatory aspects of waste disposal/recycling in Chapter 10. Guidelines on the design and operation of each of the organic waste recycling technologies are tabulated, while case studies of successful waste recycling programs are given. To enhance understanding of the organic waste recycling subjects, exercises are given at the end of each chapter for students to solve both the theoretical and practical problems.

Students majoring in environmental science and engineering in their third and fourth years of undergraduate study, and/or in their graduate study, should find this book useful for their courses, term projects and research, although prior knowledge in environmental chemistry and biology may be necessary. In addition, this book could serve as a reference text for policy-makers, planners and professionals working in the environmentally related areas.

In the preparation of the second edition, I am indebted to my students and colleagues who have given critical, but constructive, comments on the book. The following persons kindly read and contributed to the revision: Professor P. Y. Yang of the University of Hawaii at Manoa, on Chapters 4 and 7; Professor C. K. Lin and Mr. Yang Yi of the Aquaculture Program, AIT, on Chapters 5 and 6; Dr. Preeda Parkpian of the Natural Resource Conservation Program, AIT, on Chapter 8; Dr. Lim Poh Eng of the Universiti Sains Malaysia, on Chapter 9; and Dr. B. N. Lohani of the Asian Development Bank, Manila, the Philippines, on Chapter 10. Mr. Rong Xiang, my research associate, did an excellent work in assisting me in the overall revision and setting up of the exercises. Ms. Prapassorn Pokasoowan of the Environmental Engineering Program, AIT, provided administrative support in the book preparation.

I would like to express my sincere appreciation to the AEON Group Environment Foundation, Japan, for their generous support of the AEON Group Chair endowed to me during the period of 1991–95. The Wiley staff in Strasbourg and Chichester gave kind assistance to me during the whole period in this book revision. The moral support I have received from Nantana, Jim and Jeed made this project possible.

Chongrak Polprasert
October 1995

Abbreviations and symbols

AFP	advanced facultative pond
AIT	Asian Institute of Technology, Bangkok, Thailand
AIWPS	advanced integrated wastewater pond system
AMP	advanced maturation pond
APU	aquatic processing unit
ASP	advanced settling pond
atm	atmosphere (pressure)
AU	animal unit
BARC	Beltsville aerated rapid composting
BOD_5	5-day biochemical oxygen demand
BOD_L	ultimate biochemical oxygen demand
BTU	British thermal unit
C	carbon
cal.	calorie
CEC	cation exchange capacity
CH_4	methane gas
Cl	chloride
cm	centimeter
C/N	carbon/nitrogen ratio
CO	carbon monoxide
CO_2	carbon dioxide
COD	chemical oxygen demand
DAF	dissolved air flotation
DDT	dichloro-diphenyl-trichloro-ethane, $(ClC_6H_4)_2CHCCl_3$
DO	dissolved oxygen
DO_d	dissolved oxygen at dawn
e or exp	exponential
EC	electrical conductivity
FAO	United Nations Food and Agriculture Organization
FCR	food conversion ratio
FFB	fresh fruit bunches (referring to palm oil wastewater)
ft	foot
FWS	free water surface (constructed wetlands)
g	gram
μg	microgram $= 10^{-6}$ gram
gal.	gallon
H	hydrogen
h	hour

H_2	hydrogen gas
H_2S	hydrogen sulfide gas
ha	hectare
HRAP	high-rate algal pond
HRT	hydraulic retention time
in.	inch
I_i or I_o	light intensity at the water surface
IRCWD	International Reference Centre for Waste Disposal, Switzerland
I_z	light intensity at the water depth z
J	joule
K	potassium
kcal	kilocalorie
kg	kilogram
kJ	kilojoule
km	kilometer
KVIC	Khadi and Village Industries Commission (India)
L	liter
LAS	linear alkylbenzene sulfonates (major compounds in detergent)
lb	pound
LDP	limiting design parameter
LPG	liquefied petroleum gas
m	meter
meq	milli-equivalent (referring to concentration of an ion)
mg	milligram
mgad	million gallons per acre per day
mgd	million gallons per day
min	minute
mL	milliliter
mm	millimeter
μm	micrometer
MPN	most probable number
mV	millivolt
N	nitrogen
NAS	U.S. National Academy of Sciences
NH_3	un-ionized ammonia
NH_3–N	total ammonia (or $NH_3 + NH_4^+$)
NH_4^+	ammonium ion
nm	nanometer $= 10^{-9}$ m
NO_2^-	nitrite
NO_3^-	nitrate
NO_x	nitrogen oxides in various oxidation states
O	oxygen
OF	overland flow process (referring to land treatment)
Org-N	organic nitrogen
p	ratio of oxygen produced and algae synthesized
P	phosphorus
PCBs	polychlorinated biphenyls

ppm	part per million
psi	pound per square inch (referring to pressure)
PVC	polyvinyl chloride
RCRA	Resource Conservation and Recovery Act of the U.S.A.
RI	rapid infiltration process (referring to land treatment)
RMP	red mud plastic
rpm	revolution per minute
S	sulfur
s	second
SAR	sodium adsorption ratio
SD	stocking density of fish
SF	subsurface flow (constructed wetlands)
SO_x	sulfur oxides in various oxidation states
SR	slow rate (irrigation) process (referring to land treatment)
SS	suspended solids
STP	standard temperature (0 °C) and pressure (1 atm)
TKN	total Kjeldahl nitrogen
TLW	total live weight of animal
TN	total nitrogen
TOC	total organic carbon
ton	metric tonne = 1,000 kg
TP	total phosphorus
TS	total solids
TSS	total suspended solids
TVS	total volatile solids
TWW	total wet weight
UASB	upflow anaerobic sludge blanket digester
UMER	unit mass emission rate (referring to tapioca starch wastewater)
U.N.	United Nations
UNEP	United Nations Environment Program
UNESCO	United Nations Education, Science and Culture Organization
U.S. EPA	United States Environmental Protection Agency
UV	ultraviolet
VFA	volatile fatty acids
VIP	ventilated improved pit
VSS	volatile suspended solids
W	watt
WHO	World Health Organization
WSP	waste stabilization pond
θ_c	mean cell residence time
°C	degree centigrade (temperature)
°F	degree Fahrenheit (temperature)
%	percent

1

Introduction

1.1 PROBLEMS AND NEED FOR WASTE RECYCLING

A significant challenge confronting engineers and scientists in developing countries is the search for appropriate solutions to the collection, treatment, and disposal or reuse of domestic waste. The technologies of waste collection and treatment that have been taught to civil engineering students and practiced by professional engineers for decades are, respectively, the waterborne sewerage and conventional waste treatment systems such as activated sludge and trickling filter processes. However, the above systems do not appear to be applicable or effective in solving the sanitation and water pollution problems in developing countries. Supporting evidence for the above statement is the result of a United Nations (UN) survey on water supply and sanitation coverage for the Decade 1981–90 and their projection for the year 2000 (Najlis and Edwards, 1991). As shown in Table 1.1, although the percentages of population served with adequate water supply and sanitation increased during the past decade, due to rapid population and urban growth, these percentages for the urban areas are expected to decrease in the year

Table 1.1 Water supply and sanitation coverage

	Percent covered in year		
Region	1980	1990	2000
Asia and Pacific			
Urban water	73	77	71
Rural water	28	67	99
Urban sanitation	65	65	58
Rural sanitation	42	54	65
Global totals			
Urban water	77	82	77
Rural water	30	63	89
Urban sanitation	69	72	67
Rural sanitation	37	49	58

Source: Adapted from Najlis and Edwards (1991)

2000. The same trends are observed both for the Asia and Pacific region and globally.

Sanitation conditions in both urban and rural areas need to be much improved as large percentages of the population still and will lack these facilities (Table 1.1). There are approximately 25,000 people per day (or 9.1 million per year) who die from preventable water-related diseases alone, or in conjunction with malnutrition (Dale, 1979).

Polprasert and Edwards (1981) cited several reasons for the failure to provide sewerage to the population of the cities of the developing countries. The construction of sewerage systems implies large civil engineering projects with high investment costs. These projects are ill-suited to incremental implementation in densely built-up cities and usually involve long planning methods, which can take up to a decade to implement. Meantime the problem has once more outstripped the solution.

Conventional waste treatment is rarely linked to waste reuse, such as irrigation, fertilization, or aquaculture. Thus it does not generate either income or employment, both high priorities in developing countries.

Most obviously, sewers are simply too expensive. The cost of sewers and sewage treatment is high by the standards of the richest countries in the world. Not only are many of the cities in the developing countries larger, but an unprecedented number of people must be provided with hygienic sanitation in an extremely short time. The developed countries had the luxury of almost a century to build their sanitation systems, the developing countries must do it in a decade, on a larger scale, often with water shortages, in extremely densely populated cities, and sometimes with a lower level of technological development than existed in Europe and North America at the turn of the twentieth century. And this must be done at a cost that is affordable today.

Bangkok city, the capital of Thailand, is a typical case study to show the difficulty of implementing a sewerage scheme in a newly industrialized country. In Bangkok, excreta disposal is generally by septic tank or cesspool; other wastewaters from kitchens, laundries, bathrooms, etc. (gray water) are discharged directly into nearby storm drains or canals. Because the Bangkok subsoil is impermeable clay, overflows from septic tanks and cesspools normally find their way into the canals and storm drains, resulting in serious water pollution and a health hazard to the people. A master plan of sewerage, drainage, and flood protection for Bangkok city was completed in 1968; the required facilities to serve approximately 1.5 million people would cost about US$110 million. By the year 2000 the entire program would serve about 6 million people at a cost in excess of US$500 million. The proportions of cost by facility were: sewerage 35 percent, drainage 27 percent, and flood protection 38 percent (Lawler and Cullivan, 1972). At present (1995), Bangkok city is in the process of constructing five central wastewater treatment facilities to serve a population of 2.5 million at the cost of US$800 million. The facilities, employing primary and secondary treatment processes, are expected to be in full operation by the year 2000.

Besides the sanitation problem, our energy needs have also grown exponentially, corresponding with human population growth and technological advancements (Figures 1.1 and 1.2). Although the energy needs have been met adequately by the discovery of fossil fuel deposits, these deposits are limited in quantity and exploration and production costs to make them commercially available are high. The worldwide energy crisis in the 1970s is an example reminding us of the need for resources conservation and the need to develop additional energy sources, e.g. through waste recycling.

Another concern for rapid population growth is the pressure exerted on our fixed arable land area on earth. Table 1.2 projects the ratio of arable land area over world population in the year 2063 to be less than half of that in the year 2000. There is an obvious need to either control population growth or produce more resources for human needs.

Organic wastes such as human excreta, wastewater, and animal wastes contain energy which may be recovered by physical, chemical, and biological techniques, and combinations of these. Incineration and pyrolysis of sewage sludge are examples of physical and chemical methods of energy recovery from municipal and agricultural solid wastes respectively; however, these methods involve very high investment and operation costs, which are not yet economically viable. The treatment and recycling of organic wastes can be most effectively accomplished by biological processes, employing the activities of microorganisms such as bacteria, algae, fungi, and other higher life forms. The by-products of these biological processes include compost fertilizer, biogas, and protein biomass. Because the growth of organisms (or the efficiency of organic waste treatment/recycling) is temperature-dependent, areas having hot climates should be most favorable for implementation of waste recycling schemes. However, waste recycling is applicable to temperate-zone areas also, where successful results from several projects, from which many design criteria were derived, are presented in this book.

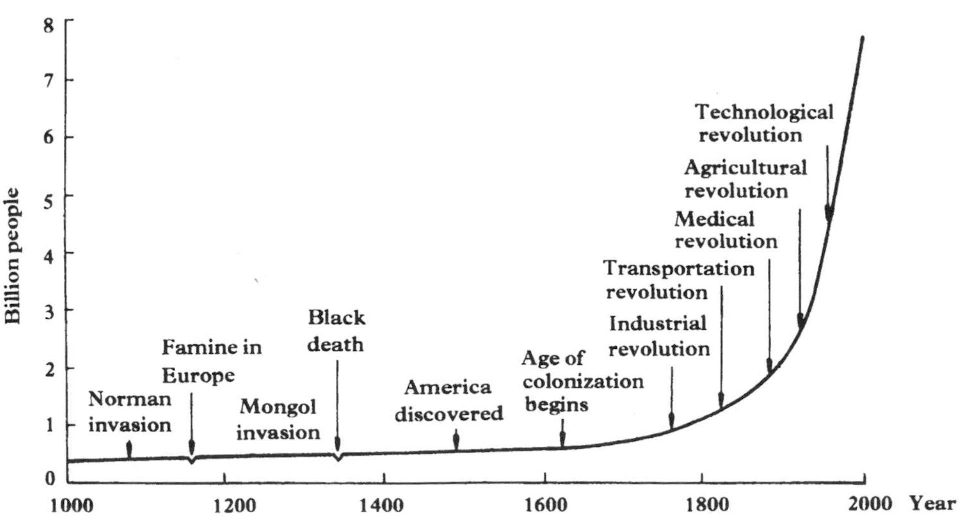

Figure 1.1 World population increases with human advances in science and technology (Taiganides, 1978)

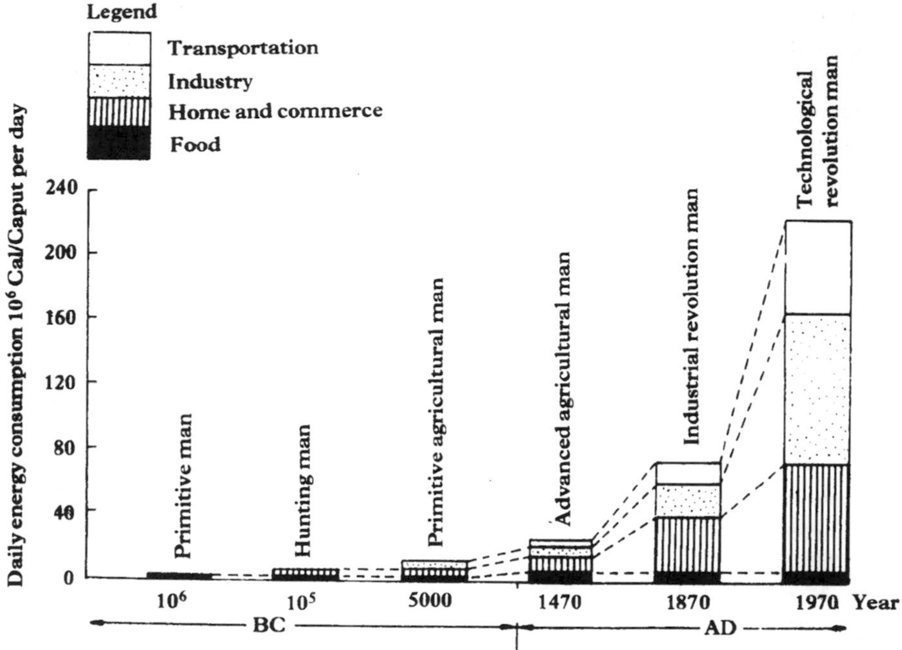

Figure 1.2 Energy consumption increases and changes in use pattern with human advances in science and technology (Taiganides, 1978)

Table 1.2 Population growth and arable land

Year	Estimated population (billion)	Arable land (ha/capita)
2000	6	1
2020	10	0.6
2063	22	0.3
2100	?	0.1–0.2

Note: Earth total surface is 51 billion ha. Only 6 billion ha is arable land suitable for crop production (adapted from Oswald, 1995)

It is therefore evident that technologies of waste management which are simple, practical, and economical for use should be developed, and they should both safeguard public health and reduce environmental pollution. With the current energy crisis, and as one of the greatest assets in tropical areas—where most developing countries are located—is the production of natural resources, the concept of waste recycling rather than simply waste treatment has received wide attention. A combination of waste treatment and recycling such as through biogas (methane) production, composting, or aquaculture, besides increasing energy or food production, will, if carried out properly, reduce pollution and disease transfer

(Rybczynski *et al.*, 1978). Waste recycling also brings about a financial return on the biogas, compost, and algae or fish which may be an incentive for the local people to be interested in the collection and handling of wastes in a sanitary manner.

The technologies to be discussed in this book apply mainly to human waste (i.e. excreta, sludge, nightsoil, or wastewater), animal wastes, and agro-industrial wastewaters whose characteristics are organic in nature.

1.2 OBJECTIVES AND SCOPE OF ORGANIC WASTE RECYCLING

The objectives of organic waste recycling are to treat the wastes and to reclaim valuable substances present in the wastes for possible reuse. These valuable substances include carbon (C), nitrogen (N), phosphorus (P), and other trace elements. The characteristics and significance of these substances are described in Chapter 2.

The possible methods of organic waste recycling are as follows.

1.2.1 Agricultural reuse

Organic wastes can be applied to crops as fertilizers or soil conditioners. However, direct application of raw wastes containing organic forms of nutrients may not yield good results since crops normally take up the inorganic forms of nutrients such as nitrate (NO_3^-) and phosphate (PO_4^{3-}). Bacterial activities can be utilized to break down the complex organic compounds into simple organic, and finally inorganic, compounds. The technologies of composting and aerobic or anaerobic digestion are examples in which organic wastes are stabilized and converted into products suitable for reuse in agriculture. The use of untreated wastes is undesirable from a public health point of view because of the occupational hazard to those working on the land being fertilized, and the risk that contaminated products of the reuse system may subsequently infect humans or other animals contacting or eating the products.

Wastewater that has been treated (e.g. by sedimentation and/or biological stabilization) can be applied to crops or grasslands through sprinkling or soil infiltration. The application of sludge to crops and forest lands has been practiced in many parts of the world.

1.2.2 Biogas production

Biogas, a by-product of anaerobic decomposition of organic matter, has been considered as an alternative source of energy. The process of anaerobic decomposition takes place in the absence of oxygen. The biogas consists mainly of methane (about 65 percent), carbon dioxide (about 30 percent), and trace amounts of ammonia, hydrogen sulfide, and other gases. The energy of biogas originates mainly from methane (CH_4) whose calorific value is 1,012 BTU/ft^3 (or 9,005 $kcal/m^3$) at 15.5 °C and 1 atmospheric pressure. The approximate calorific value of biogas is 500–700 BTU/ft^3 (4,450–6,230 $kcal/m^3$).

For small-scale biogas digesters (1–5 m^3 in size) situated at individual households or farm lands, the biogas produced is used basically for household cooking, heating, and lighting. In large wastewater treatment plants the biogas produced from anaerobic digestion of sludge is frequently used as fuel for boiler and internal combustion engines. Hot water from heating boilers may be used for digester heating and/or building heating. The combustion engines fueled by the biogas can be used for wastewater pumping, and have other miscellaneous uses in the treatment plants or in the vicinity.

The slurry or effluent from biogas digesters, though still polluted, is rich in nutrients and is a valuable fertilizer. The normal practice is to dry the slurry, and subsequently spread it on land; it can also be used as fertilizer for fish ponds, although little work in this area has been conducted to date. There are, however, potential health problems with biogas digesters in the handling and reuse of the slurry; it should be treated further, such as through prolonged drying or composting prior to being reused.

1.2.3 Aquacultural reuses

Three main types of aquacultural reuses of organic wastes in hot climates involve the production of micro-algae (single-cell protein), aquatic macrophytes, and fish. Micro-algal production normally utilizes wastewater in high-rate photosynthetic ponds. Although the algal cells produced during wastewater treatment contain about 50 percent protein, their small size, generally less than 10 μm, has caused some difficulties for the available harvesting techniques which as yet are not economically viable. Aquatic macrophytes such as duckweeds, water lettuce, or water hyacinth grow well in polluted waters, and, after harvesting, can be used as animals feed supplements or in producing compost fertilizer.

There are basically three techniques for reusing organic wastes in fish culture: by fertilization of fish ponds with human or animal manure; by rearing fish in effluent-fertilized fish ponds; or by rearing fish directly in waste stabilization ponds. Fish farming is considered to have a great potential in developing countries because fish can be easily harvested and have a high market value. However, to safeguard public health in those countries where fish are raised on wastes, it is essential to have good hygiene in all stages of fish handling and processing, and to ensure that fish are consumed only after being well cooked.

1.2.4 Indirect reuse of wastewater

The discharge of wastewater into rivers or streams can result in the self-purification process in which the microbial activities (mainly those of bacteria) decompose and stabilize the organic compounds present in the wastewater. Therefore, at a station downstream and far enough from the point of wastewater discharge, the river water can be reused in irrigation or as a source of water supply for communities located downstream. Figure 1.3 depicts typical patterns of dissolved oxygen (DO) sag along distance of flow of a stream receiving organic waste discharge. Type 1 pollution occurs when the organic waste load is mild and little DO is utilized by the bacteria in waste decomposition. At higher organic waste load (Type 2 pollution), more oxygen is utilized by the bacteria, causing a greater DO sag and consequently a longer recovery time or distance of flow before the DO reaches the normal level again. Type 3 pollution has an overloading of organic waste into the stream, resulting in anaerobic conditions (zero DO concentration). This is detrimental to aquatic organisms; the recovery time for DO will be much longer than that of Types 1 and 2 pollution. Although DO is an indicator of stream recovery from pollution discharge, other parameters such as the concentrations of pathogens and toxic compounds should be taken into account in the reuse of stream water.

Similarly, raw or partially treated wastewater can be injected into wells upstream and, through the processes of filtration, straining, and some microbial activities, good-quality water can be obtained from wells downstream. The subject of groundwater recharge is discussed in Chapter 8.

1.3 INTEGRATED TECHNOLOGIES

Depending on local conditions, the above-mentioned technologies can be implemented individually or in combination. For optimum use of resources the integration of various waste recycling technologies in which the wastes of one process serve as the raw material for another should be considered. In these integrated systems animal, human, and agricultural wastes are all used to produce food, fuel, and fertilizers. The conversion processes are combined and balanced to

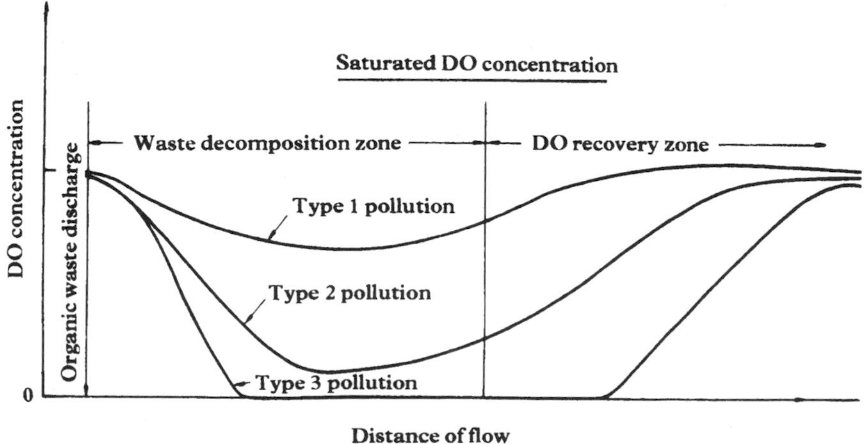

Figure 1.3 DO sag profiles in polluted stream

minimize external energy inputs and maximize self-sufficiency. The advantages of the integrated system include (NAS, 1981):

increased resource utilization,
maximized yields,
expanded harvest time based on diversified products,
marketable surplus, and
enhanced self-sufficiency.

Some of the possible integrated systems of organic waste recycling are shown in Figure 1.4. In scheme (a) organic waste such as excreta, animal manure, or sewage sludge is the raw material for the composting process; the composted product then serves as fertilizers for crops or as soil conditioner for infertile soil. Instead of composting, scheme (b) has the organic waste converted into biogas, and the digested slurry serves as fertilizer or feed for crops or fish ponds, respectively. Schemes (c)–(f) generally utilize organic waste in liquid form and the biomass yields such as weeds, algae, crops, and fish can be used as food or feed for other higher life forms, including human beings, while treated wastewater is discharged to irrigated land. The integrated systems that combine several aspects of waste recycling at small- and large-scale operations have been tested and/or commercially implemented in both developing and developed countries. Some examples of the above systems that have been in successful operation on a commercial scale are described below.

1.3.1 Kamol Kij Co. Rice Mill Complex and Kirikan Farm, Thailand (Ullah, 1979)

The Rice Mill Complex (area 18 ha) and Kirikan Farm (area 81 ha), owned by the same company, are located in Pathumthani Province, about 30 km north of Bangkok. Figure 1.5 shows the recycling of by-products or wastes generated from the rice milling units which produce about 500 tons/day of parboiled and polished rice from purchased paddy. Rice husks are burned to produce the energy needed for parboiling, paddy drying, and oil extraction. The husk ash is mixed with clay to make bricks, and the white ash from the kiln, containing about 95 percent silica, is sold for use in making insulators and abrasives.

Fine bran from the rice mill is passed to the oil extraction plant to produce crude vegetable oil about 20 percent in concentration; this crude oil is sold to a vegetable oil refinery factory to produce edible vegetable oil. The defatted rice bran is used as animal feed at Kirikan Farm.

Kirikan Farm, one of the biggest integrated farms in Thailand, maintains livestock, fish, and crops, as shown in Figure 1.6. There are approximately 7,000 ducks, 6,500 chickens, and 5,000 pigs being raised there, using feed from the rice mill defatted bran and other feed components. The farm could sell about 430,000 duck eggs and 1.2 million chicken eggs in a year.

The chicken pens are located above the pigsties so that the waste food and chicken droppings are consumed by the pigs. The manure from pigs, chickens and

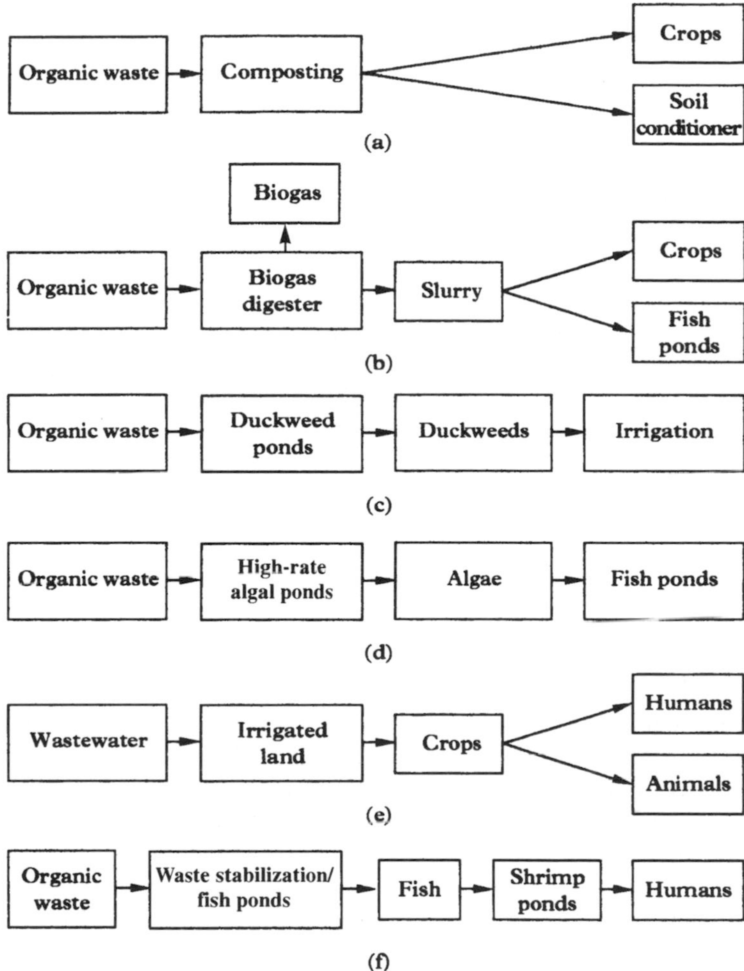

Figure 1.4 Some integrated systems of organic waste recycling

ducks is fed to fish ponds. Part of the pig manure is used as influent to the biogas digester to produce biogas for use in the farm. The fish ponds are cultured with *Pangasius,* tilapia, and *Clarias* and good fish yields are reported. The crops cultivated on the farm are sugar cane, potatoes, beans, bananas, mangoes and some vegetables. These crops are fertilized with the fish pond water and the biogas digester slurry.

1.3.2 Maya Farms, the Philippines

The 24-ha Maya Farms complex maintains 15,000 pigs and markets nearly twice that number annually (NAS, 1981). The biogas plants have been established primarily to control the pollution from its integrated piggery farms, meat

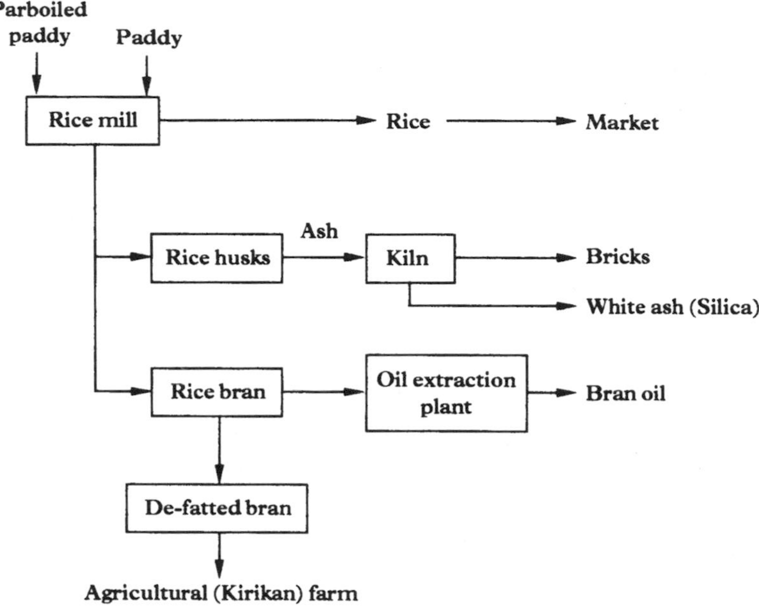

Figure 1.5 Kamol Kij Co. Rice Mill Complex

processing, and canning operations. There are various kinds of biogas digesters, but the large-scale, plug-flow digesters are part of the waste recycling system. About 7.5 tons of pig manure are fed daily into the three biogas digesters; the size of each digester is 500 m^3 and each produces approximately 400 m^3 of biogas daily. The biogas replaces liquefied petroleum gas (LPG) for cooking and other heating operations in the canning plant. It also replaces steam for heating the scalding tank in the slaughterhouse, and the cooking tank in the meat processing plant; and replaces the electric heaters in the drying rooms. The farms use biogas to run the gas refrigerators and electric generators, and converted gas engines replace the electric motors for deepwell pumping (Maramba, 1978).

Digested slurry from the above biogas digesters is conditioned in a series of waste stabilization ponds and used as a partial ration (10 percent) in the pig feed. Treatment of the sludge supernatant (liquid portion) is through a combination of planktonic algae (*Chlorella*) and fish (tilapia) ponds (Diaz and Golueke, 1979).

1.3.3 Werribee Farm, Australia (Barnes, 1978)

The Werribee Farm was established at the inception of the Melbourne Sewerage System, and has been in operation since 1897. It is situated in an agricultural area 33 km from Melbourne and has a frontage of 21 km to Port Phillip Bay. Of the Melbourne sewage flow received at the Farm, 70 percent is municipal waste with the remainder coming from trade and industrial wastes. Table 1.3 shows the

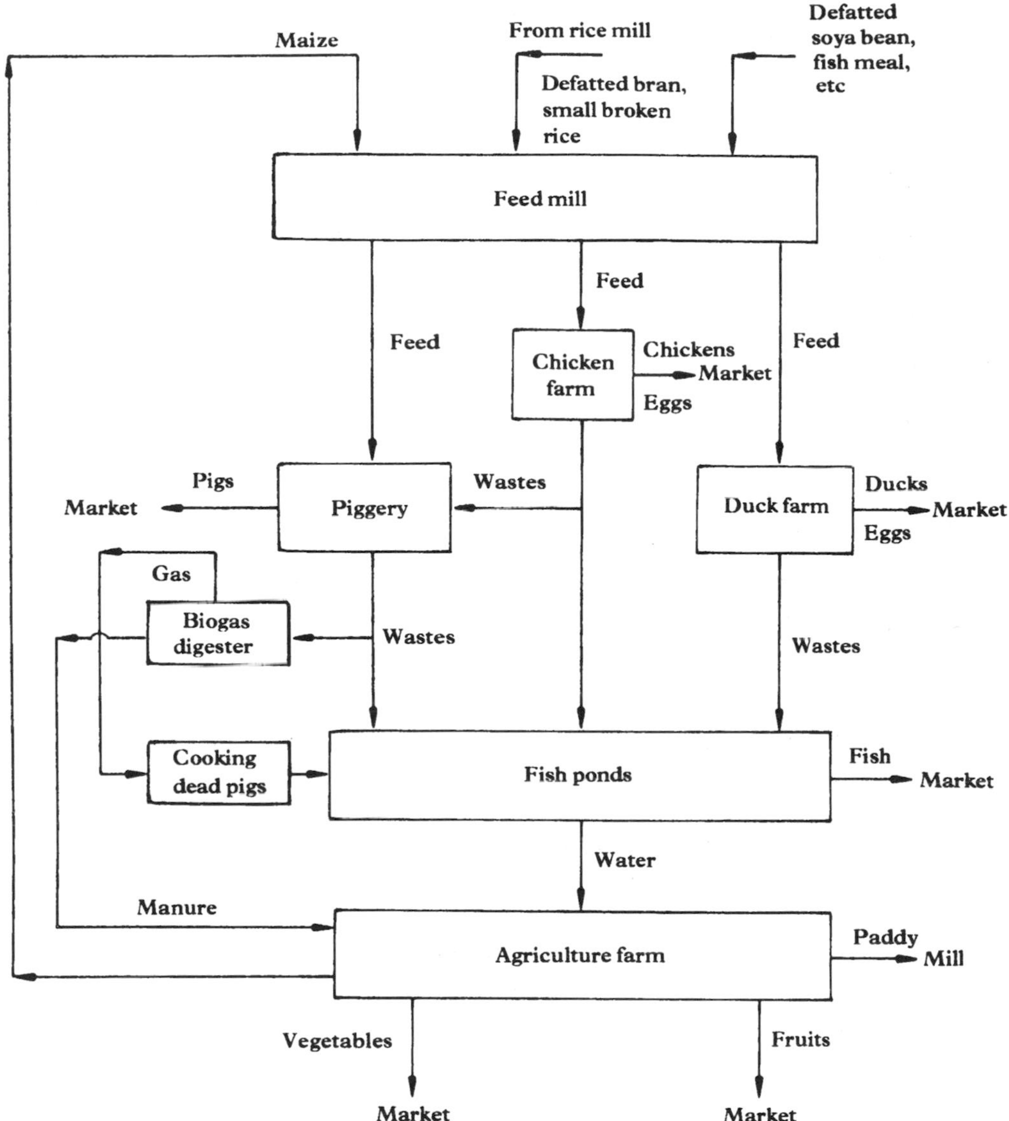

Figure 1.6 Flow diagram of Kirikan Farm

infrastructure and facilities for sewage treatment, and data on livestock being raised at the farm.

Figure 1.7 shows the sewage purification and waste recycling schemes in operation at Werribee Farm, consisting of three treatment processes, namely waste stabilization ponds (or lagooning), land filtration (or irrigation), and grass filtration (or overland flow). Sewage is primarily treated through a series of waste stabilization ponds. Sewage irrigation through land filtration and cattle grazing of grass are conducted during the summer period. The annual gross return from the sale of livestock is of the order of Australian $1 million. Grass filtration of sewage

Table 1.3 Facts about the Board of Works Farm—Werribee (Bremner and Chiffings, 1991)

Area	10,850 ha
Road constructed	231 km
Fencing erected	Approx. 2,024 km
Channels constructed	853 km
Drains constructed	666 km
Annual rainfall	530 mm
Annual evaporation	1,350 mm
Average daily flow	470,000 m^3/day
Sewage BOD	510 mg/L
Sewage SS	450 mg/L
Area used for purification of sewage	
Land filtration	3,830 ha
Grass filtration	1,520 ha
Lagoon treatment	1,500 ha
Livestock	
Cattle on farm	12,000
Cattle bred on farm during year	4,500
Sheep on farm	25,000–70,000

BOD = Biochemical oxygen demand
SS = Suspended solids

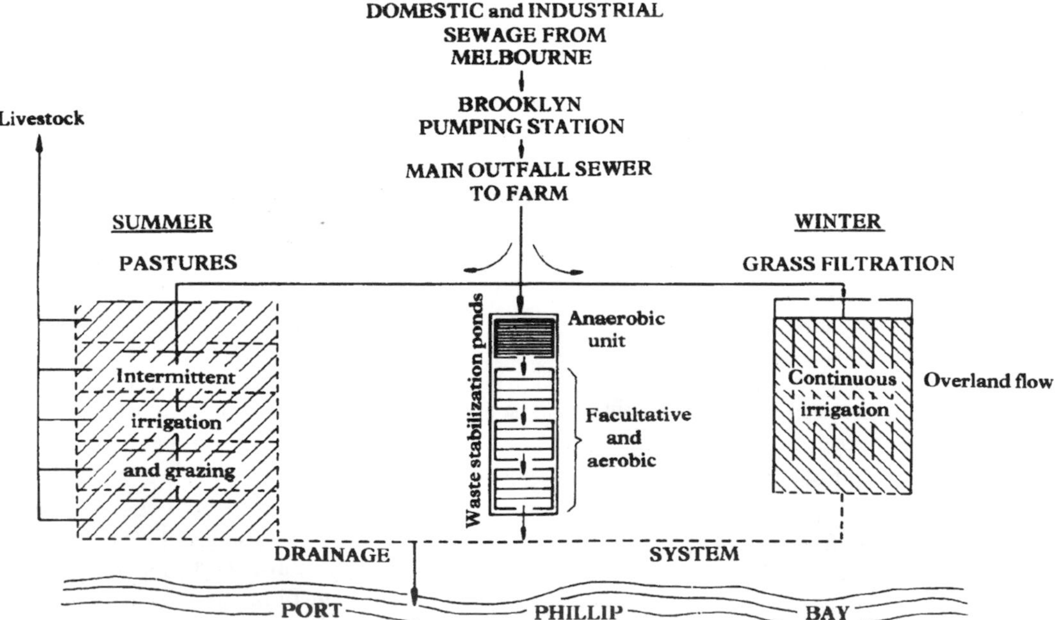

Figure 1.7 Sewage purification and waste recycling—Werribee Farm (Barnes, 1978)

is employed during the winter season. A description of the methods of land treatment of wastewater is given in Chapter 8. The final effluent from the farm is discharged into the sea at Port Phillip Bay without causing any serious environmental impact to the marine ecosystem. The Werribee Farm treatment complex supports an abundance of wildlife, including over 250 species of birds and several species of fauna and flora (Bremner and Chiffings, 1991).

1.4 FEASIBILITY AND SOCIAL ACCEPTANCE OF WASTE RECYCLING

The feasibility of a waste recycling scheme depends not only on technical aspects, but also on social, cultural, public health, and institutional considerations. Although waste recycling has been in practice successfully in many countries (both developed and developing), as cited in the examples in section 1.3, a large number of people still lack understanding and neglect the benefits to be gained from these waste recycling schemes. Waste recycling should not aim at producing only food or energy. In considering the cost-effectiveness of a waste recycling scheme, unquantifiable benefits to be gained from pollution control and public health improvement should be taken into account. Because human excreta and animal manure can contain several pathogenic microorganisms, the recycling of these organic wastes has to be undertaken with great care; the public health aspects of waste recycling technology are discussed in the following chapters. Institutional support and cooperation from various governmental agencies in the promotion, training, and maintenance/monitoring of waste recycling programs are also essential for success.

As the success of any program depends greatly on public acceptance, the communities and people concerned should be made aware of the waste recycling programs to be implemented, their processes, and advantages and drawbacks. A public opinion survey was conducted in 10 communities in Southern California, U.S.A. (Stone, 1976) to assess social acceptability of water reuse. For lower-contact uses (such as in irrigating parks/golf courses, factory cooling, toilet flushing, and scenic lakes), public attitudes are largely accepting; besides, treatment costs are generally low due to the requirement for a lesser degree of treatment, and adverse impacts on public health are minimized. In contrast, the reuse of wastewater for body contact uses (such as in boating/fishing, beaches, bathing, and laundry) produced more neutral or negative attitudes, while human consumptive reuses (such as in food canning, cooking, and drinking) were not acceptable to the people surveyed.

The assessment of public acceptance for wastewater reuse has not been undertaken or is rarely, if at all, conducted in developing countries. Because several countries such as China, India, and Indonesia have been recycling either human or animal wastes for centuries, and due to socio-economic constraints, the social acceptability for wastewater reuses should be more positive than that in developed countries.

REFERENCES

Barnes, F. B. (1978). Land treatment and irrigation of wastewaters. Paper presented at Preconference short course, *International Conference on Water Pollution Control in Developing Countries*, 17 pp. Asian Institute of Technology, Bangkok.

Bremner, A. J. and Chiffings, A. W. (1991). The Werribee treatment complex—an environmental perspective. *J. Aust. Water Wastewater Assoc.*, **19**, 22–4.

Dale, J. T. (1979). World Bank shifts focus on third world sanitation projects. *J. Water Pollut. Control Fed.*, **51**, 662–5.

Diaz, L. F. and Golueke, C. G. (1979). How Maya farms recycle wastes in the Philippines. *Compost Sci./Land Util.*, **20**, 32–3.

Lawler, J. C. and Cullivan, D. E. (1972). Sewerage, drainage, flood protection—Bangkok, Thailand. *J. Water Pollut. Control Fed.*, **44**, 840–6.

Maramba, F. D., Sr (1978). *Biogas and Waste Recycling—the Philippine Experience*. Regal Printing Co., Manila.

Najlis, P. and Edwards, A. (1991). The International Water Supply and Sanitation Decade in retrospect and implications for the future. *Natural Resources For.*, **15**, 110–7.

NAS (1981). *Food, Fuel and Fertilizer from Organic Wastes*. National Academy Press, Washington, D.C.

Oswald, W. J. (1995). Ponds in the twenty-first century. *Water Sci. Tech.*, **31**, 1–8.

Polprasert, C. and Edwards, P. (1981). Low-cost waste recycling in the tropics. *Biocycle, J. Waste Recycling*, **33**, 30–5.

Rybczynski, W., Polprasert, C., and McGarry, M. G. (1978). *Low-cost Technology Options for Sanitation*, IDRC-102e, International Development Research Centre, Ottawa, Canada (republished by the World Bank, Washington, DC, 1982).

Stone, R. (1976). Water reclamation—technology and public acceptance. *J. Env. Eng. Div.*—ASCE, **102**, 581–94.

Taiganides, E. P. (1978). Energy and useful by-product recovery from animal wastes. In *Water Pollution Control in Developing Countries* (eds. E. A. R. Ouana, B. N. Lohani, and N. C. Thanh), pp. 315–23. Asian Institute of Technology, Bangkok.

Ullah, M. W. (1979). A case study of an integrated rice mill farm complex. M. Eng. thesis no. AE 79–8, Asian Institute of Technology, Bangkok.

WHO (1976). Community water supply and excreta disposal in developing countries: review of progress. *World Health Stat. Rep.*, **29**, 544–631.

PROBLEMS

1.1 List the major types of organic wastes available in your own country and give examples of organic waste recycling practices being undertaken there.

1.2 Based on the data given in Table 1.1, give reasons why the percent coverage of urban and rural sanitation in the year 2000 is not much higher than in the year 1990. Suggest measures to increase these coverage percentages.

1.3 Find out the current practices of agricultural fertilization in your country with respect to the use of chemical fertilizer. How can you convince the farmers to use more organic waste fertilizer?

1.4 Conduct a survey, through interviews or questionnaire, about the public acceptance of the following methods of organic waste recycling: composting, biogas production, algal production, fish production, and crop irrigation. Based on the survey results, rank the public preference of these waste recycling methods and discuss the results.

1.5 You are to visit an agro-based industry or farm and investigate the present methods of waste treatment/recycling being undertaken there. Determine the potential or improvements that can be made at the industry or farm on the waste management aspects.

2

Characteristics of organic wastes

Almost all kinds of organic waste can be recycled into valuable products according to the technologies outlined in Chapter 1. In designing facilities for the handling, treatment, and disposal or reuse of these wastes, knowledge of their nature and characteristics is essential for proper sizing and selection of a suitable process. This chapter will describe characteristics of organic wastes generated from human, animal, and some agro-industrial activities. Pollution caused by these organic wastes, and possible diseases associated with the handling and recycling of both human and animal wastes are described. A section on waste minimization and clean technology is presented to emphasize the current trend of waste management. The analysis of physical, chemical, and biological characteristics of the organic wastes is according to the procedures outlined in *Standard Methods for the Examination of Water and Wastewater* (APHA, AWWA, WPCF, 1993) and *Official Methods of Analysis of the Association of Official Analytical Chemists* (AOAC, 1990); while the significance of these characteristics for waste treatment and recycling can be found in *Chemistry for Environmental Engineering* (Sawyer *et al.*, 1994) and *Wastewater Engineering: Treatment, Disposal and Reuse* (Metcalf and Eddy, 1993).

2.1 HUMAN WASTES

Excreta is a combination of feces and urine, normally of human origin. When diluted with flushing water or other gray water (such as from washing, bathing, and cleansing activities), it becomes domestic sewage or wastewater. Another type of human waste, called solid waste, refers to the solid or semi-solid forms of waste discarded as useless or unwanted. It includes food wastes, rubbish, ashes, and residues, etc.; in this case, the food wastes, which are mostly organic, are suitable for recycling.

The quantity and composition of human excreta, wastewater, and solid wastes vary widely from location to location depending upon, for example, food diet, socio-economic factors, weather, and water availability. Therefore generalized data from the literature may not be readily applicable to a specific case and, wherever possible, field investigation at the actual site is recommended prior to the start of facility design.

2.1.1 Human excreta

Literature surveys by Feachem *et al.* (1983) found the quantity of feces production in some European and North American cities to be between 100 and 200 g (wet weight) per capita daily, while that in developing countries is between 130 and 520 g (wet weight) per capita daily. Most adults produce between 1 and 1.3 kg urine, depending on how much they drink, and on the local climate. The water content of feces varies with the fecal quantity generated, being between 70 and 85 percent. The composition of human feces and urine is shown in Table 2.1. The solid matter of feces is mostly organic, but its carbon/nitrogen (C/N) ratio is only between 6 and 10, which is lower than the optimum C/N ratio of 20–30 required for effective biological treatment. If such processes as composting and/or anaerobic digestion are to be employed for excreta treatment, other organic matters high in C content need to be added to raise the C/N ratio. Garbage (food wastes), rice straw, water hyacinth, and leaves are some easily available C compounds used to mix with excreta. A person normally produces from 25 to 30 g of BOD_5 daily through excreta.

In areas where sewerage systems are not available, excreta is commonly treated by on-site methods such as septic tanks, cesspools, or pit latrines. Periodically (about once in every 1–5 years) septage, or the sludge produced in septic tanks and cesspools, needs to be removed so that it does not overflow from the tanks to clog the soakage pits (Figure 2.1) or the drainage trenches. (A soakage pit and/or drainage trench is a unit into which septic tank/cesspool overflow flows, and from where it seeps into the surrounding soil, where the soil microorganisms will biodegrade its organic content.) The most satisfactory method of septage removal

Table 2.1 Composition of human feces and urine[a]

	Feces	Urine
Quantity (wet) per person per day	100–400 g	1.0–1.31 kg
Quantity (dry solids) per person per day	30–60 g	50–70 g
Moisture content	70–85%	93–96%
Approximate composition (percent dry weight)		
Organic matter	88–97	65–85
Nitrogen (N)	5.0–7.0	15–19
Phosphorus (as P_2O_5)	3.0–5.4	2.5–5.0
Potassium (as K_2O)	1.0–2.5	3.0–4.5
Carbon (C)	44–55	11–17
Calcium (as CaO)	4.5	4.5–6.0
C/N ratio	~6–10	1
BOD_5 content per person per day	15–20 g	10 g

[a] Adapted from Gotaas (1956) and Feachem *et al.* (1983)

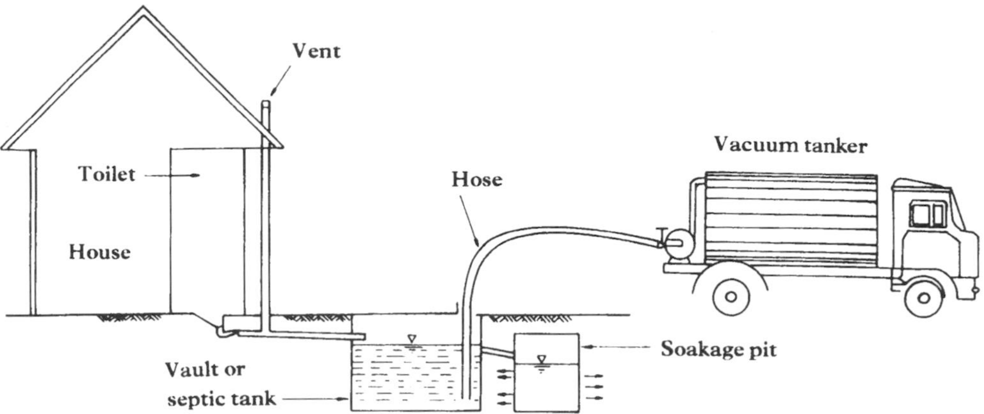

Figure 2.1 Vacuum tanker removing septage

is to use a vacuum tanker (size about 3–10 m^3) equipped with a pump and a flexible suction hose (Figure 2.1). If vacuum tankers are not available, the septage has to be manually collected by shovel and buckets; in this case the laborer who does the work can be subjected to disease contamination from the septage, and the practice is considered to be unaesthetic.

Septage is characterized by a high solid and organic content, large quantities of grit and grease, a great capacity to foam upon agitation, and poor settling and dewatering characteristics. A highly offensive odor is often associated with brown to black septage. The composition of septage is highly variable from one location to another. This variation is due to several factors, including: the number of people utilizing the septic tank and their cooking and water use habits; tank size and design; climatic conditions; septage pumping frequency; and the use of ancillary appliances such as kitchen waste grinders and washing machines.

Table 2.2 summarizes septage characteristics in the U.S.A. and Europe/Canada as reported in the literature. The last column of this table shows suggested design values of septage characteristics for use as guidelines in the design and operation of septage handling and treatment facilities. Characteristics of some septage in Asia are shown in Table 2.3.

Brandes (1978) reported that longer detention time of septage in the tanks contributes to better decomposition of organic materials and, consequently, to smaller amounts of septage pumped out per year. He found the septage accumulation rate for the residents of Ontario, Canada, which is applicable for septage disposal and treatment planning, to be approximately 200 L per capita yearly. Septage accumulation rates under Japanese conditions were estimated to be 1–1.1 L per capita daily (or 365–400 L per capita yearly) (Pradt, 1971). However, field investigations at the actual site are strongly recommended prior to the inception of detailed planning and design of septage treatment facilities. Because of its concentrated characteristics, septage needs to be properly collected and treated prior to disposal. On the other hand, its concentrated form would be advantageous for reclaiming the valuable nutrients contained in it.

Table 2.2 Physical and chemical characteristics of septage, as found in the literature, with suggested design values[a,b] (U.S. EPA, 1984)

Parameter	United States				Europe/Canada				EPA mean	Suggested design value
	Average	Minimum	Maximum	Variance	Average	Minimum	Maximum	Variance		
TS	34,106	1,132	130,475	115	33,800	200	123,860	619	38,800	40,000
TVS	23,100	353	71,402	202	31,600	160	65,570	422	25,260	25,000
TSS	12,862	310	93,378	301	45,000	5,000	70,920	14	13,000	15,000
VSS	9,027	95	51,500	542	29,900	4,000	52,370	13	8,720	10,000
BOD_5	6,480	440	78,600	179	8,343	700	25,000	36	5,000	7,000
COD	31,900	1,500	703,000	469	28,975	1,300	114,870	88	42,850	15,000
TKN	588	66	1,060	16	1,067	150	2,570	17	677	700
NH_3–N	97	3	116	39	—	—	—	—	157	150
Total P	210	20	760	38	155	20	636	32	253	250
Alkalinity	970	522	4,190	8	—	—	—	—	—	1,000
Grease	5,600	208	23,368	112	—	—	—	—	9,090	8,000
pH	—	1.5	12.6	8	—	5.2	9.0		6.9	6.0
LAS	—	110	200	2	—	—	—		157	150

[a] Values expressed as mg/L, except for pH

[b] The data presented in this table were compiled from many sources. The inconsistency of individual data sets results in some skewing of the data and discrepancies when individual parameters are compared. This is taken into account in offering suggested design values

Table 2.3 Characteristics of septage in Asia[a]

	Japan[b]	Bangkok, Thailand[c]
pH	7–9	7–8
TS	25,000–32,000	5,000–25,400
TVS	—	3,300–19,300
TSS	18,000–24,000	3,700–24,100
VSS	50–70% of TSS	3,000–18,000
BOD_5	4,000–12,000	800–4,000
COD	8,000–15,000	5,000–32,000
Total N	3,500–7,500	—
NH_3–N	—	250–340
Total P	800–1,200	—
Total coliform, no./100 mL	10^6–10^7	10^6–10^8
Fecal coliform, no./100 mL	—	10^5–10^7
Bacteriophages, no./100 mL	—	10^3–10^4
Grit (%)	0.2–0.5	—

[a] Values expressed as mg/L, except for pH and those specified
[b] Data from Magara *et al.* (1980)
[c] Data from Arifin (1982) and Liu (1986)

Human excreta deposited in pit latrines normally stays there under anaerobic conditions for 1–3 years prior to being dug out for possible reuse as a soil conditioner or fertilizer. The rather long period of anaerobic decomposition in pit latrines will cause the excreta to be well stabilized and most pathogens to be inactivated.

2.1.2 *Wastewater*

Urban cities in developed countries and many cities in developing countries have sewerage systems to carry wastewater from households and buildings to central treatment plants. This wastewater is a combination of excreta, flushing water, and other gray water or sullage, and is much diluted depending on the per capita water uses. According to White (1977), the volume of water used ranges from a daily mean consumption per person of a few liters to about 25 L for rural consumers without tap connections or standpipes. Consumption is 15–90 L for those with a single tap in the household, and 30–300 L for those with multiple taps in the house.

It should be noted that householders with per capita water consumption less than 100 L/day may produce wastewater containing very high solids content, which could possibly cause sewer blockage. The strength of a wastewater depends mainly on the degree of water dilution, which can be categorized as strong, medium, or weak, as shown in Table 2.4. These wastewater characteristics can vary widely with local conditions, hour of the day, day of the week, seasons, and types of sewers (either separate or combined sewers where storm water is included). Table 2.4 shows that domestic wastewaters generally contain sufficient amount of

Table 2.4 Typical characteristics of domestic wastewater (all values are expressed in mg/L)

	Concentration		
Parameters	Strong	Medium	Weak
BOD_5	400	220	110
COD	1,000	500	250
Org-N	35	15	8
NH_3–N	50	25	12
Total N	85	40	20
Total P	15	8	4
Total solids	1,200	720	350
Suspended solids	350	220	100

Data from Metcalf and Eddy, 1991; reproduced by permission of McGraw-Hill Book Co.

nutrients (based on $BOD_5/N/P$ ratio) suitable for biological waste treatment and recycling where microbial activities are employed.

In sewerless areas, where septic tanks or cesspools are employed for wastewater treatment, the hydraulic retention time (HRT) designed for these units is only about 1–3 days to remove the settleable solids and retain the scum. Because of short HRT, the effluent from a septic tank or septic tank overflow is an obnoxious liquor, containing high concentrations of organic matter, nutrients, and enteric microorganisms. The septic tank effluent is normally treated through a subsurface soil absorption system or soakage pits as shown in Figure 2.1 (Polprasert and Rajput, 1982). Where land is not available for the treatment of septic tank effluent, the effluent can be transported via small-bore sewers to a central wastewater treatment/recycling plant. (Small-bore sewers have diameters smaller than conventional sewers and carry only liquid effluents from septic tanks or aqua privies.) The characteristics of septic tank effluent are more or less similar to those of the wastewater (Table 2.4), but with a lower solid content.

2.1.3 *Solid wastes*

Solid wastes generated from human activities include those from residential, commercial, street sweepings, institutional, and industrial categories. Table 2.5 is a comparative analysis of solid waste characteristics of some developing and developed countries. The percentage of food wastes (organic matter) were 36–70 for the developing countries (Thailand, India, and Egypt) which were a few-fold higher than those of the developed countries (U.K. and U.S.A.). On the other hand, proportions of paper and cardboard were found to be higher in the solid wastes of the developed countries. The quantity of solid waste generation is generally correlated with the per capita income and affluence—i.e. the higher the income the greater the amount of solid waste generated. The global range of solid waste generation is 0.2–3 kg per capita daily (Pickford, 1977). Solid waste generation rates of less than 0.4 kg per capita daily can be applied to some

Table 2.5 Comparative solid wastes analysis

	Bangkok, Thailand[a]	Calcutta, India[b]	Cairo, Egypt[b]	U.K.[c]	U.S.A.[d]
Food wastes (organic)	39.2	36	70	17.6	9.0
Paper and cardboard	13.6	3	10	36.9	40.0
Metals	1.9	1	4	8.9	9.5
Glass	1.1	1	2	9.1	8.0
Textiles	4.8	4	2	2.4	2.0
Plastics and rubber	14.5	1	1	1.1	7.5
Miscellaneous, incombustible	3	50	10	21.9	4.0
Miscellaneous, combustible	21.9	4	1	3.1	20.0
Bulk density, kg/L	0.28	—	—	0.16	0.18–0.41

[a] Units in percentage weight (Lohani and Boonthanon, 1985)
[b] Units in percentage dry weight (Cook and Kalbermatten, 1982)
[c] Units in percentage wet weight (Department of Environment, 1971)
[d] Units in percentage weight (Tchobanoglous *et al.*, 1993)

cities in India, while generation rates greater than 2 kg per capita daily are applicable to several cities in the U.S.A. (Pickford, 1977; Cook and Kalbermatten, 1982).

As the food waste portion of the solid wastes is suitable for recycling (e.g. through composting), it should ideally be stored and collected separately for this purpose. However, for convenient and practical reasons the food waste is normally collected together with other kinds of waste to be processed and treated at a solid waste treatment plant. If composting is to be employed as a means to stabilize and produce fertilizer from the solid wastes, several methods of solid waste processing have to be utilized to separate out the food waste. These processing methods include mechanical size reduction and component separation, the details of which can be found in the books *Integrated Solid Wastes Management* (Tchobanoglous *et al.*, 1993) and *Handbook of Solid Waste Management* (Wilson, 1977). The subject of composting is described in Chapter 3 of this book.

2.2 ANIMAL WASTES

The amount and composition of animal wastes (feces and urine) excreted per unit of time also vary widely. They depend on various factors such as the total live weight of the animal (TLW), animal species, animal size and age, feed and water intake, climate, and management practices, etc. For design of facilities for animal waste collection and treatment, measurements and samples should be taken at the farm site or (if the farm is not built) at similar sites. For planning purpose, Taiganides (1978) suggests that the general guideline values given in Table 2.6 may be used. Young animals excrete more waste per unit of TLW than mature animals. The quantities of wastewater or wastes to be handled would, in general, be larger than those given in Table 2.6 due to the addition of dilution water, washwater, moisture-absorbing materials, and litter, etc.

Table 2.6 Bioengineering parameters of animal wastes (Taiganides, 1978)

Parameter	Symbol	Units	Pork pigs	Laying hens	Feedlot beef	Feedlot sheep	Dairy cattle
Wet waste	TWW	% TLW/day	5.1	6.6	4.6	3.6	9.4
Total solids	TS	% TWW	13.5	25.3	17.2	29.7	9.3
		% TLW/day	0.69	1.68	0.70	1.07	0.89
Volatile solids	TVS	% TS	82.4	72.8	82.8	84.7	80.3
		% TLW/day	0.57	1.22	0.65	0.91	0.72
Biochemical oxygen demand	BOD_5	% TS	31.8	21.4	16.2	8.8	20.4
		% TVS	38.6	29.4	19.6	10.4	25.4
		% TLW/day	0.22	0.36	0.13	0.09	0.18
COD/BOD_5 ratio	COD/BOD_5	Ratio	3.3	4.3	5.7	12.8	7.2
Total nitrogen	N	% TS	5.6	5.9	7.8	4.0	4.0
		% TLW/day	0.039	0.099	0.055	0.043	0.036
Phosphate	P_2O_5	% TS	2.5	4.6	1.2	1.4	1.1
		% TLW/day	0.017	0.077	0.008	0.015	0.010
Potash	K_2O	% TS	1.4	2.1	1.8	2.9	1.7
		% TLW/day	0.010	0.035	0.013	0.031	0.015

Table 2.7 shows the approximate weights of animals, the quantity of waste produced and their BOD_5 values. The annual production of nutrients from animal wastes is given in Table 2.8. During storage of animal wastes a considerable portion of N which exists in the form of ammonia (NH_3) can be lost through NH_3 volatilization.

For systems handling animal wastes as solids (> 30 percent TS), N losses will range from 20 percent in deep lagoons to 55 percent for open feedlots. For animal waste liquid handling systems (< 12 percent TS), N losses can range from about 25 percent in anaerobic lagoons to 80 percent for aerated systems (Taiganides, 1978). P and K are physically and chemically less mobile than N. However, when applied to land the actual amounts of nutrients available to crops can be much less than those shown in Table 2.8 for reasons such as nutrient losses through soil leaching and the inability of crops to utilize the nutrients effectively, etc.

2.3 AGRO–INDUSTRIAL WASTEWATERS

In this section the general processes, and the quantity and characteristics of waste generation of six types of agro-industries, namely tapioca, palm oil, sugar cane, brewery, slaughterhouse, and fruit and vegetables are described. The processing diagrams for each agro-industry include the sources of waste generation, which should be useful in helping environmental engineers and scientists to understand the waste characteristics and in investigating means of reducing waste generation in these industries.

2.3.1 Tapioca industry

Tapioca, also known as cassava or manioc, is grown in most tropical areas of the world. The root of the plant contains approximately 20 percent starch in a cellulose matrix. Tapioca starch is in particular demand for sizing paper or fibers, and is also used in the food industry.

Tapioca products include pellets, chips, and flour. The production of pellets has been increasing steadily, due to increased demand for pellets as animal feeds from European countries (Unkulvasapaul, 1975).

Production of pellets and chips is not a water-using process and hence causes no water pollution. On the other hand, the production of tapioca flour requires large quantities of water and the resulting wastewaters are highly polluting. About 5–10 m^3 of water is used to process 1 ton of input root, or about 30–50 m^3 of water is used per ton of starch produced. The wastewaters produced are organic in nature and highly variable in quantity and quality. They are characterized by high BOD_5 and SS values, with low pH and few nutrients (Jesuitas, 1966).

Table 2.7 Waste production by various animals (adapted from Lohani and Rajagopal, 1981)

Animal	Average weight of animal, in pounds (kg)	Total waste[a] in pounds/head per day[b] (kg/head per day)	BOD_5 in pounds/ head per day[b] (kg/head per day)
Beef cattle	800 (363)	40–60 (18–27)	1.0–1.5 (0.45–0.68)
Dairy cattle (milk cows, replacement heifers, breeding stock)	1,300 (590)	96 (44)	2.0 (0.9)
Swine	100 (45)	—	—
Chickens			
Broilers[c]	—	0.05 lb/lb per day	0.0044 lb/lb per day
Laying hens	—	0.059 lb/lb per day	0.0044 lb/lb per day
Sheep and lambs	—	15.5 (7.0)	0.35 (0.16)
Turkeys	15 (6.8)	0.90 (0.41)	0.05 (0.023)
Ducks	3.5 (1.59)	—	0.011–0.065
Horses[c] (pleasure, farm racing)	—	82 (37)	0.8 (0.36)

[a] Total excreta including feces and urine
[b] All units for waste and BOD_5 production are in pounds (kilograms)/head per day except for those for chickens (broilers and laying hens) in pounds/pound of bird per day as indicated
[c] Manure for broilers and horses is mixed with bedding material or litter

Table 2.8 Annual production (kg nutrient per year) of fertilizer nutrients from average weight animals and per AU[a] (Taiganides, 1978)

		Dairy cow		Feedlot beef		Pork pig		Laying hen		Feedlot sheep	
Nutrient	Symbol	400 kg cow	Per AU	300 kg steer	Per AU	55 kg pig	Per AU	2 kg hen	Per AU	50 kg sheep	Per AU
Total nitrogen	TN	53	66	60	100	8	71	0.7	181	8	78
Phosphates	P	14	18	9	15	3	31	0.6	141	3	27
Potash	K	22	27	14	24	2	18	0.3	64	6	57

[a] Animal unit (AU): animals whose total live weight is 500 kg

TAPIOCA PROCESSING

The main products of tapioca roots are pellets, chips, and flour. Most of the tapioca products (about 90 percent by weight) are exported in the form of pellets.

The manufacture of tapioca products is seasonal: plants begin processing in June and production increases steadily to a peak period between September and January, then falls off gradually and comes to a halt in April. A few larger plants operate throughout the year.

Chip and pellet production

Chips are manufactured by chopping the tapioca roots and then spreading them out on large concrete pads for drying. They are either transported directly to market or are pelletized before shipping overseas.

Pellets are produced by pressing chips into a cylindrical shape under high pressure and raised temperature; a small quantity of waste pulp from the starch plant is also added for adhesive purposes. Sometimes an even smaller quantity of rice bran is added to improve the nutritional value of the pellets.

Flour production

Tapioca starch is produced in two grades by two types of processes. The final quality of starch is, however, similar.

First-grade tapioca processes A typical flow diagram of the first-grade starch plant is illustrated in Figure 2.2. Roots transported to the plant should be processed within 24 h to avoid degradation of the starch. First, the sand on the roots is removed by dry rasping in a revolving drum and the peel is then removed by mechanical tumbling in a wash basin, from which the root washwater is derived. The roots are then mechanically crushed, releasing the starch granules from their surrounding cellulose matrix. Most of the cellulose material is removed by centrifugal means in a jet extractor and then by continuous centrifugation. The cellulose material or pulp is sold as poultry feed, provided it is fresh; or dewatered, dried and sold as animal feed. After primary centrifugation the starch milk is passed through a series of three sieves decreasing in pore size to assist in separating the starch from the small amount of pulp remaining. The recovered pulp is recycled to the jet extractor and the processed starch milk is led to a second centrifuge, from which wastewater is derived and by which a more concentrated starch is produced. After dewatering to a paste-like substance in a basket centrifuge, the product is spray-dried and packed.

As shown in Figure 2.2, there are three main sources of wastewaters, namely that from the root washer, centrifuge 1, and centrifuge 2. These wastewaters are normally combined and discharged. The wastewaters from centrifuge 1 and centrifuge 2 are sometimes called separator wastewater.

Second-grade tapioca process Second-grade tapioca plants are labor-intensive, employing simple processes with little mechanization, and are mostly small private-enterprise operations. A typical process flow diagram is shown in Figure 2.3. The roots are washed in a wooden tank with revolving paddles; sand and clay particles, as well as some peel, are removed at this step. The washed roots are conveyed to the rasper, followed by filtration through nylon mesh supported by a large cylindrical drum. The starch is sprayed through and the pulp is slowly drawn off and collected for dewatering. The starch milk is then released into large concrete settling basins. After 24 h settlement the supernatant is removed by decanting. The surface of the starch cake on the bottom is washed, the starch is then resuspended and pumped to a second sedimentation basin. After 24 h the

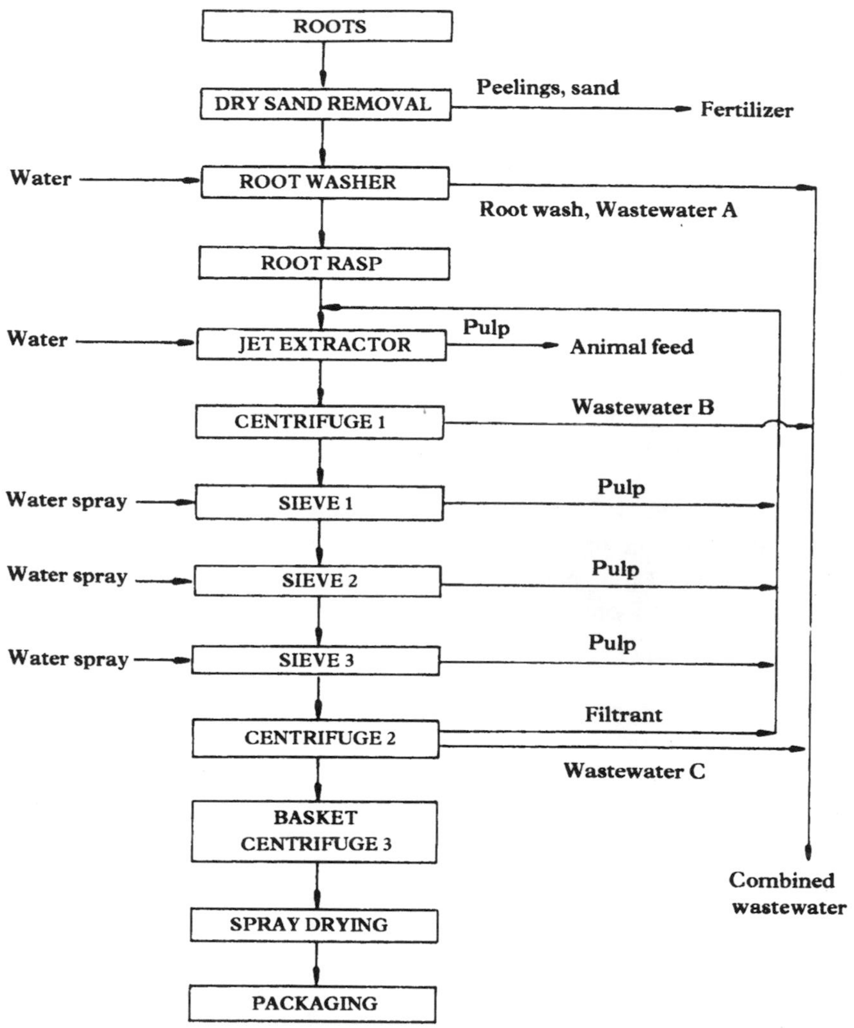

Figure 2.2 First-grade tapioca processing (Thailand) (Thanh, 1984)

supernatant is decanted and the surface is then washed again. The starch is then removed in large cakey chunks and spread on a heated concrete pad to dry. After drying the starch is packed. The supernatant and surface washwaters from the first and second settling basins are discharged or directed to a third settling basin. Where a third settling basin is available, the supernatant and the surface washwaters are allowed to settle for 24 h before the supernatant is decanted and discharged. The bottom sediment is dredged about once every 2 months; the sediment is resuspended twice again, as mentioned above, and the starch thus recovered is sold as a lower-grade starch.

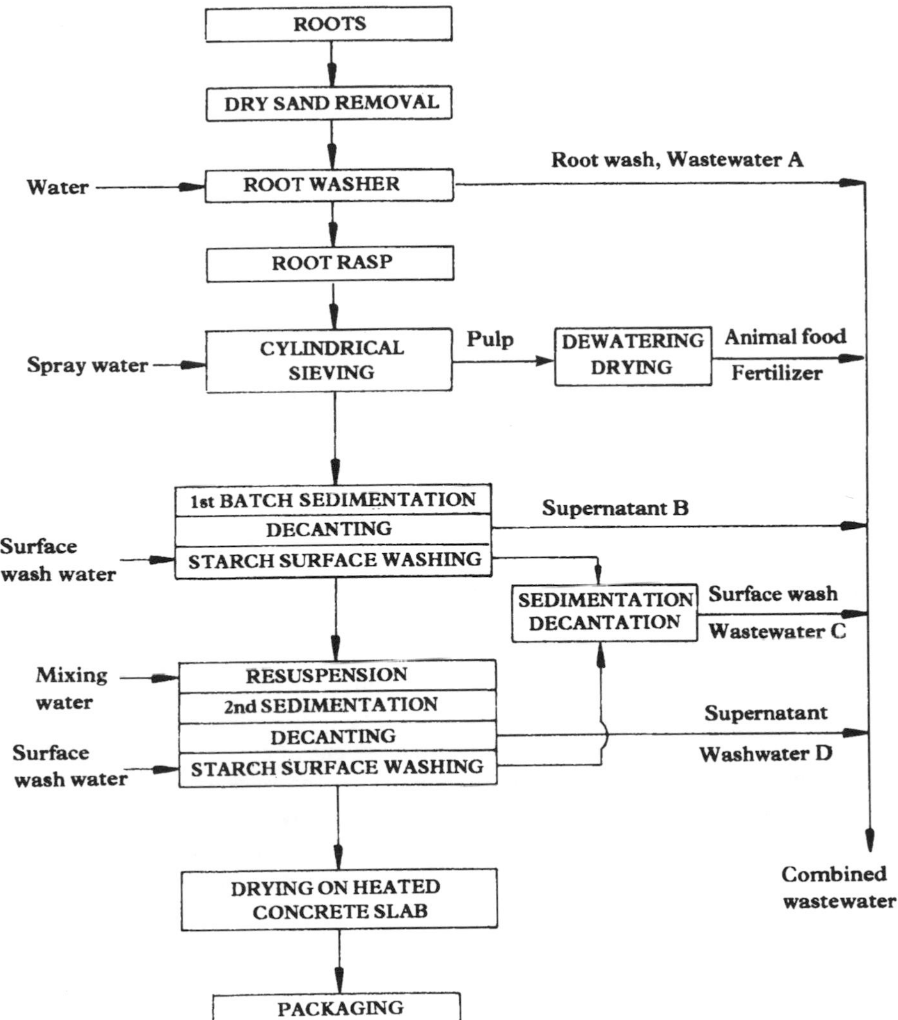

Figure 2.3 Second-grade tapioca processing (Thailand) (Thanh, 1984)

TAPIOCA STARCH WASTEWATER CHARACTERISTICS

The combined wastewater from tapioca starch production is composed mainly of root washwater and either the starch supernatant decanted from sedimentation basins or the separator wastewater, depending upon whether a second-grade or first-grade starch factory is being considered. First-grade and second-grade factories in Thailand commonly process in the order of 200 and 30 tons of tapioca root per day, respectively, and release wastewaters with unit mass emission rates (UMER values) as shown in Table 2.9. Designations A, B, C, and D refer to wastewater sources shown in Figures 2.2 and 2.3.

Table 2.9 Unit mass emission rate (UMER) values of tapioca starch wastewater in g/kg root (Thanh, 1984)

	Wastewater				
Characteristics	A	B	C	D	Combined
First-grade factory (Figure 2.2)					
COD	0.69	55.2	6.0	—	61.9
BOD_5	0.33	26.5	2.9	—	29.8
Suspended solids	0.54	10.5	3.7	—	15.7
Dissolved solids	4.91	46.3	9.6	—	60.8
Second-grade factory (Figure 2.3)					
BOD_5	1.8	44.8	2.1	2.1	48.7
Total solids	3.1	102.8	2.2	2.2	109.2
Suspended solids	3.1	83.5	0.6	0.6	87.2

The characteristics of tapioca starch wastewaters are summarized in Table 2.10. Root washwater contains high settleable solids, mainly sand and clay particles from the raw roots. The combined waste is acidic in nature, its pH ranging from 3.8 to 5.2, resulting from the addition of sulfuric acid in the extraction process and also from the release of some prussic acid by the tapioca root. Tapioca starch wastewaters are highly organic but have relatively low N and P concentrations. The ratio of soluble BOD_5 to soluble COD in the settled separator waste is 0.6–0.8, indicating that the waste is biologically degradable. It is likely that biological treatment methods will be most economical for this organic waste. The high BOD_5 and COD concentrations suggest that anaerobic biological processes, as the first-stage treatment, will be effective for organic reduction.

2.3.2 *Palm oil industry*

Palm oil is basically a vegetable oil used mainly for human consumption. It is a semi-solid, edible oil, extracted from the pulpy portion of the fruit wall of the palm fruit. The palm plant seems to have originated in the Guinea coast of West Africa and is now Malaysia's second most important crop. Almost all the palm oil produced in Malaysia and Thailand is used for margarine and other edible purposes.

EXTRACTION PROCESS

A general process flow diagram of a palm oil mill is shown in Figure 2.4. The freshly cut fruit bunches, delivered to the mill, are first loaded in cages mounted on rails and run directly into a horizontal sterilizer where steam is used to heat the fruits to about 140 °C at a pressure of 2.5–3.5 kg/cm^2 (35–45 psi) for 50–75 min.

Table 2.10 Characteristics of tapioca starch wastewaters (after Thongkasame, 1968; Unkulvasapaul, 1975)

Characteristics, mg/L (except as indicated)	Second-grade starch wastewater				First-grade starch wastewater		
	Washwater		Supernatant		Plant A		Plant B
	Range	Mean	Range	Mean	Separator wastewater	Washwater	Combined wastewater
BOD_5	300–2,490	1,190	1,720–6,820	4,150	3,000–4,000	200–1,700	5,550–7,400
COD	610–6,110	2,700	4,700–10,000	7,580	3,100–13,900	2,000–4,850	13,300–19,500
Suspended solids	290–4,240	1,880	470–1,710	850	1,480–8,400	400–6,100	1,970–3,850
NH_3–N	0–7.8	1.9	1.2–35.0	16.0	0–4.7	0–1.1	0
Org-N	0–67	32	4–109	70	19–39	14–18	86–115
Phosphorus	0–6.0	3.6	0–10.5	6.6	5.6–8.5	1.2–1.3	0
Total dissolved solids	30–6,960	2,750	3,890–16,390	9390	—	—	—
pH	3–4.6	4.2	2.6–4.0	3.4	3.4–4.2	4.2–7.1	3.8–5.2
DO	0–4.9	2.7	0–2.7	1.0	0	0.6–5.3	0
Acidity	—	—	—	—	670–860	20–220	135–1,010
Settleable solids	—	—	—	—	60–200	10–100	50–115
Temperature (°C)	—	—	—	—	28–33	28–31	30–31

The purpose of this sterilization is to deactivate the enzymes responsible for the breakdown of oil into free fatty acids and to loosen the fruits from the stalks. Thereafter, the sterilized bunches are fed into a rotary-drum threshing or stripping machine which separates the fruit from the bunches. The empty bunches drop into a conveyor belt that carries them into an incinerator and are burnt into ash, whereas the loose fruits are converted into a homogeneous oily mash by a series of rotating arms (digester). The digested mash is then led into a press for the extraction of crude oil. Nuts must not be broken at this stage of the process.

Nuts and fresh fiber are then separated to recover the nuts which are subsequently cracked to produce kernels for sale. The pressed fiber and some of the shells are usually burnt as fuel in the steam-raising boiler.

The extracted crude oil, consisting of a mixture of oil, water, and some fine solid materials, is passed through a vibrating screen to remove the solid matter. Hot water is often added in this operation. A clarification tank is used to separate the oil by gravity, and the oily sludge settles at the bottom. The clarified oil is further purified in a vacuum dryer prior to being pumped to a storage tank. The oil sludge, after straining and desanding, is centrifuged to recover the oil, which is returned to the clarifier. The sludge is discharged into an oil trap, where further oil is recovered by heating the sludge with steam prior to discharge of the sludge to a waste treatment facility.

With respect to Figure 2.4, the sources of wastewater and sludge generated during palm oil milling are given below:

1. sterilizers' condensates, cleansing of the sterilizers and floor washing at the sterilizer station;
2. floor washing and desanding of the press station;
3. steam condensates;
4. steam condensates;
5. hydrocyclones discharge;
6. steam condensates;
7. turbines cooling water and steam condensates;
8. boilers blowdown;
9. overflow and backwash water of the water-softening plant;
10. floor washing of the oil room;
11. }
12. } wastewaters discharged from various units in the oil room;
13. }
14. overflow from the vacuum dryers;
15. oil trap discharge.

A general mass balance of various products generated from a palm oil mill plant is shown in Figure 2.5. It is seen that of the 100 tons fresh fruit bunches (FFB) processed, about 21 tons of palm oil and 6 tons of palm kernel will be produced.

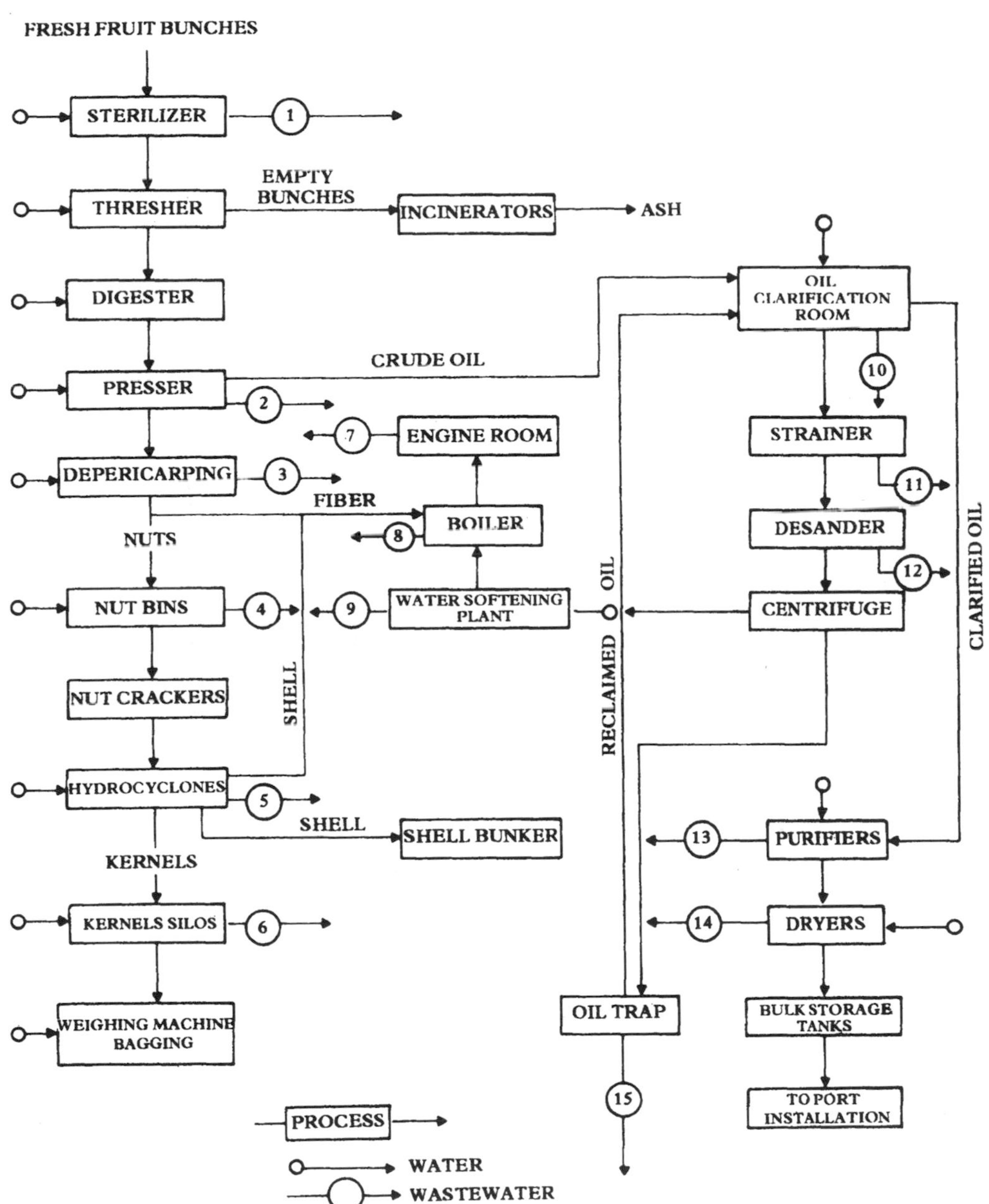

Figure 2.4 General process diagram of a palm oil mill (Thanh *et al.*, 1980)

Other miscellaneous by-products can be used as animal feed, fertilizer, and boiler fuel.

PALM OIL MILL WASTEWATER CHARACTERISTICS

Thanh *et al.* (1980) conducted a study on wastewater characteristics of four palm oil mills in Malaysia and Thailand. The data of wastewater flows per ton of FFB processed were classified into 50, 70, 80, and 90 percentiles. Table 2.11 shows the percentiles of wastewater flows in m^3 per ton of FFB measured at some sampling

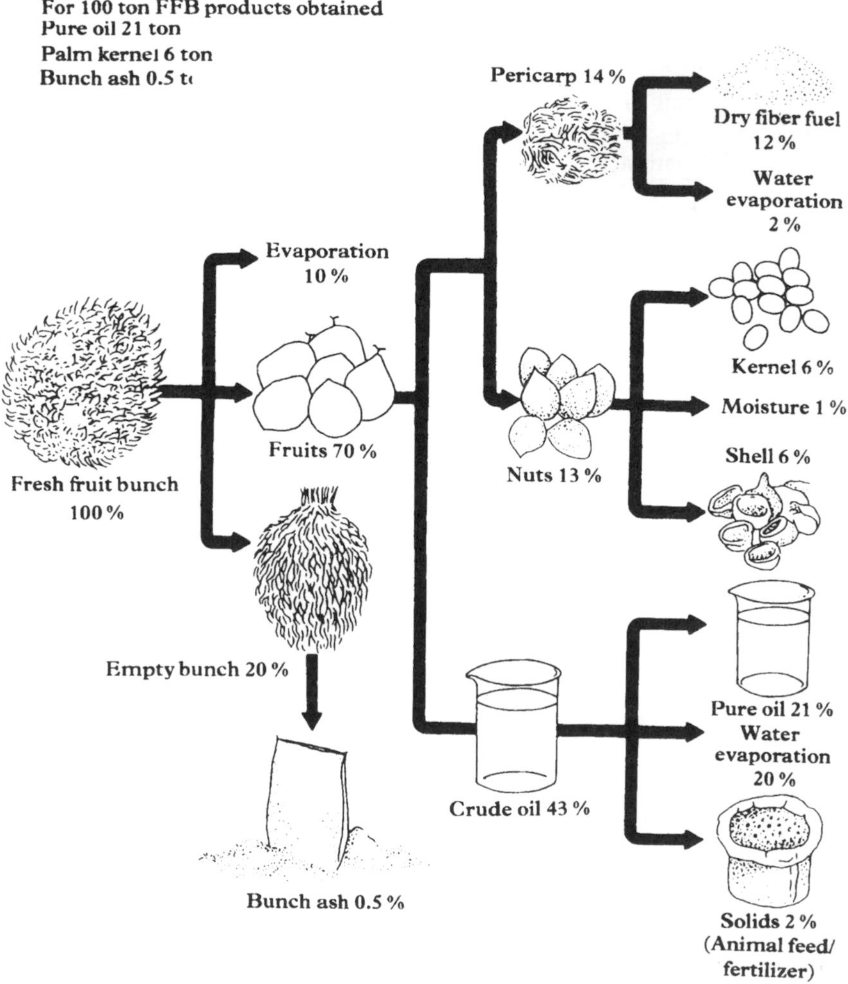

Figure 2.5 Product mass balance of a typical palm oil mill (Muttamara *et al.*, 1987)

Table 2.11 Probability of wastewater flows per ton of FFB processed at different measuring stations of palm oil mills (Thanh *et al.*, 1980)

Palm oil mill	Measuring stations (see Figure 2.4)	Flow in m^3/ton FFB (m^3/h)			
		50 Percentile[a]	70 Percentile[a]	80 Percentile[a]	90 Percentile[a]
Plant A (Malaysia)	Oil clarification room	1.04 (22.0)	1.14 (24.8)	1.21 (26.5)	1.31 (29.0)
	Boiler house and hydrocyclones	0.49 (10.5)	0.57 (12.5)	0.62 (13.8)	0.70 (16.0)
	Sterilizer building and press station	0.76 (15.8)	0.84 (17.5)	0.89 (18.8)	0.96 (20.6)
	Combined flow	2.32 (47.0)	2.50 (49.8)	2.60 (51.5)	2.75 (54.0)
Plant B (Malaysia)	Oil clarification room	1.25 (15.5)	1.50 (18.0)	1.68 (19.5)	1.94 (22.0)
	Boiler house, sterilizer building and hydrocyclones	1.90 (21.0)	2.21 (23.7)	2.43 (25.7)	2.80 (28.5)
	Combined flow	3.23 (37.0)	3.72 (42.0)	4.10 (46.0)	4.60 (51.0)
Plant C (Malaysia)	Sterilizer building	0.11 (2.2)	0.15 (2.7)	0.17 (3.1)	0.22 (3.8)
	Oil clarification room	1.30 (22.5)	1.60 (27.0)	1.83 (31.0)	2.21 (36.0)
	Boiler house	0.52 (8.7)	0.61 (10.2)	0.67 (11.2)	0.76 (13.0)
	Hydrocyclones	0.27 (4.7)	0.34 (5.9)	0.38 (5.8)	0.45 (6.5)
	Combined flow	2.25 (36.0)	2.60 (41.5)	2.85 (45.0)	3.25 (50.0)
Plant D (Thailand)	Oil clarification room	0.89 (6.1)	1.01 (8.0)	1.09 (9.2)	1.11 (11.8)
	Sterilizer building and oil clarification room	1.09 (8.9)	1.21 (10.5)	1.29 (12.0)	1.39 (13.0)
	Miscellaneous sources	0.64 (4.9)	0.80 (5.9)	0.87 (6.5)	1.00 (7.8)
	Combined flow	1.73 (13.8)	2.01 (16.4)	2.16 (18.5)	2.39 (20.8)

[a] If a 50 percentile flow is used in the design of wastewater treatment facility, it means that 50 percent of the time the flow of that particular waste stream will be equal or less than the stated figures

stations of the palm oil mills. The characteristics of these wastewaters at 50 percentile values are given in Table 2.12.

It appears from these data that wastewaters from the oil clarification room and sterilizer building are the most concentrated, containing high concentrations of COD, BOD, solids, and oil and grease. Palm oil mill wastewaters generally have pH below 5, and their temperature can range from 30 to 70 °C. The main solid wastes produced from palm oil milling are nuts, fibers, and shells. The nuts are processed to make kernels, and fibers and shells are used as boiler fuel, as explained earlier.

2.3.3 Sugar cane industry

RAW SUGAR CANE MANUFACTURING PROCESS

A flow diagram of the sugar cane manufacturing process is shown in Figure 2.6. The first step in raw sugar processing is juice extraction, carried out by crushing

Table 2.12 Fifty percentile values of palm oil wastewater characteristics at different sampling stations (all values are in mg/L, except pH which is dimensionless and temperature is in °C) (Thanh *et al.*, 1980)

Palm oil mill	Measuring stations	COD	BOD[a]	BOD[a]/COD	TS	SS	Total P	Total N	Oil and grease	pH	Temperature (°C)
Plant A (Malaysia)	Oil clarification room	64,000	30,000	0.47	45,000	28,000	285	490	18,500	4.0–4.3	46–77
	Boiler house and hydrocyclones	1,860	1,050	0.56	1,300	850	20.5	14.5	800	4.7–5.9	30–65
	Sterilizer building and press station	10,300	5,500	0.53	6,000	1,250	42	60	1,100	4.3–4.7	32–77
	Combined flow	28,500	14,000	0.49	23,000	10,000	163	265	9,600	4.3–4.7	32–75
Plant B (Malaysia)	Oil clarification room	45,000	16,800	0.37	31,000	20,000	230	450	11,500	3.9–4.8	36–53
	Boiler house, sterilizer building and hydrocyclones	2,800	1,600	0.57	2,180	680	28	39	—	4.9–6.6	30–51
	Combined flow (1 + 2)	18,300	10,000	0.55	15,500	7,500	135	230	8,200	4.1–6.3	30–51
Plant C (Malaysia)	Sterilizer building	52,500	27,000	0.51	38,500	9,000	320	590	6,100	4.0–4.5	30–88
	Oil clarification room	46,000	20,000	0.43	33,500	18,400	297	590	6,800	4.0–4.4	36–57
	Boiler house	88	40	0.45	210	54	1.0	1.2	—	—	31–78
	Hydrocyclones	3,600	1,950	0.54	2,600	2,000	23	26	1,600	5.5–6.2	31–70
Plant D (Thailand)	Sterilizer building and clarification room	61,000	28,500	0.46	47,500	31,000	330	720	86,000	4.5–4.9	45–49
	Miscellaneous sources	750	460	0.61	840	360	73	15	180	5.6–7.1	31–33

[a] For plants in Malaysia, BOD was measured at 30 °C for 3 days. For plants in Thailand, BOD was measured at 20 °C for 5 days

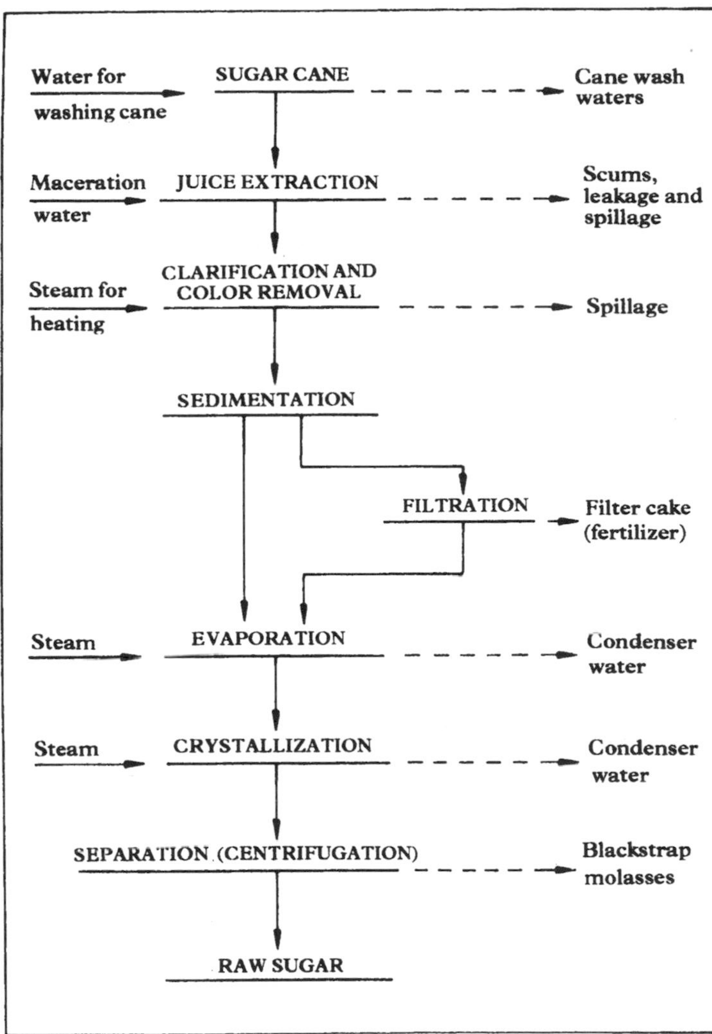

Figure 2.6 Diagrammatic flow sheet of the process of raw sugar manufacture (adapted from Meade and Chen, 1977)

the cane between a series of rollers under pressure. To aid in the extraction of the juice, sprays of water or thin juice are directed on the blanket of bagasse (fibrous part of the cane) as it emerges from each mill unit to leach out sugar. The final bagasse from the last roller contains the unextracted sugar, the woody fiber, and 40–50 percent water. This final bagasse is used as fuel or as material for wallboard and paper manufacture.

The extracted juice is acidic, turbid, and dark green in color. It is treated with chemicals such as lime, sulfur dioxide, carbon dioxide, and phosphate, and heated for clarification. This treatment has the effect of precipitating suspended solids and some impurities, which are allowed to settle, and color removal. The clear juice is then filtered through vacuum filters. The filter press juice returns to the clarification process or goes directly to clarified juice. The press cake is discarded or returned to the fields as fertilizer.

The clarified juice contains approximately 80 percent water. Two-thirds of this water is evaporated in vacuum multiple-effect evaporators. Crystallization takes place in a single-effect vacuum pan where the syrup is evaporated until saturated with sugar. It is at this point that the sugar first appears as fine crystals, which are then built up to the size required for the final product. The mixture of crystals and syrup is then concentrated to a dense mass (massecuite) and the contents of the pan are discharged into a mixer or crystallizer.

The massecuite from the mixer or crystallizer is drawn into revolving machines called centrifugals. The sugar crystals are retained and the mother liquor, molasses, passes through. The final molasses or blackstrap is a heavy viscous material containing approximately one-third reducing sugars, and the remainder ash, organic non-sugars, and water. It may be used as cattle feed or as a raw material for distilleries.

SUGAR INDUSTRY WASTEWATER EFFLUENTS

The sugar industry uses large amounts of water, mainly for cane washing, condensing of vapor, boiler and feed water, and other miscellaneous uses. Sources of wastewater and sludge generation during the manufacturing of sugar are given in Figure 2.6. There are air emissions from the sugar manufacturing process which include a variety of pollutants such as particulates, SO_x, NO_x, CO, and hydrocarbons; these pollutants have to be reduced to the required air emission standards set by the appropriate regulatory agency.

The characteristics of wastewater from a sugar mill in Pakistan are given in Table 2.13. Its pH varied from 6.7 to 10 and the COD range was 600–3,700 mg/L. The wastewater contained rather high solid concentrations in which the ranges of total solids and volatile solids were 4,520–10,790 and 1,820–4,600 mg/L, respectively. Similar data from a sugar mill in the U.S.A. are given in Table 2.14, for information.

2.3.4 Brewing industry

THE BREWING PROCESS

The principal stages in brewing are mashing, boiling, and fermentation with subsequent packaging into bottles and casks followed by cooling, clarification, and pasteurization.

During mashing, the mixture of finely ground malt and hot water undergoes enzymatic changes, whereby the starch is converted into sugar and dextrins and the protein into amino acids and polypeptides. The soluble product from mashing is

Table 2.13 Wastewater characteristics of Crescent Sugar Mills, Pakistan (Qureshi, 1977) (all values are in mg/L, except pH)

Characteristics	Range	Fifty percentile	Ninety percentile
pH	6.7–10.0	8.15	9.7
Alkalinity	260–490	370	490
Total solids	4,520–10,790	6,500	10,300
Fixed solids	1,850–6,150	4,000	6,100
Volatile solids	1,825–4,600	2,700	4,500
COD	607–3,680	2,030	3,500
Nitrogen	15–20	33	—
Phosphorus	6.7–11.25	9.25	—

Table 2.14 Wastewater characteristics of a sugar mill plant in the U.S.A. (Oswald *et al.*, 1973)

Parameter	Units	Value	Variance
Total nitrogen (N)	mg/L	16.4	—
Ammonia (N)	mg/L	6.3	—
Nitrate (N)	mg/L	2.6	—
Chlorides	mg/L	400[a]	—
Sulfate	mg/L	210.0	—
Alkalinity ($CaCO_3$)	mg/L	538.0[b]	—
Sulfide	mg/L	0.68	—
Phosphate (P)	mg/L	3.4	—
Calcium	mg/L	178.0	—
Magnesium	mg/L	66.0	—
Sodium	mg/L	222.0[c]	—
Potassium	mg/L	88.0[c]	—
BOD_5 (unfiltered)	mg/L	930.0	—
COD (unfiltered)	mg/L	1,601.0	—
COD (filtered)	mg/L	1.195.0	—
Suspended solids	mg/L	1,015	11,269
Suspended volatile	mg/L	360	7,140
Suspended ash	mg/L	655	6,348
Dissolved solids	mg/L	2,209	6,016
Dissolved volatile	mg/L	1,139	2,764
Dissolved ash	mg/L	1,070	7,754
Total solids	mg/L	3,224	14,979
Total volatile	mg/L	1,499	4,456
Total ash	mg/L	1,725	8,328
Sugar	mg/L	1.25	0.25
Dissolved oxygen	mg/L	0.00	—
pH		7.1	1.8
Light penetration	cm	45.6	17.8
Specific conductance	μmhos/cm	300	299

[a] Based on specific conductivity
[b] By difference
[c] Single values

known as sweet wort, and this is subsequently boiled with hops in a metal vessel. Boiling destroys the enzymes and, at the same time, extracts resins from the hops to give a bittering effect. The wort then goes to a cooling process in the cool-ship, yeast is added and, during the resultant fermentation, the sugars present are converted into alcohol and CO_2. The nitrogenous material and phosphates in the wort are also utilized by the yeast for growth and fermentation. Then the beer is stored in lager tanks for a period of time, then filtered and pasteurized before being bottled or canned. A schematic diagram of this brewing process is given in Figure 2.7.

SOURCES AND CHARACTERISTICS OF WASTEWATERS

The sources of wastewaters that contain very large amounts of suspended and dissolved solids are: washing from cool-ships, lager tanks, and fermentation tanks. These wastewaters contain excessive amounts of beer, malt, and yeast which possess a very high COD, sometimes as high as 20,000 mg/L, amounting to about 10 percent of the total waste discharged. Brewery wastewaters normally have an amber color and a rich grainy smell of malt.

According to Lovan and Foree (1971), brewery wastewaters have been found to be highly contaminated with soluble organics, low in nutrients and pH, and they have a high temperature. The COD content can vary from as much as 24,000 mg/L to as little as 6,000 mg/L. The ratio of BOD_5/COD varies between 0.2 and 0.9. Table 2.15 gives the characteristics of a brewery wastewater in Thailand.

2.3.5 Slaughterhouse

This section will describe waste generation from abattoirs, meat processing plants, and poultry processing plants.

Table 2.15 Characteristics of brewery wastewater (Tantideeravit, 1975)

Parameter	Value
BOD_5 (mg/L)	1,800
COD (mg/L)	3,100
Total solids (mg/L)	1,750
Suspended solids (mg/L)	800
pH	7.5
Alkalinity (mg/L as $CaCO_3$)	160

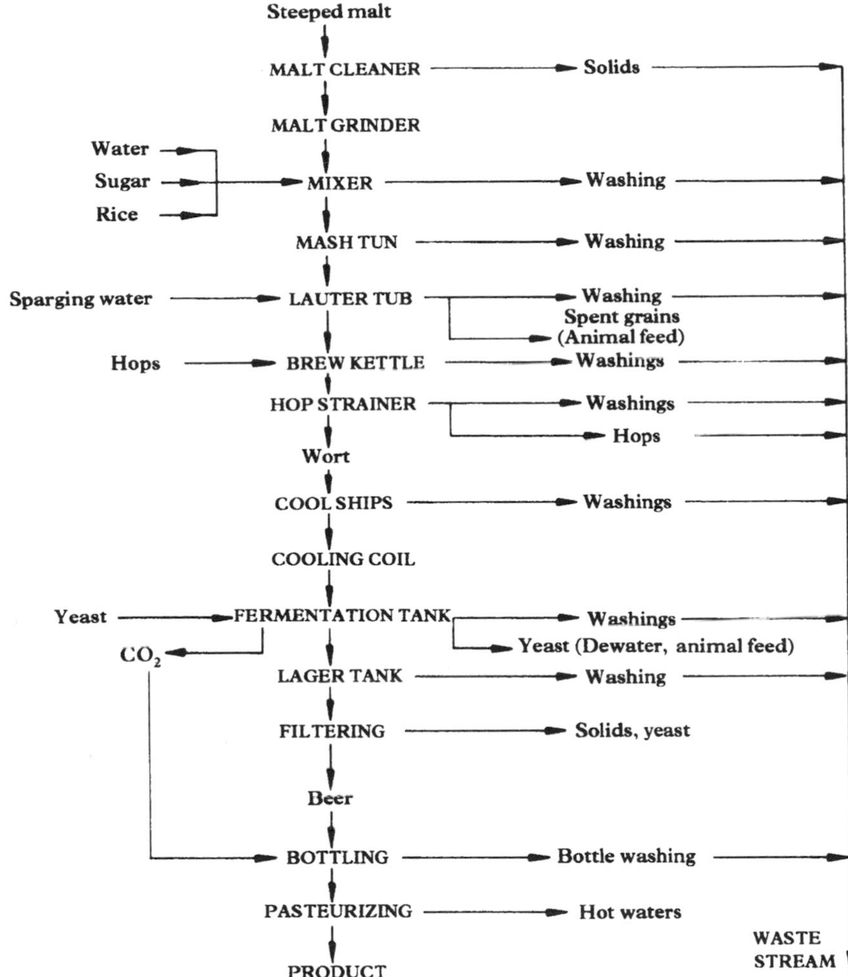

Figure 2.7 Brewing processes (Tantideeravit, 1975)

ABATTOIRS (GRIFFITHS, 1981)

Lairage/stockyard

Lair refers to animals' lying-places or cages. The effluent arising is dung and urine—contaminated by straw. Straw is separated by a conventional trap and the effluent is fairly weak due to frequent washing-down. This effluent is not usually kept separate unless fat is to be recovered from other effluents.

Table 2.16 shows the analysis of wastes from one Chicago stockyard. Another study of these same wastes showed a volume of 623,000 gallons (165 m^3) per day

Table 2.16 Stockyard wastes (U.S. PHS, 1943)

Characteristics	Concentration (mg/L)
Total suspended solids	173
Volatile suspended solids	132
Organic nitrogen	11
Ammonia nitrogen	8
BOD_5	64

for a 27 acre (11 ha) section of the yard and an average BOD_5 of 100 mg/L (U.S. PHS, 1943).

Slaughter: bleeding and carcass separation

Figure 2.8 is a schematic diagram of a slaughterhouse operation. After killing, the carcasses are bled, the hide is stripped, the abdominal mass is removed, and the carcass is split into smaller pieces. At some stage washing takes place, which gives rise to dirt, dung, grit, etc. in the effluent. Blood is normally kept separate to reduce the polluting load in the effluent and because it can be a valuable by-product.

The abdominal mass is removed and sorted for collection by the pharmaceutical and by-products industries.

The paunch (stomach) is removed, and it is important to keep the contents out of the normal effluent stream (to simplify subsequent treatment). The contents, partly digested food, are handled separately; they are best kept as dry as possible and disposed of on land as fertilizer.

The cutting down of the carcass involves blood, bone dust, tissue, and fat contaminating the washwater, some of which can be caught in traps, but this is the main pollutant which has to be dealt with.

Meat processing (Griffiths, 1981)

Meat processing plants are completely different from abattoirs. There is no paunching, no blood, no gut; but there are a lot more meat and fat wastes plus all sorts of additives, e.g. pastry, soya, sauces, spices, preservatives, colorings, etc.

The major problem is the flow of the effluent stream not being constant. Little water is used while production is under way, but when a batch is complete a lot of water is used to wash down. The additives vary and so does production. In order to treat the effluents properly it is necessary in each case to study and to understand the variations in the process and the waste characteristics.

Poultry processing (Griffiths, 1981)

Owing to the large demand for fowls, poultry processing is now a high-volume operation which has developed from its original farmhouse background. The basic procedures for poultry processing are shown in Figure 2.9.

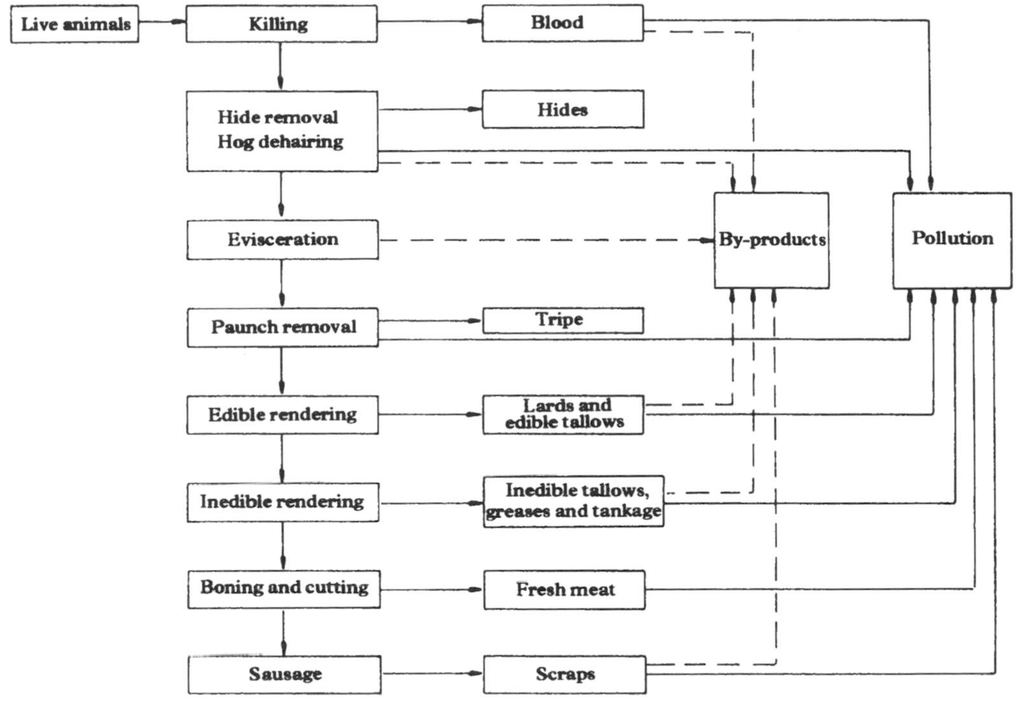

Figure 2.8 Fundamental processes in a slaughterhouse (Nemerow, 1971) (© 1971, Addison-Wesley Publishing Co., Reading, Mass.)

The birds are hung by the feet on a conveyor. They are then killed and bled. The blood is taken away and, if recovered, used for human or pet foods (it is of less value than beef blood).

The next stage is feather plucking. This is done by a series of flaps and flails in a tunnel, under a hot-water spray. There are two ways of disposing of the feathers. They can be taken away by water in a flume or taken away by a conveyor. The feathers can then be reused as a material for mattresses.

Next the heads, feet, and innards are removed. The removal of innards is done by hand, and they are disposed of either by a wet or dry viscera line. In a dry line innards are removed and taken to a disposal skip by conveyor or by vacuum. In a wet line they are taken by a water flume to be separated later.

The final stages are washing using a fixed amount of water per fowl, then spin chilling and freezing or washing (with the addition of chlorine for disinfection), drying and vacuum sealing.

The following are possible causes of pollution in the poultry processing effluent:

1. blood—if not recovered: this has a high BOD_5;
2. feather water and flume: if a wet flume is used this can have a high proportion of dissolved or emulsified fats;

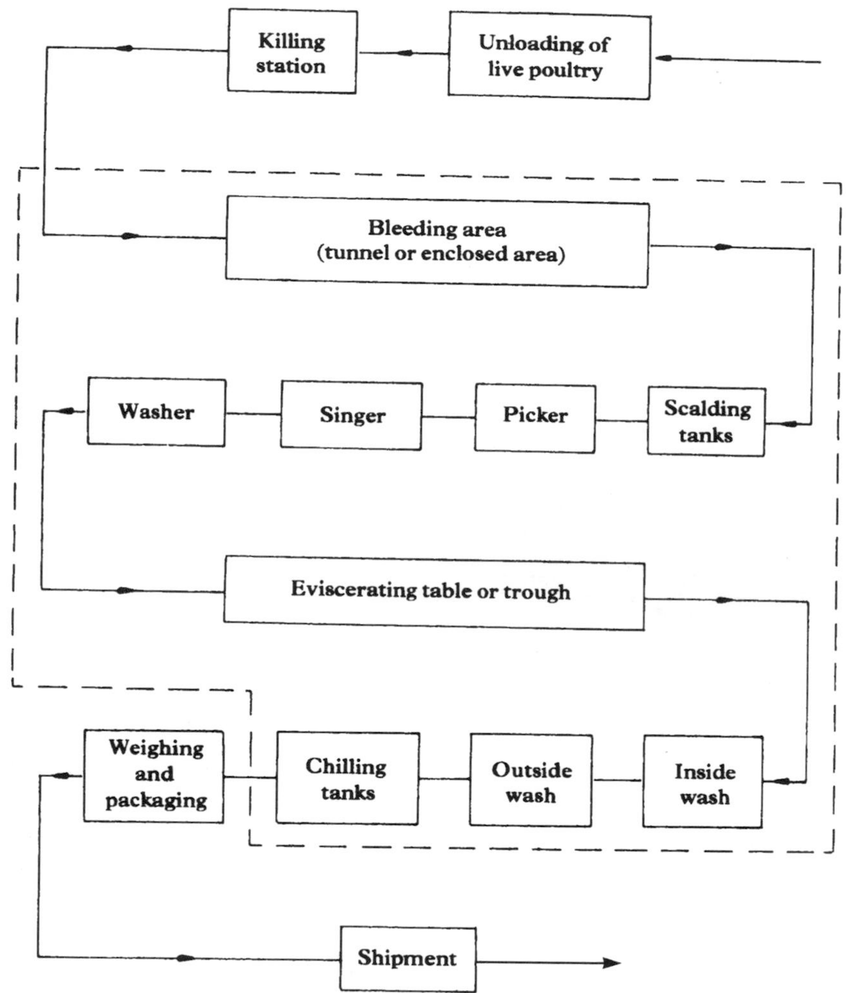

Figure 2.9 Major steps in poultry processing (Nemerow, 1971) (© 1971, Addison-Wesley Publishing Co., Reading, Mass.)

3. wet viscera: requires screening to remove the solid matter;
4. wash and spin chiller discharge: note that chlorine is introduced in the washwater.

Characteristics of slaughterhouse and meat-packing wastes

Effluents from slaughterhouses and meat-packing houses are usually heavily polluted with solids, floatable matter (fat), blood, manure, and a large variety of proteinaceous compounds. The composition depends very much on the type of production and facilities. The main wastes originate from lairage, killing, hide removal or dehairing, paunch handling, rendering, trimming, processing, and clean-up operations. The BOD_5 and solids concentrations in the plant effluent depend on in-plant control of water use, by-products recovery, waste separation at the source, and plant management.

When considering effluent disposal the first consideration is water consumption. The range of wastewater production from abattoirs is 5–15 m^3/ton live weight. For the poultry industry a typical discharge lies between 10 and 55 L per bird (Brolls and Broughton, 1981).

In terms of BOD_5, typical loads from abattoirs and meat processing are 6–12 kg/ton live weight at U.K. abattoirs, usually at a concentration of 1,000–3,000 mg/L. From poultry packing the effluent is normally about half the strength of the abattoir wastewater, with a BOD_5 content of 13–23 kg/1,000 birds (Brolls and Broughton, 1981).

Some characteristics of slaughterhouse wastewaters, as reported by Brolls and Broughton (1981), are given in Table 2.17, and those of meat-packing wastes are given in Table 2.18. Table 2.19 shows the wastewater characteristics of a poultry processing plant in Thailand.

2.3.6 *Fruits and vegetables industry*

PROCESSING OPERATIONS AND WASTE GENERATION

Unit operations

The processing steps for five typical commodities, namely peaches (to exemplify fruits), peas and corn (common vegetables), beets (peeled root crops), and tomato products (pulped commodities), are outlined in Figure 2.10. The principal steps where water (or steam) is used and where solid and dissolved residue are generated are indicated.

The containers of fruits or vegetables are dumped into the first-stage washer or flume or on to belts. The products are conveyed by flumes, pipes, belts, elevators, or other conveyors between processing steps. Fluming water, generally reused, and small flows of water to belts, graders, and other equipment for lubrication and sanitation are not noted in the flow diagrams.

Table 2.17 Characteristics of slaughterhouse wastes

Item	Lower range	Upper range
BOD_5, mg/L	360–1,880	900–2,600
Suspended solids, mg/L	489–1,450	600–1,800
Organic N, mg/L	36–510	—
Grease, mg/L	88–440	200–800
Chlorides, mg/L, as Cl	190–5,690	—
pH	6.5–8.4	6.8
COD, mg/L	—	2,000–4,900
Flow, L/1,000 kg live weight	—	1,690–7,920

From Brolls and Broughton (1981); reproduced by permission of Elsevier Science Publishers Ltd

Table 2.18 Characteristics of meat packing wastes[a]

Item	Average	Range
BOD_5, mg/L	1,240	600–2,720
COD, mg/L	2,940	960–8,290
Organic N, mg/L	85	22–240
Grease, mg/L	1,010	250–3,000
Suspended solids, mg/L	1,850	300–4,200
Volatile suspended solids, mg/L	92	80–97

[a] Composite samples from 16 plants
From Brolls and Broughton (1981); reproduced by permission of Elsevier Science Publishers Ltd

Table 2.19 Characteristics of poultry processing wastewater (Visittavanich, 1987)

Parameters	Mean	Range
COD, mg/L	2,110	1,260–3,260
BOD_5, mg/L	1,010	520–1,740
TKN, mg/L	150	90–200
TS, mg/L	1,940	1,290–2,640
Total P, mg/L	14	12–17
pH	7.1	6.7–7.7

Waste generation by unit operations

Much of the total waste flows from many commodities, especially from citrus, comes from relatively clean water used in cooling, condensing, and concentrating. The segregation of this water for reuse is practiced to some extent, and should become almost universal. Other large waste flows are from washing tomatoes, peeling potatoes, peeling peaches, washing potatoes, cutting corn, cutting and pitting peaches, washing and blanching corn. Large quantities of BOD_5 are generated in those processes. Citrus by-products recovery also generates very large amounts of BOD_5 and SS. Other large sources of SS are peeling potatoes, cutting corn, and washing tomatoes.

Peeling Table 2.20 shows a summary of estimated pollution loads in the effluent from peeling fruits and vegetables, reported by various investigators. Peel wastes comprise a high percentage of the total pollution loads in the effluent of fruit and vegetable plants.

Blanching Table 2.21 shows examples of losses from vegetables during blanching expressed in BOD_5, COD, and SS. The volume of effluent from blanchers is generally relatively small, but the concentrations of suspended and

Water	Operation	Peaches	Peas	Corn	Beets	Tomato products	Type of waste generated: Solid	Soluble	Soil
	Dry dump	X	X	X	(X)*		●		
→	Water dump				(X)	X	●	●	●
	Air cleaner		X	X			●		
	Trash eliminator	X					●		
	Husker			X			●		
	Sorting/trimming	X	X	X	X	X	●	●	
→	Washer	X	X	X	X	X	●	●	●
→	Grader/sizer	X	X		X		●	●	
	Cutter	X		X			●	●	
→	Peeler/rinse	X			X	(X)	●	●	
→	Blancher/rinse		X	(X)	(X)			●	
	Pulper/finisher	(X)				(X)	●		
	Slicer/dicer	(X)			(X)		●	●	
→	Evaporator	(X)				(X)	●	●**	

* Optional or alternative operations in ( ).

** Relatively clean water

Figure 2.10 The use of water and generation of wastes in typical unit operations of fruit and vegetable processing (U.S. EPA, 1971)

total solids can be high. These wastewaters contain a high organic content (in terms of BOD_5 and COD) which can be treated or recycled by biological processes.

2.4 POLLUTION CAUSED BY HUMAN WASTES AND OTHER WASTEWATERS

Because human and animal wastes are organic in nature, when being discharged into a stream or lake they will serve as food for heterotrophic bacteria. Bacterial reactions will decompose the organic compounds to simple, inorganic end-products with the production of energy for cell synthesis as shown in eqs 2.1 and 2.2 (Metcalf and Eddy, 1991):

Oxidation or dissimilatory process

$$\underset{\text{(organic matter)}}{\text{(COHNS)}} + O_2 + \text{aerobic bacteria} \rightarrow CO_2 + NH_3 + \text{other end-products} + \text{energy} \qquad (2.1)$$

Table 2.20 Characteristics of wastewater from peeling fruits and vegetables[a] (U.S. EPA, 1971)

	BOD_5		SS
Product	kg/ton	Percentage of plant waste stream	Rate
Beets (blancher/peeler)	88	84	100 kg/h
Peaches (rinse after peeling)		40	
Peaches	27 (COD)		5 kg/ton
Peaches	4–6		2.5–4 kg/ton
Potatoes (peeler)	9		41 kg/ton
Potatoes	17–36		34–49 kg/ton
Potatoes (peeler, chips)	1		2 kg/ton
Potatoes (peeler/dehydration)	9		41 kg/ton
Potatoes	15		
Potatoes (lye peel)	84	89	
Potatoes (dry caustic peel)	12	80	
Potatoes (lye peel)	171		
Potatoes (infrared peel)	118		
Potatoes (steam peel)	118		
Potatoes (infrared peel)	91		
Peach/tomato (flume after peel)		1	383 mg/L
Peach/tomato (rinse after peel)		19	1,230 mg/L
Peach/tomato (rinse after peel)		20	
Tomato (lye peel)		39	
Tomato (lye peel)		35	

[a] Kilograms of BOD_5 or suspended solids per ton of raw product, or per hour of operation

Table 2.21 Pollution loads in effluents from water blanching of vegetables[a] (U.S. EPA, 1971)

Vegetable	Effluent flow (L/h)	BOD_5 (kg/ton)	COD (kg/ton)	SS (kg/ton)
Beets (and peeler)	3,013	88 (85)[b]	147 (83)	109 (55)
Carrots (and peeler)	1,937	44 (65)	90 (67)	153 (64)
Corn	62	277 (kg/day)	390 (kg/day)	65 (kg/day)
Corn	523	11.2 (16)	14 (18)	3 (12)
Peas	294	3,500 ppm in effluent		
Potatoes	580	24	26	17
Potatoes	531	10	15	11
Potatoes (and peeler)	2,118	84 (89)	127 (86)	82 (37)

[a] BOD_5, COD, and SS in kg per ton of raw product except as noted
[b] Percentage of total effluent pollution load in parentheses

Synthesis or assimilatory process

$$(COHNS) + O_2 + \text{aerobic bacteria} + \text{energy} \rightarrow \underset{\text{(new bacterial cells)}}{C_5H_7O_2N} \quad (2.2)$$

Under anaerobic conditions (i.e. in the absence of oxygen), anaerobic bacteria will decompose the organic matter as follows:

$$(COHNS) + \text{anaerobic bacteria} \rightarrow CO_2 + H_2S + NH_3 + CH_4 + \text{other end-products} + \text{energy} \quad (2.3)$$

$$(COHNS) + \text{anaerobic bacteria} + \text{energy} \rightarrow \underset{\text{(new bacterial cells)}}{C_5H_7O_2N} \quad (2.4)$$

Note: $C_5H_7O_2N$ is a common chemical formula used to represent bacterial cells.

In the absence of organic matter the bacteria will undergo endogenous respiration or self-oxidation, using their own cell tissue as substrate:

$$C_5H_7O_2N + 5O_2 \rightarrow 5CO_2 + NH_3 + 2H_2O + \text{energy} \quad (2.5)$$

The compounds CO_2 and NH_3 are nutrients for algae. With sufficient sunlight, algal photosynthesis will occur (Oswald and Gotaas, 1956):

$$NH_3 + 7.62CO_2 + 2.53H_2O \xrightarrow[\text{algae}]{\text{sunlight}} \underset{\text{(new algal cells)}}{C_{7.62}H_{8.06}O_{2.53}N} + 7.62O_2 \quad (2.6)$$

Another equation to incorporate P compound in the algal photosynthesis is (Oswald, 1988):

$$106CO_2 + 236H_2O + 16NH_4^+ + HPO_4^{2-} \xrightarrow[\text{algae}]{\text{sunlight}} C_{106}H_{181}O_{45}N_{16}P + 118O_2 + 171H_2O + 14H^+ \quad (2.7)$$

In a natural water course receiving low organic loading, the oxygen produced in eq. 2.6 can be used by the bacteria in eqs 2.1 and 2.2, and the cycle is repeated. This cycle, called 'algal–bacterial symbiosis', is a natural phenomenon occurring in a water body that receives a low organic loading, and these algal–bacterial reactions are in a state of dynamic equilibrium. Parts of the algal and bacterial cells serve as food for protozoa and some small fish, which are further consumed by big fish, and finally humans.

The discharge of untreated wastes into a receiving water body will create a biologically unbalanced condition. At high organic loading the bacteria will require more oxygen for their oxidation and synthesis (eqs 2.1 and 2.2), resulting in the depletion of oxygen in the water body which is detrimental to all aquatic life (Figure 1.3). Although algal photosynthesis produces some oxygen (eq. 2.6 or 2.7), at night when there is no sunlight, algae respire or use oxygen, causing a further depletion of oxygen in the water. Eventually the water will become anaerobic (without oxygen), dark in color, and only anaerobic microorganisms and certain types of worms can live there. Besides this organic pollution, the water body would become very unaesthetic to nearby inhabitants, and would lower the environmental quality of its surrounding areas.

However, knowledge of the pollution effects as cited above has, in turn, been used for the design of technologies and facilities for waste treatment and recycling. For example, the reactions of eqs 2.1 and 2.2 are basically used in the design and operation of aerobic waste treatment facilities such as activated sludge, trickling filtration, and composting. Algal–bacterial symbiosis is the basis for the design of facultative waste stabilization ponds and high-rate algal ponds (for the production of algal protein biomass). Equation 2.3 is the reaction where biogas (CH_4 and CO_2) is produced. Application of this basic information to waste recycling is described in Chapters 3–9.

2.5 DISEASES ASSOCIATED WITH HUMAN AND ANIMAL WASTES

There are various kinds of enteric microorganisms present in human excreta and animal manure; some of these are pathogenic and some are not. They can be classified into such major groups as bacteria, viruses, protozoa, and helminths. Some of the important enteric pathogens commonly found in human excreta and wastewater, the diseases they cause, modes of transmission, and geographical distribution are listed in Table 2.23. Feachem *et al.* (1983) estimated the possible outputs of some pathogens in the feces and wastewater of a tropical community as shown in Table 2.22. It is obvious from these tables that human wastes (excreta and wastewater) are a potential public health hazard, being the beginning of the transmission route of many diseases. The engineering profession responsible for the collection, transport, treatment and disposal or reuse of these wastes must be aware of the potential infectivity and transmission of these diseases, so as to be able to ensure that these pathogens do not pose an actual threat to human health. Details of these diseases, their modes of transmission and die-offs can be found in *Sanitation and Disease* by Feachem *et al.* (1983).

Animals and their manure can also be sources of disease transmission to human beings. The term 'zoonoses' is defined as those diseases and infections which are naturally transmitted between human and other vertebrate animals. There are two groups of zoonoses. In the first group, animals act as hosts alternative to humans, for example the protozoon *Balantidium* (Table 2.23). In this case both human excreta and animal manure have to be properly controlled to avoid transmission of the disease. *Taenia* (Table 2.23) is an example of the second group of zoonoses in which the animals (either cows or pigs) are the intermediate step in the disease transmission. Inactivation of *Taenia* ova in the feces or consumption of well-cooked beef or pork will prevent the transmission of taeniasis; therefore the control of either human excreta or animal manure would be sufficient in controlling the spread of the second group of zoonoses. Some of the zoonoses relating to viral, rickettsial, and bacterial diseases are given in Table 2.24, and those relating to protozoan and helminthic infections are shown in Table 2.25.

The detection and identification of pathogenic microorganisms are generally difficult, time-consuming, and expensive. For routine analyses or monitoring,

Table 2.22 Possible output of selected pathogens in the feces and sewage of a tropical community of 50,000 in a developing country

Pathogen	Prevalence of infection in country (percentage)[a]	Average number of organisms per gram of feces[b]	Total excreted daily per infected person[c]	Total excreted daily by town	Concentration per liter in town sewage[b]
Viruses					
Enteroviruses[d]	5	10^6	10^8	2.5×10^{11}	5,000
Bacteria					
Pathogenic *E.coli*[e]	?	10^8	10^{10}	?	?
Salmonella spp.	7	10^6	10^8	3.5×10^{11}	7,000
Shigella spp.	7	10^6	10^8	3.5×10^{11}	7,000
Vibrio cholerae	1	10^6	10^8	5×10^{10}	1,000
Protozoa					
Entamoeba histolytica	30	15×10^4	15×10^6	2.25×10^{11}	4,500
Helminths					
Ascaris lumbricoides	60	10^4[f]	10^6	3×10^{10}	600
Hookworms[g]	40	800[f]	8×10^4	1.6×10^9	32
Schistosoma mansoni	25	40[f]	4×10^3	5×10^7	1
Taenia saginata	1	10^4	10^4	5×10^8	10
Trichuris trichiura	60	2×10^3[f]	2×10^5	6×10^9	120

? Uncertain

Note: This table is hypothetical, and the data are not taken from any actual, single town. For each pathogen, however, the figures are reasonable and congruous with those found in the literature. The concentrations derived for each pathogen in sewage are in line with higher figures in the literature, but it is unlikely that all these infections at such relatively high prevalences would occur in any one community

[a] The prevalences given in this column refer to infection and not to morbidity

[b] It must be recognized that the pathogens listed have different abilities to survive outside the host, and that the concentrations of some of them will rapidly decline after the feces have been passed. The concentrations of pathogens per liter in the sewage of the town were calculated by assuming that 100 liters of sewage are produced daily per capita, and that 90 percent of the pathogens do not enter the sewers or are inactivated in the first few minutes after the excretion

[c] To calculate this figure it is necessary to estimate a mean fecal weight for those people infected. This must necessarily be the roughest of estimates because of the age-specific fecal weights and the age distribution of infected people in the community. It was assumed that people over 15 years old excrete 150 g daily, and that people under 15 excrete, on average, 75 g daily. It was also assumed that two-thirds of all infected people are under 15. This gives a mean fecal weight for infected individuals of 100 g

[d] Includes polio-, echo-, and coxsackieviruses

[e] Includes enterotoxigenic, enteroinvasive, and enteropathogenic *E.coli*

[f] The distribution of egg output from people infected by these helminths is extremely skewed; a few people excrete very high egg concentrations

[g] *Ancylostoma duodenale* and *Necator americanus*

From Feachem *et al.*, 1983; reproduced with permission of the World Bank

Table 2.23 Enteric pathogens in human excreta and wastewater (Gaudy and Gaudy, 1980; Feachem *et al.*, 1983)

Pathogen	Disease	Transmission[a]	Geographical distribution
Bacteria			
Vibrio cholerae	Cholera	Person → person	Asia, Africa, some parts of Europe
Salmonella typhi	Typhoid fever	Person (or animals) → person	Worldwide
Other Salmonellae	Various enteric fevers (often called paratyphoid), gastroenteritis, septicemia (generalized infections in which organisms multiply in the bloodstream)		
Shigella dysenteriae and other species	Bacterial dysentery	Person → person	Worldwide
Pathogenic *Escherichia coli*	Diarrhea	Person → person	Worldwide
Viruses			
Poliovirus	Poliomyelitis	Person → person	Worldwide
Coxsackievirus (some strains)	Various, including severe respiratory disease, fevers, rashes, paralysis, aseptic meningitis, myocarditis	Person → person	Worldwide
Echovirus	Various, similar to coxsackievirus	Person → person	Worldwide
Reovirus	Various, some similar to coxsackievirus (evidence is not definitive except in experimental animals)	Person (or animals) → person	Worldwide
Adenovirus	Respiratory and gastrointestinal infections	Person → person	Worldwide

Hepatitis	Infectious hepatitis (liver malfunction), also may affect kidneys and spleen	Person → person	Worldwide
Protozoa			
Giardia lamblia	Diarrhea (intestinal parasite)	Person (or animals) → person	Worldwide
Entamoeba histolytica	Amoebic dysentery, infections of other organs	Person → person	Worldwide
Balantidium coli	Dysentery, intestinal ulcers	Person (or animals) → person	Worldwide
Helminths			
Ascaris lumbricoides	Ascariasis (roundworm)	Person → soil → person	Worldwide
Ancylostoma duodenale	Hookworm	Person → soil → person	Mainly in warm, wet climates
Schistosoma	Schistosomiasis or bilharziasis	Person → aquatic snail → person	Africa, Asia, South America
Clonorchis sinensis	Liver fluke	Person (or animals) → aquatic snail → fish → person	Southeast Asia
Taenia saginata	Beef tapeworm	Person → cow → person	Worldwide
Taenia solium	Pork tapeworm	Person → pig (or person) → person	Worldwide
Trichuris trichiura	Whipworm	Person → soil → person	Worldwide

[a] Transmission mode is generally through the fecal–oral route, i.e. the excreted disease may be ingested by other persons or the excreted disease go through some development stages in the environment (water, soil, or animals) and infect other persons. Some virus diseases may be transmitted by the oral–oral route, e.g. through throat secretion. For some helminthic infections such as *Ancylostoma* and *Schistosoma*, the infective larvae will penetrate the skin of a person directly

Table 2.24 Epidemiological aspects of some of the zoonoses

Disease	Causative organism	Principal animals affected	Geographical distribution	Probable vector or means of spread
1. *Virus diseases*				
Arthropod-borne infections:				
Japanese B encephalitis	Virus	Birds, horses, cattle, pigs	Eastern Asia and East Indies	Various *Culex* mosquitoes
St. Louis encephalitis	Virus	Birds, fowl	North America	Various mosquitoes
Tick-borne spring–summer group including louping ill, Russian spring–summer encephalitis, Omsk hemorrhagic fever, Kyasanur forest disease	Virus	Birds, mammals, cattle, sheep, goats	Europe and Asia	Wood tick, *Ixodes persulcatus* Pasture tick, *Ixodes ricinus*
South African diseases				
Rift Valley fever	Virus	Sheep, goats, cattle, related wild animals	South and central Africa	Various mosquitoes
Wesselsbron fever	Virus	Sheep	South Africa	Various mosquitoes
Middleburg disease	Virus	Sheep	South Africa	Various mosquitoes
Contagious ecthyma	Virus	Sheep, goats	Europe and North America	Occupational exposure
Cowpox	Virus	Cattle, horses	Worldwide, especially where smallpox exists	Contact exposure
Pseudo-cowpox	Virus	Cattle	Worldwide	Occupational exposure
Psittacosis–ornithosis	Virus	Birds; related virus found in cattle and cats	Worldwide	Contact and occupational exposure
Foot and mouth diseases	Virus	Cattle, pigs, related species	Europe, Asia, Africa, and South America	Contact exposure (humans are quite resistant)
Influenza and parainfluenza including Type A (Shope and Prague) and Sendai (Type D)	Virus	Pigs, rodents	Asia and Europe	Contact exposure

Lymphocytic choriomeningitis	Virus	Rodents, pigs, dogs	Worldwide	Virus contaminates food and environment
Newcastle disease	Virus	Fowl	Worldwide	Occupational exposure
Vesicular stomatitis	Virus	Pigs, cattle, horses	North and South America, Europe	Contact exposure and possibly insect bites
2. *Rickettsial diseases*				
Q fever	*Coxiella burnetii*	Sheep, cattle, goats, fowl, wild birds, mammals	Worldwide	Mainly airborne, although milk may be a vehicle and occasionally ticks
3. *Bacterial diseases*				
Anthrax	*Bacillus anthracis*	Cattle, sheep, goats	Worldwide	Occupational exposure, ingestion of contaminated meat Occasionally airborne or biting insects
Brucellosis	*Brucella abortus* *Br. suis*	Cattle, sheep, goats, pigs	Worldwide	Occupational exposure
	Br.melitensis			Ingestion of contaminated milk products and other foods
Bacterial food poisoning	Various bacteria and their toxins including *Salmonella*, *Staphylococcus*, and *Clostridia*	Cattle, pigs, fowl	Worldwide	Occasionally airborne Ingestion
Colibacillosis	*Escherichia* spp. Arizona group	Cattle, pigs, fowl	Worldwide	Ingestion

(*continues*)

Table 2.24 Epidemiological aspects of some of the zoonoses (*continued*)

Disease	Causative organism	Principal animals affected	Geographical distribution	Probable vector or means of spread
3. *Bacterial diseases (continued)*				
Erysipeloid	*Erysipelohrix rhusiopathiae*	Pigs, fowl, fish	Worldwide	Occupational contact
Leptospirosis	*Leptospira* spp.	Rodents, dogs, pigs, cattle	Worldwide	Occupational contact, immersion exposure and ingestion
Listeriosis	*Listeria monocytogenes*	Rodents, sheep, cattle, pigs	Worldwide	Unknown
Melioidosis	*Pseudomonas pseudomallei*	Rodents, sheep	Asia and North America	Exposure and ingestion
Pasteurellosis	*Pasteurella multocida*	Mammals, birds	Worldwide	Exposure and ingestion
Pseudotuberculosis	*Pasteurella pseudo-tuberculosis*	Guinea pigs, other rodents	Worldwide	Occupational exposure
Salmonellosis	*Salmonella* spp.	Cattle, pigs, fowl	Worldwide	Ingestion, airborne and contact
Staphylococcus	*Staphylococcus* spp.	Cattle, pigs	Worldwide	Ingestion and contact
Streptococcus	*Streptococcus* spp.	Cattle, fowl	Worldwide	Ingestion and contact
Tuberculosis	*Mycobacterium tuberculosis*		Worldwide	Ingestion, inhalation and occupational exposure
	var. *bovis*	Cattle		
	var. *hominis*	Monkeys		
	var. *avium*	Fowl		
Vibriosis	*Vibrio fetus*	Cattle, sheep	Europe, North and South America	Unknown

From Steele (1966); reproduced by permission of the Food and Agriculture Organization of the United Nations

Table 2.25 Partial list of parasitic zoonoses

Disease in humans	Causative organism	Vertebrate animals principally involved
1. *Protozoan infections*		
Amoebiasis	*Entamoeba histolytica*	Nonhuman primates, dogs
Babesiosis	*Babesia divergens*	Cattle
	Babesia microti	Voles, mice
Balantidiasis	*Balantidium coli*	Swine, rats, nonhuman primates
Giardiasis	*Giardia* spp.	Beavers
Sarcosporidiosis intestinal	*Sarcocystis miesheriana* (syn. *S. suihominis;* =*Isospora hominis proparte*)	Pigs
	Sarcocystis hominis (syn. *S. bovihominis;* =*Isospora hominis proparte*)	Cattle
Toxoplasmosis[a]	*Toxoplasma gondii*	Cats, mammals, birds
Trypanosomiasis[a]		
African	*Trypanosoma rhodesiense*	Game animals, cattle
American	*Trypanosoma cruzi*	Dogs, cats, pigs, other small mammals
2. *Helminthic infections*		
Trematode infections:		
Amphistomiasis	*Gastrodiscoides hominis*	Swine
Cercarial dermatitis	Schistosomatids	Birds, mammals
Clonorchiasis[a]	*Clonorchis sinensis*	Dogs, cats, swine, wild mammals, fish
Dicrocoeliasis	*Dicrocoelium dendriticum*	Ruminants
	Dicrocoelium hospes	Ruminants
Fascioliasis	*Fasciola hepatica*	Ruminants
	Fasciola gigantica	Ruminants
Fasciolopsiasis[a]	*Fasciolopsis buski*	Swine, dogs
Heterophyasis and other heterophids	*Heterophyes heterophyes*	Cats, dogs, fish
Metagonimiasis	*Metagonimus yokogawi*	Cats, dogs, fish
Opisthorchiasis[a]	*Opisthorchis felineus*	Cats, dogs
	Opisthorchis viverrini and occasionally other *Opisthorchis* species	Cats, wildlife, fish
Schistosomiasis	*Schistosoma japonicum*[a]	Wild and domestic mammals
	Schistosoma haematobium	Rodents
	Schistosoma mansoni	Baboons, rodents
	Schistosoma mekongi	Dogs, monkeys
	Schistosoma mattheej and occasionally other schistosomes	Cattle, sheep, antelopes

(*continued*)

Table 2.25 Partial list of parasitic zoonoses (*continued*)

Disease in humans	Causative organism	Vertebrate animals principally involved
Cestode infections:		
Coenuriasis	*Taenia multiceps*	Sheep, ruminants, pigs
Diphyllobothriasis[a]	*Diphyllobothrium latum*	Fish, carnivores
Taeniasis[a]	*Taenia saginata*	Cattle
Taeniasis-cysticercosis[a]	*Taenia solium*	Swine
Nematode infections:		
Ancylostomiasis	*Ancylostoma ceylanicum* and other *Ancylostoma* species	Dogs
Ascariasis	*Ascaris suum*	Swine
Capillariasis		
Intestinal	*Capillaria philippinensis*	Fish
Dioctophymiasis	*Dioctophyme renale*	Fish, dogs, mustelids
Gongylonemiasis	*Gongylonema* species	Ruminants, rats
Mammomonogamiasis	*Mammomonogamus laryngeus*	Ruminants
Strongyloidiasis	*Strongyloides stercoralis*	Dogs, monkeys
	Strongyloides fuelleborni	Dogs, monkeys
Thelaziasis	*Thelazia* species	Dogs, ruminants
Trichinellosis[a]	*Trichinella spiralis* and other *Trichinella* species	Swine, rodents, wild carnivores, marine mammals
Trichostrongyliasis	*Trichostrongylus colubriformus* and other *Trichostrongylus* species	Ruminants

From WHO (1979); reproduced by permission of the World Health Organization, Geneva. This list is not comprehensive and is confined to those diseases in which the animal link in the chain of infection to humans is considered to be important, although not always essential.

[a] Diseases or causative organisms of particular importance over large areas

fecal indicator microorganisms are the preferred microbes analyzed. An ideal indicator should be non-pathogenic, easily detected and counted, present where fecal pathogens are present but in higher numbers, and a member of the normal flora of the intestinal tract of healthy persons.

Some of the common fecal indicators for bacteria are fecal coliforms, fecal streptococci, *Clostridium perfringens*, and *Pseudomonas aeruginosa*. Bacteriophages or coliphages have been employed as indicators for enteric viruses. It should be noted that, at present, definite relationships between the die-offs of the indicator microorganisms and those of the pathogenic microorganisms are not well established. For example, in a sludge composting unit the absence of fecal coliforms does not necessarily mean that other enteric bacteria will be dead. Therefore, appropriate indicator microorganisms should be selected for a specific case or a treatment/reuse method being employed. Because they are the most

Table 2.26 Relative health risks from use of untreated excreta and wastewater in agriculture and aquaculture

Class of pathogen	Relative amount of excess frequency of infection or disease
1. Intestinal nematodes: *Ascaris*, *Trichuris*, *Ancylostoma*, *Necator*	High
2. Bacterial infections: bacterial diarrheas (e.g. cholera), typhoid	Lower
3. Viral infections: viral diarrheas, hepatitis A	Least
4. Trematode and cestode infections: schistosomiasis, clonorchiasis, taeniasis	From high to nil, depending upon the particular excreta use practice and local circumstances

From IRCWD (1985); reproduced with permission

hardy and resistant of all helminth pathogens, viable *Ascaris* ova have been recommended to be the best pathogen indicator for non-effluent wastes (such as nightsoil, the contents of pit latrines, and septage (Feachem *et al.*, 1983). Liquid or effluent wastes are normally treated by waste stabilization ponds and/or other conventional waste treatment processes including sedimentation. Under satisfactory operation most helminth ova would be removed by sedimentation, while bacteria and viruses are still carried over with the effluents. In this case the use of indicators for either fecal bacteria or viruses would be appropriate. Methods to enumerate common fecal indicators for bacteria and viruses can be found in *Standard Methods* (APHA, AWWA, WPCF, 1993). The detection of helminth ova can be conducted according to the procedures outlined in Mara and Cairncross (1989).

In the design and operation of a waste treatment/recycling system the engineer/scientist in charge has to ensure that the public health risks are kept to a minimum. Each country, state, or province normally develops its own standards of microbiological quality to be used in the disposal or reuses of wastewater and sludge. A report published by the International Reference Centre for Wastes Disposal (IRCWD, 1985), and concurred by the World Health Organization (Mara and Cairncross, 1989), proposed the relative health risks from the use of human excreta and wastewater, as shown in Table 2.26. The microbiological quality guidelines for wastewater agricultural irrigation are recommended as given in Table 2.27, while those for aquacultural use of wastewater and excreta are given in Table 2.28. Because of their relatively high health risks, almost complete removal of the ova of intestinal nematodes is emphasized. A fecal coliform concentration of 1,000 (geometric mean) per 100 mL or less is recommended for wastewater to be used in unrestricted irrigation and aquaculture. In areas where there is a high prevalence, or an outbreak of a particular disease, attention should be given to the detection of the infectious agent present in the waste to be recycled, or the waste recycling practice is terminated until the disease is under control.

As the organic wastes described in sections 2.1, 2.2 and 2.3 would normally contain low concentrations of heavy metals (and other toxic organic compounds),

Table 2.27 Tentative microbiological quality guidelines for treated wastewater reuse in agricultural irrigation[a]

Reuse process	Intestinal nematodes[b] (arithmetic mean no. of viable eggs per litre)	Fecal coliforms (geometric mean no. per 100 mL)
Restricted irrigation[c]		
Irrigation of trees, industrial crops, fodder crops, fruit trees[d] and pasture[e]	$\leqslant 1$	Not applicable[c]
Unrestricted irrigation		
Irrigation of edible crops, sports fields, and public parks[f]	$\leqslant 1$	$\leqslant$ 1,000[g]

[a] In specific cases local epidemiological, sociocultural, and hydrogeological factors should be taken into account, and these guidelines modified accordingly
[b] *Ascaris*, *Trichuris*, and hookworms
[c] A minimum degree of treatment equivalent to at least a 1-day anaerobic pond followed by a 5-day facultative pond or its equivalent is required in all cases
[d] Irrigation should cease 2 weeks before fruit is picked, and no fruit should be picked off the ground
[e] Irrigation should cease 2 weeks before animals are allowed to graze
[f] Local epidemiological factors may require a more stringent standard for public lawns, especially hotel lawns in tourist areas
[g] When edible crops are always consumed well-cooked, this recommendation may be less stringent
From IRCWD (1985); reproduced with permission

Table 2.28 Tentative microbiological quality criteria for the aquacultural use of wastewater and excreta (Mara and Cairncross, 1989)

Reuse process	Viable trematode eggs[a] (arithmetic mean number per liter or kg)	Fecal coliforms (geometric mean number per 100 mL or per 100 g)[b]
Fish culture	0	$< 10^4$
Aquatic macrophyte culture	0	$< 10^4$

[a] *Clonorchis*, *Fasciolopsis*, and *Schistosoma*. Consideration should be given to this guideline only in endemic area
[b] This guideline assumes that there is a one $\log_{10}$ unit reduction in fecal coliforms occurring in the pond, so that in-pond concentrations are $< 1{,}000$ per 100 mL. If consideration of pond temperature and retention time indicates that a higher reduction can be achieved, the guideline may be relaxed accordingly

the health risks with respect to heavy metal contamination resulting from the practice of organic waste recycling should not be of a great concern. However, there is a possibility of an accumulation of these toxic compounds (even though at trace concentrations) in the biological food chain and in soils receiving wastewater and sludge for a long period of time. This long-term accumulation of toxic compounds can lead to a problem of bio-magnification where organisms in the upper part of the food chain (e.g. animals or human beings) that feed on the contaminated crops could have toxic compounds concentrations several times greater than those present in the soil or crops. Information about standards or guidelines of toxic compounds concentrations in wastewater and sludge to be reused in agriculture or aquaculture are available for most developed countries in North America and Europe; this information will be referred to in Chapters 8 and 9 (Land treatment of wastewater and sludge, respectively). A description on toxicity of chemicals to crops, animals and humans is given in section 9.4.

2.6 WASTE MINIMIZATION AND CLEAN TECHNOLOGY

As one of the objectives of waste recycling is to minimize the pollution effects caused by organic wastes, the practice of waste minimization and clean technology by industries or any waste generators is to be encouraged.

Waste minimization is defined as the reduction, to the extent feasible, of pollutant waste that is generated or subsequently treated, stored or disposed of. It is realized through the implementation of clean technologies. Waste minimization includes any source reduction or recycling activities undertaken by a waste generator that results in the reduction of total volume or quantity of waste. Waste minimization has the following advantages (U.S. EPA, 1988):

1. save money by reducing waste treatment and disposal costs, raw material purchases, and other operating costs;
2. meet state and national waste minimization policy goals;
3. reduce potential environmental liabilities;
4. protect public health and worker health and safety;
5. enhance public image; and
6. protect the environment.

Clean technology, also known as low and non-waste technology or low-emission technology, refers to any technical measures taken in the various industries to reduce, or even eliminate at source the production of any nuisance, pollution or waste, and to help save raw materials, natural resources (Anonymous, 1989). It is an outcome of:

1. increasingly stringent environmental regulations;
2. high investment and operational costs of traditional end-of-pipe waste treatment facilities; and
3. increasing costs and/or scarcity of water, energy and raw materials.

It is generally known that source reduction and recycling are the most viable pollution prevention techniques, preceding waste treatment and disposal. Reducing the amount of waste generated at the source will often be more cost-effective than recycling, as waste as a raw material or product will require time and money to recover. Therefore, recycling should be considered only when all source reduction options have been investigated and implemented. Some descriptions about source reduction and waste recycling are given below.

2.6.1 Source reduction

Source reduction refers to any practice which reduces the amount of any pollutant or contaminant entering any waste stream or otherwise released into the environment prior to recycling, treatment or disposal (Holmes *et al.*, 1993). By avoiding the generation of wastes, source reduction eliminates the problems associated with the handling and disposal of wastes. A wide variety of facilities can adopt procedures to minimize the quantity of waste generated. Many source reduction options involve a change in procedural or organizational activities, rather than a change in technology. For this reason, these options tend to affect the managerial aspect of production and usually do not demand large capital and time investments. Figure 2.11 shows the various options of source reduction which can be categorized as good operating practices, changes in technology, input materials, or products.

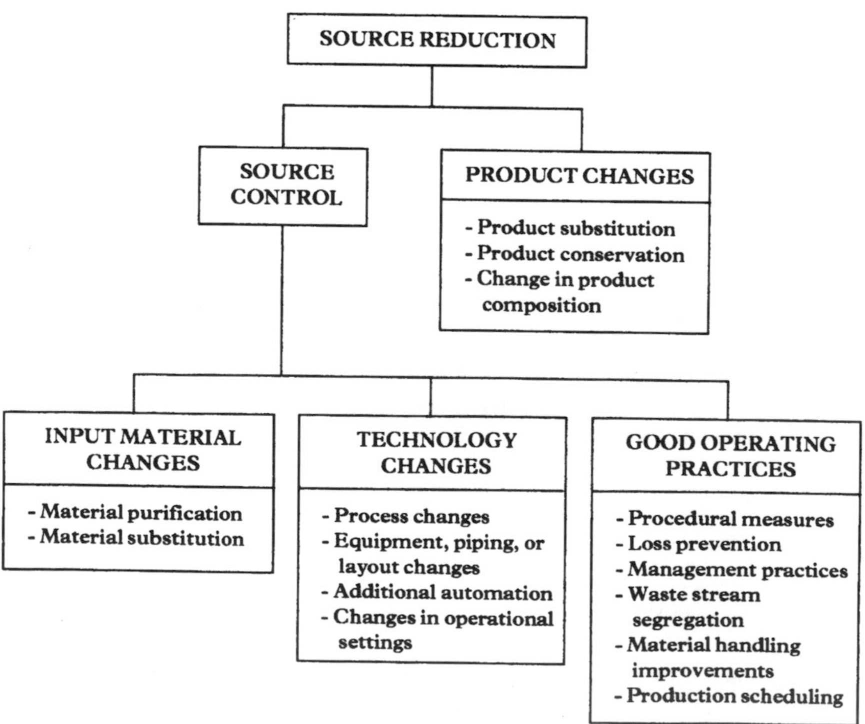

Figure 2.11 Source reduction (U.S. EPA, 1988)

GOOD OPERATING PRACTICES

Good operating practices are procedural, administrative, or institutional measures that an industry can use to minimize waste. Good operating practices apply to the human aspect of manufacturing operations which can often be implemented with little cost and, therefore, have a high return on investment. Good operating practices include the following (U.S. EPA, 1988):

1. waste minimization programs;
2. management and personnel practices, including employee training, incentives and bonuses, and other programs that encourage employees to conscientiously strive to reduce waste;
3. material handling and inventory practices, including programs to reduce loss of input materials due to mishandling, expired shelf life of time-sensitive materials
4. loss prevention by avoiding leaks from equipment and spills;
5. waste segregation;
6. cost accounting practices, including programs to allocate waste treatment and disposal costs directly to the departments or groups that generate waste, rather than charging these costs to general factory overhead accounts; and
7. production scheduling.

TECHNOLOGY CHANGES

Technology changes are oriented toward process and equipment modifications to reduce waste, primarily in a production setting. Technology changes include the following:

1. changes in the production process;
2. equipment, layout, or piping changes;
3. use of automation; and
4. changes in process operating conditions, such as flow rate, temperatures, pressures, and residence times.

Example: Technology changes The U.S.DA-Magnuson Infrared Antipollution Peeling Process uses infrared energy at 900 °C (1650 °F) to condition the surfaces of potatoes treated with strong sodium hydroxide solutions. The peel can then be removed mechanically by soft rubber scrubbing rolls rather than by water as is done in conventional caustic peeling. A final spray rinse using low volumes of water removes residual peel fragments and excess sodium hydroxide. Direct comparison of this process with conventional peeling has demonstrated that the strength of the waste discharged has been reduced by 40 percent (Jones, 1973).

INPUT MATERIAL CHANGES

Input material changes accomplish waste minimization by reducing or eliminating the pollutant materials that enter the production process. Also, changes in input materials can be made to avoid the generation of pollutant wastes within the production processes. Input material changes include: (i) material purification; and (ii) material substitution.

PRODUCT CHANGES

Product changes are performed by the manufacturer of a product with the intent of reducing waste resulting from a product's use. Product changes include:

1. product substitution;
2. product conservation; and
3. changes in product composition.

2.6.2 Waste recycling

A material is 'recycled' if it is used, reused, or reclaimed. Recycling through use and/or reuse involves returning waste material either to the original process as a substitute for an input material, or to another process as an input material. Recycling through reclamation is the processing of a waste for recovery of a valuable material or for regeneration. Recycling techniques may be performed on-site or at an off-site facility designed to recycle the waste. Recycling of wastes can provide a very cost-effective waste management alternative. This option can help eliminate waste disposal costs, reduce raw material costs, and provide income from salable waste. The technologies described in Chapters 3–9 are mainly for waste recycling purposes.

REFERENCES

Anonymous (1989). Clean technology policy of the European Economic Community. *UNEP: Industry and Environment*, **12**, 11–13.

AOAC (1990). *Official Methods of Analysis of the Association of Official Analytical Chemists*, 15th edition. Association of Official Analytical Chemists, Arlington, VA.

APHA, AWWA, WPCF (1993). *Standard Methods for the Examination of Water and Wastewater*, 18th edition. American Public Health Association, Washington, D.C.

Arifin, B. (1982). Design criteria development of septage recycling ponds. Master's thesis no. EV-82-24. Asian Institute of Technology, Bangkok.

Brandes, M. (1978). Accumulation rate and characteristics of septic tank sludge and septage. *J. Water Pollut. Control Fed.*, **50**, 936–43.

Brolls, E. K. and Broughton, M. (1981). The treatment of effluents arising from the animal by-products industry. In *Food Industry Wastes: Disposal and Recovery* (eds. A. Herzka and R. G. Booth), pp. 184–203. Applied Science Publishers, London.

Cook, D. B. and Kalbermatten, J. M. (1982). Prospects for resource recovery from urban solid wastes in developing countries. In *Proceedings Regional Seminar on Solid Waste Management* (ed. G. Tharun), pp. 267–73. Asian Institute of Technology, Bangkok.

Department of Environment (1971). *Refuse Disposal*. Report of the working party on refuse disposal. HMSO, London.

Feachem, R. G., Bradley, D. J., Garelick, H., and Mara, D. D. (1983). *Sanitation and Disease—Health Aspects of Excreta and Wastewater Management*. Wiley, Chichester.

Gaudy, A. and Gaudy, E. (1980). *Microbiology for Environmental Scientists and Engineers*. McGraw-Hill, New York.

Gotaas, H. B. (1956). *Composting: Sanitary Disposal and Reclamation of Organic Wastes*. WHO Monograph Series No. 31. World Health Organization, Geneva.

Griffiths, A. (1981). Meat and animal by-products: processes and effluents. In *Food Industry Wastes: Disposal and Recovery* (eds. A. Herzka and R. G. Booth), pp. 171–83. Applied Science Publishers, London.

Holmes, G., Singh, B. R., and Theodore, L. (1993). *Handbook of Environmental Management and Technology*. Wiley, New York.

IRCWD (1985). The Engelberg report: health aspects of wastewater and excreta use in agriculture and aquaculture. International Reference Centre for Waste Disposal, *IRCWD News*, **23**, 11–18.

Jesuitas, E. P. (1966). An investigation of tapioca wastes. AIT thesis, No. 136. Asian Institute of Technology, Bangkok.

Jones, H. R. (1973). *Waste Disposal Control in the Fruit and Vegetable Industry*. Noyes Data Corporation, Park Ridge, New Jersey.

Liu, C. L. (1986). Anaerobic lagoon treatment of septage. Master's thesis No. EV-86-15. Asian Institute of Technology, Bangkok.

Lohani, B. N. and Boonthanon, S. (1985). Comparative study of the properties of Bangkok solid waste and compost. Research report No. 183. Asian Institute of Technology, Bangkok.

Lohani, B. N. and Rajagopal, K. (1981). Human and animal waste management strategies in developing countries. *Environmental Sanitation Review*, No. 4/5. Environmental Sanitation Information Center, Asian Institute of Technology, Bangkok.

Lovan, C. R. and Foree, E. G. (1971). The anaerobic filter for the treatment of brewery press liquor waste. *Proc. Industrial Waste Conference, Purdue University*, pp. 1074–86. W. Lafayette, Indiana, U.S.A.

Magara, Y., Sugito, D., and Yagome, K. (1980). Design and performance of night soil treatment plants. *J. Water Pollut. Control Fed.*, **52**, 914–22.

Mara, D. D. and Cairncross, A. M. (1989). *Guidelines for the Safe Use of Wastewater and Excreta in Agriculture and Aquaculture*. World Health Organization, Geneva.

Meade, G. P. and Chen, J. C. P. (1977). *Cane Sugar Handbook*, 10th edition. Wiley-Interscience, New York.

Metcalf and Eddy, Inc. (1991). *Wastewater Engineering: Treatment, Disposal, Reuse*, 3rd edition. McGraw-Hill, New York.

Muttamara, S., Vigneswaran, S., and Shin, H. S. (1987). *Palm Oil Mill Effluent Management in Thailand: A Preliminary Study*. Final report submitted to UNDP/FAO. Asian Institute of Technology, Bangkok.

Nemerow, N. L. (1971). *Liquid Wastes of Industry—Theories, Practices and Treatment*. Addison-Wesley, Reading, Massachusetts.

Oswald, W. J. (1988). Micro-algae and waste-water treatment. In *Micro-algal Biotechnology* (eds. M. A. Borowitzka and J. Borowitzka), pp. 305–28. Cambridge University Press, Cambridge, U.K.

Oswald, W. J. and Gotaas, H. B. (1955). Photosynthesis in sewage treatment. *Trans. Am. Soc. Civ. Eng.*, **122**, 73–97.

Oswald, W. J., Tsugita, R. A., Golueke, C. G., and Cooper, R. C. (1973). *Anaerobic–Aerobic Ponds for Beet Sugar Waste Treatment*. EPA-R2-73-025. Environmental Protection Agency, Washington, D.C.

Pickford, J. (1977). Solid waste in hot climates. In *Water, Wastes and Health in Hot*

Climates (eds. R. Feachem, M. McGarry and D. Mara), pp. 320–44. Wiley, Chichester.
Polprasert, C. and Rajput, V. S. (1982). Septic tanks and septic systems. *Environmental Sanitation Review*, No. 7/8. Environmental Sanitation Information Center, Asian Institute of Technology, Bangkok.
Pradt, L. A. (1971). Some recent developments in night soil treatment. *Water Res.*, **5**, 507–21.
Qureshi, M. S. (1977). A study of cane sugar waste treatment. AIT thesis no. 1092. Asian Institute of Technology, Bangkok.
Sawyer, C. N., McCarty, P. L., and Parkin, G. F. (1994). *Chemistry for Environmental Engineering*, 4th edition. McGraw-Hill, New York.
Steele, J. H. (1966). Animal disease and human health. *Freedom from Hunger Campaign Basic Study No. 3*. Food and Agriculture Organization of the United Nations, Rome.
Taiganides, E. P. (1978). Energy and useful by-product recovery from animal wastes. In *Water Pollution Control in Developing Countries* (eds. E. A. R. Ouano, B. N. Lohani and N. C. Thanh), pp. 315–23. Asian Institute of Technology, Bangkok.
Tantideeravit, T. (1975). Evaluation and treatment of brewery wastewaters. AIT thesis no. 835. Asian Institute of Technology, Bangkok.
Tchobanoglous, G., Theisen, H., and Eliassen, R. (1993). *Integrated Solid Wastes Management—Principles and Management Issues*. McGraw-Hill, New York.
Thanh, N. C. (1984). Pollution control and management of agro-industrial wastes. *Proceedings: UNEP/ESCAP/FAO Workshop on Agricultural and Agro-Industrial Residue Utilization in the Asian and Pacific Region*, pp. 93–164. Pattaya, Thailand.
Thanh, N. C., Muttamara, S., and Lohani, B. N. (1980). Palm oil wastewater treatment study in Malaysia and Thailand. IDRC Final Report, No. 114. Asian Institute of Technology, Bangkok.
Thongkasame, C. (1968). Anaerobic treatment of tapioca starch waste. AIT thesis no. 288. Asian Institute of Technology, Bangkok.
Unkulvasapaul, Y. (1975). Evaluation and treatment of wastes from the tapioca starch industry. AIT thesis no. 836. Asian Institute of Technology, Bangkok.
U.S. EPA (1971). *Liquid Wastes from Canning and Freezing Fruits and Vegetables*. Program No. 12060 EDK. United States Environmental Protection Agency, Washington, D.C.
U.S. EPA (1984). *Handbook — Septage Treatment and Disposal*. EPA-625/6-84-009. United States Environmental Protection Agency, Cincinnati, Ohio.
U.S. EPA (1988). *Waste Management Opportunity Assessment Manual*, EPA/625-7-88/033. United States Environmental Protection Agency, Cincinnati, OH.
U.S. PHS (1943). *Industrial Waste Guide to the Meat Industry*. American Public Health Service, Washington, D.C.
Visittavanich, D. (1987). The physicochemical treatment of poultry processing wastewater with regionally available chemicals. AIT thesis no. EV-87-21. Asian Institute of Technology, Bangkok.
White, A. U. (1977). Patterns of domestic water use in low-income communities. In *Water, Wastes and Health in Hot Climates* (eds. R. G. Feachem, M. G. McGarry, and D. D. Mara), pp. 96–112. Wiley, London.
Wilson, D. G. (ed.) (1977). *Handbook of Solid Waste Management*. Van Nostrand Reinhold, New York.
World Health Organization (1979). *Parasitic Zoonoses*. Report of a WHO Expert Committee with the participation of FAO. Technical Report Series 637. World Health Organization, Geneva.

EXERCISES

2.1 You are to visit an agro-based industry or an animal farm to determine the waste quantity and characteristics. Compare the results obtained with the data presented in Chapter 2.

2.2 From the data given in Table 2.1, derive the following information for your hometown:

a Daily generation of human excreta both in wet weight and in volume, assuming a density of 1.05 kg/L.

b Daily production of total solids from the excreta.

c Daily production of total N from the excreta.

d Daily production of BOD_5 from the excreta.

e C/N ratio of the excreta.

2.3 A chicken farm has 5,000 laying hens in stock, each with an average live weight (LW) of 1.5 kg. Use the data in Table 2.6 to estimate the following:

a Daily production of chicken wastes in wet weight.

b Daily production of total solids from the chicken farm.

c Daily production of volatile solids.

d Daily production of total N.

2.4 Based on the microbiological guidelines given in Tables 2.27 and 2.28, you are to survey a waste recycling project (either in agriculture or aquaculture) and collect the wastewater for analysis of fecal coliform and nematode egg contents.

From the results obtained, discuss the possible health impacts of this waste recycling practice.

2.5 Discuss the extent of waste minimization that is being implemented in your country at national and local (city) levels.

3

Composting

Haug (1980) defined composting as the biological decomposition and stabilization of organic substrates under conditions which allow development of thermophilic temperatures as a result of biologically produced heat, with a final product sufficiently stable for storage and application to land without adverse environmental effects. Another definition, recently agreed in Europe, refers composting to a controlled aerobic process carried out by successive microbial populations combining both mesophilic and thermophilic activities, leading to the production of carbon dioxide (CO_2), water, minerals and stabilized organic matter (Pereira-Neta, 1987). Generally, composting is applied to solid and semi-solid organic wastes, such as nightsoil, sludge, animal manures, agricultural residues, and municipal refuse.

Aerobic composting is the decomposition of organic wastes in the presence of oxygen (air); the end-products of biological metabolism are CO_2, NH_3, water, and heat (similar to eq. 2.1). Anaerobic composting is the decomposition of organic wastes in the absence of oxygen; the end-products are methane (CH_4), CO_2, NH_3, trace amounts of other gases, and other low-molecular-weight organic acids (eq. 2.3). NH_3 is further oxidized to become nitrate (NO_3^-) by the nitrifying bacteria during the maturation or curing phase. Because it can release more heat energy, resulting in a rapid decomposition rate, aerobic composting has been the preferred process for stabilizing large quantities of organic wastes. Anaerobic composting is a slow process and can produce obnoxious odors originating from the intermediate metabolites such as mercaptans and sulfides. Depending on the methods of operation, anaerobic composting can produce temperatures near to or at thermophilic levels. Because of its simplicity, anaerobic composting has found some applications in many rural areas of developing countries in the stabilization of wastes generated from households and farms.

It should be noted that, in contrast to wastewater treatment, the terms 'aerobic' and 'anaerobic' for composting have relative meanings. They simply indicate what conditions are predominant in the process. As the compost materials are heterogeneous and bulky in character, in a compost heap there always exists 'anaerobic' composting, which is little in 'aerobic' composting but abundant in 'anaerobic' composting; and vice-versa. Some composting processes, such as composting pits being practiced in China, are aerobic at first and become anaerobic during the later stages of the composting period.

Using technology as the key, composting can be classified into 'mechanical' and 'non-mechanical' processes, or 'on-site' and 'off-site' processes. Composting can also be divided with respect to the modes of operation, i.e. batch operation and continuous or semi-continuous operation. These technologies and operational classifications of composting processes will be described in section 3.6. When temperature is the basis, composting can be divided into 'mesophilic' composting (when temperatures in the compost heap are between 25 and 40 °C) and 'thermophilic' composting (when temperatures are between 50 and 65 °C).

3.1 OBJECTIVES, BENEFITS AND LIMITATIONS OF COMPOSTING

The main purposes and advantages of composting are classified as follows:

1. *Waste stabilization.* The biological reactions occurring during composting will convert the putrescible forms of organic wastes into stable, mainly inorganic forms which would cause little further pollution effects if discharged on to land or into a water course.
2. *Pathogen inactivation.* The waste heat produced biologically during composting can reach a temperature of about 60 °C, which is sufficient to inactivate most pathogenic bacteria, viruses, and helminthic ova, provided that this temperature is maintained for at least 1 day. Therefore the composted products can be safely disposed of on land, or used as fertilizers and soil conditioners. Figure 3.1 shows the influence of time and temperature on die-off of selected pathogens in nightsoil and sludge. The higher the temperature the shorter the time required for pathogen die-off.
3. *Nutrient and land reclamation.* The nutrients (N, P, K) present in the wastes are usually in complex organic forms, difficult to be taken up by the crops. After composting, these nutrients would be in inorganic forms such as NO_3^- and PO_4^- suitable for crop uptake. The application of composted products as fertilizer to land reduces loss of nutrients through leaching because the inorganic nutrients are mainly in the insoluble forms which are less likely to leach than the soluble forms of the uncomposted wastes. In addition, the soil tilth is improved, thereby permitting better root growth and consequent ready accessibility to the nutrients (Golueke, 1982). The application of compost to unproductive soils would eventually improve the soil quality and the otherwise useless lands can be reclaimed.
4. *Sludge drying.* Human excreta, animal manure and sludge contain about 80–95 percent water, which makes the costs of sludge collection, transportation, and disposal expensive. Sludge drying through composting is an alternative in which the waste heat produced biologically will evaporate the water contained in the sludge.

A major drawback of composting concerns the unreliability of the process in providing the expected nutrient concentrations and pathogen die-offs. Because

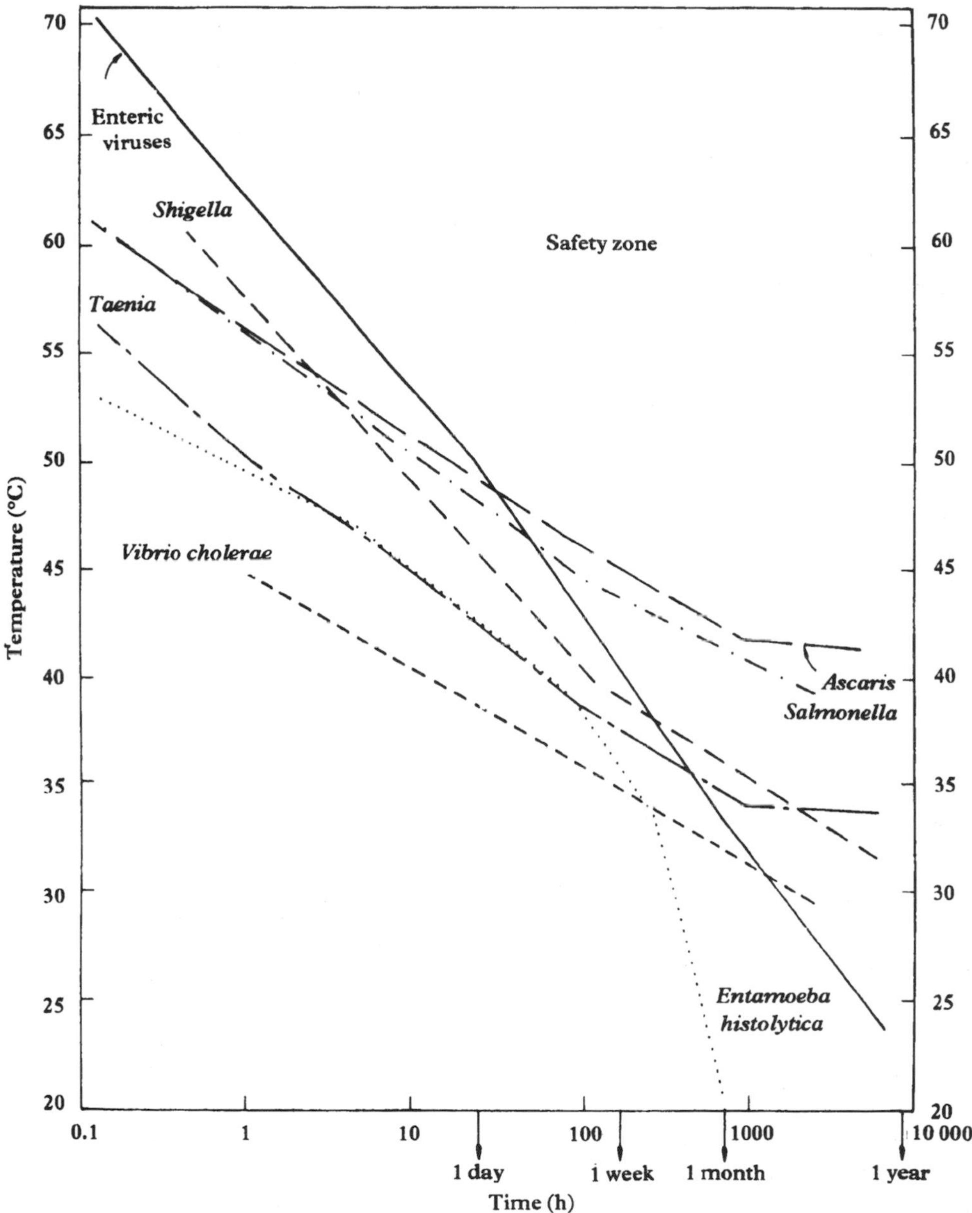

Figure 3.1 Influence of time and temperature on selected pathogens in nightsoil and sludge (Feachem *et al.*, 1983; reproduced by permission of the World Bank). *Note*: The time represents conservative upper boundaries for pathogen death—that is, estimates of the time–temperature combinations required for pathogen inactivation. A treatment process with time–temperature effect falling within the 'safety zone' should be lethal to all excreted pathogens (with the possible exception of hepatitis A virus—not included in the enteric viruses in the figure—at short retention time). Indicated time–temperature requirements are at least: 1 h at $\geqslant$ 62 °C, 1 day at $\geqslant$ 50 °C and 1 week at $\geqslant$ 46 °C

the characteristics of organic wastes can vary greatly from batch to batch, with time, climates, and modes of operation, the properties of the composted products would also vary accordingly. The heterogeneous nature of materials in the compost piles usually causes uneven temperature distribution (except in well-operated compost reactors—see section 3.6), resulting in incomplete inactivation of pathogens present in the composted materials. Other limitations of composting relate to socio-economic factors. For example, the handling of nightsoil during composting can be unappealing, unaesthetic, and obnoxious in odor. Most farmers still prefer to use chemical fertilizers because they are relatively inexpensive and, in the short term, produce reliable results on crop yields.

3.2 BIOCHEMICAL REACTIONS

Organic wastes suitable for composting vary from the highly heterogeneous materials present in municipal refuse and sludge to virtually homogeneous wastes from food processing plants. The courses of biochemically breaking down these wastes are very complex, encountering several intermediates and pathways. For example, the breaking down of proteins includes the following pathways:

Proteins → peptides → amino acids → ammonium compounds
→ bacterial protoplasm and atmospheric N or ammonia.

For carbohydrates:

Carbohydrates → simple sugars → organic acids → CO_2
and bacterial protoplasm.

The precise details of the biochemical changes taking place during the complex processes of composting are still lacking. The phases which can be distinguished in the composting processes according to temperature patterns are (Figure 3.2):

1. Latent phase, which corresponds to the time necessary for the microorganisms to acclimatize and colonize in the new environment in the compost heap.
2. Growth phase, which is characterized by the rise of biologically produced temperature to mesophilic level.
3. Thermophilic phase, in which the temperature rises to the highest level. This is the phase where waste stabilization and pathogen destruction are most effective. This biochemical reaction can be represented by eqs 2.1 and 2.3 for the cases of aerobic and anaerobic composting, respectively.
4. Maturation phase, where the temperature decreases to mesophilic and, subsequently, ambient levels. A secondary fermentation takes place which is slow and favors humification; that is, the transformation of some complex organics to humic colloids closely associated with minerals (iron, calcium, N, etc.) and finally to humus. Nitrification reactions, in which ammonia (a

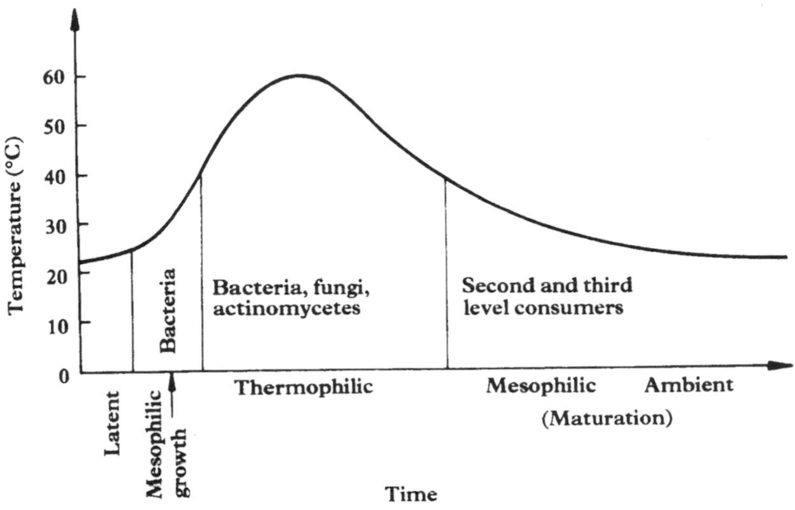

Figure 3.2 Patterns of temperature and microbial growth in compost piles

by-product from waste stabilization as shown in eqs 2.1 and 2.3) is biologically oxidized to become nitrite (NO_2^-) and finally nitrate (NO_3^-), also take place (Metcalf and Eddy, 1991):

$$NH_4^+ + \tfrac{3}{2}O_2 \xrightarrow[\text{bacteria}]{\textit{Nitrosomonas}} NO_2^- + 2H^+ + H_2O \tag{3.1}$$

$$NO_2^- + \tfrac{1}{2}O_2 \xrightarrow[\text{bacteria}]{\textit{Nitrobactor}} NO_3^- \tag{3.2}$$

Combining eqs 3.1 and 3.2, the overall oxidation reaction for nitrification is:

$$NH_4^+ + 2O_2 \rightarrow NO_3^- + 2H^+ + H_2O \tag{3.3}$$

Since NH_4^+ is also synthesized into cell tissue, a representative reaction for cell synthesis is:

$$NH_4^+ + 4CO_2 + HCO_3^- + H_2O \rightarrow C_5H_7O_2N + 5O_2 \tag{3.4}$$

The overall nitrification reaction, combining eqs 3.3 and 3.4 is:

$$22NH_4^+ + 37O_2 + 4CO_2 + HCO_3^- \rightarrow 21NO_3^- + C_5H_7O_2N + 20H_2O + 42H^+ \tag{3.5}$$

The nitrifying bacteria responsible for nitrification consist of two main groups, i.e. the *Nitrosomonas*, which converts NH_4^+ to NO_2^-, and the *Nitrobactor*, which converts NO_2^- to NO_3^-. The nitrifying bacteria have a relatively slow growth rate and are inactive at temperatures greater than 40 °C (Alexander, 1961), hence they will become active normally after the reactions of organic waste decomposition (the growth phase and thermophilic phase) are complete. As NO_3^- is the form of N which is readily available for crop uptake, the maturation phase thus becomes an essential step in composting to produce a good quality compost for use as fertilizer and soil conditioner.

At this stage the organisms classified as second- and third-level consumers (Figure 3.3), such as protozoa and beetles, will grow and will feed on the first-level consumers (e.g. bacteria, fungi, actinomycetes).

The composted products after maturation can be used as fertilizers or soil conditioners to crops. In this way the nutrients returned as compost are in the form of microbial protoplasm and/or organic compounds that break down slowly. Other nutrients present in the compost, such as nitrates, are readily available to crops.

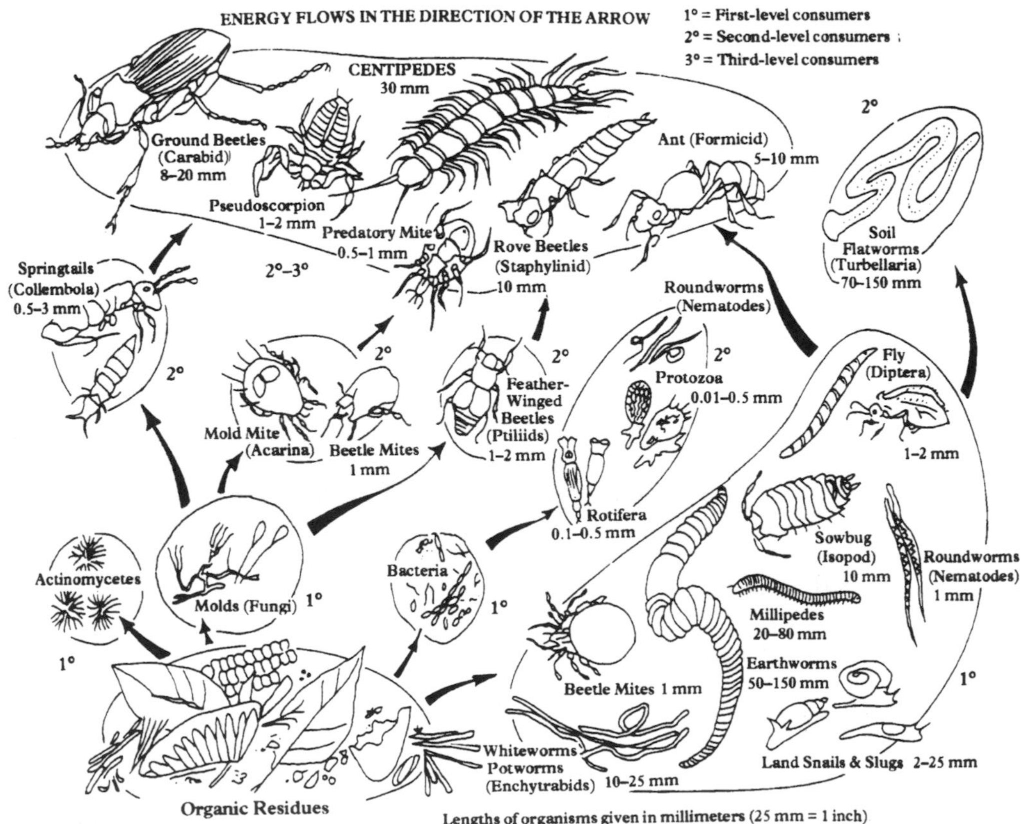

Figure 3.3 Food web of the compost pile (Dindal, 1978; reproduced by permission of the JG Press)

In the aerobic composting systems the degradation of organic matter depends on the presence of oxygen. Oxygen serves two functions in the metabolic reactions: as the terminal electron-acceptor in aerobic respiration, and as a substrate required for the operation of the class of enzymes called oxygenase (Finstein *et al.*, 1980).

Organic matter generally degrades more rapidly and more completely if oxygen is plentiful. This can be explained by the presence of the large amount of free energy produced for microbial growth where the prominent electron-acceptor is oxygen. Oxygen can be incorporated into molecules devoid of this element through the action of the widely distributed, non-substrate-specific and inducible enzymes, 'oxygenases'. This is often the first necessary step in metabolic sequences leading to the degradation of molecules resistant to biological attack. Classes of organic microcontaminants acted upon by oxygenases include saturated alkanes, aromatic hydrocarbons, and halogenated hydrocarbons; anaerobic environments lack this mechanism (Finstein *et al.*, 1980). In anaerobic composting the free energy (heat) produced is much less than that of aerobic composting, and therefore a longer time is required for organic decomposition and pathogen inactivation.

The kinetics of composting systems is a subject of vital interest to the design engineer, who must determine the type and size of composting plants and the detention time required to achieve a certain degree of organic stabilization and pathogen inactivation. Haug (1980) conceptually described the various rate-controlling phenomena occurring during aerobic composting; these include:

1. release of extracellular hydrolytic enzymes by the cell and transport of enzymes to the surface of the substrate;
2. hydrolysis of substrate molecules into lower molecular weight, soluble fractions;
3. diffusion transport of solubilized substrate molecules to the cell;
4. diffusion transport of substrate into the microbial cell, floc, or mycelia;
5. bulk transport of oxygen (usually in air) through the voids between particles;
6. transport of oxygen across the gas–liquid interface and the unmixed regions which lie on either side of such an interface;
7. diffusion transport of oxygen through the liquid region;
8. diffusion transport of oxygen into the microbial cell, floc, or mycelia; and
9. aerobic oxidation of the substrate by biochemical reaction within the organism.

In practice the design of a composting plant is based on such criteria as: type and quantity of materials to be composted; time required for waste stabilization and pathogen inactivation; degree of compost maturity; type of composting process to be employed; and area and location of the composting plant. Data from laboratory and pilot-scale investigation, together with knowledge of past experiences, greatly help in designing an efficient composting plant.

3.3 BIOLOGICAL SUCCESSION

Composting is a biological process in which organic wastes are converted into a stabilized humus by the activity of complex organisms naturally present in the wastes. These include microorganisms such as bacteria, fungi, and protozoa, and may also involve invertebrates such as nematodes, earthworms, mites, and various other organisms (Figure 3.3).

The organic wastes are initially decomposed by the first-level consumers such as bacteria, fungi (molds), and actinomycetes. Waste stabilization is accomplished mainly through the bacterial reactions. Mesophilic bacteria are the first to appear. Thereafter, as the temperature rises, thermophilic bacteria, which inhabit all parts of the compost heap, appear. Thermophilic fungi usually grow after 5–10 days of composting. If the temperature becomes too high, i.e. greater than 65–70 °C, fungi, actinomycetes, and most bacteria become inactive, and only spore-forming bacteria can develop. In the final stages, as the temperature declines, members of the actinomycetes become the dominant group which may give the heap surface a white or gray appearance.

Thermophilic bacteria, mostly *Bacillus* spp. (Strom, 1985), play a major role in the decomposition of proteins and other carbohydrate compounds. In spite of being confined primarily to the outer layers of the compost piles and becoming active only during the latter part of the composting period, fungi and actinomycetes play an important role in decomposing cellulose, lignins, and other more resistant materials, which are attacked after the readily decomposed materials have been utilized. The common species of actinomycetes are reported to be *Streptomyces* and *Thermoactinomyces*, while *Aspergillus* is the common fungus species (Strom, 1985).

After these stages the first-level consumers become the food of second-level consumers, such as mites, beetles, nematodes, protozoa, and rotifers. Third-level consumers, such as centipedes, rove beetles and ants, prey on the second-level consumers. A schematic diagram showing the growth patterns of different-level consumers, with composting time and temperature, is presented in Figure 3.2.

In order for the composting process to function effectively a suitable number of organisms must be present which are capable of attacking the types of wastes to be stabilized. These organisms are naturally present in the wastes such as nightsoil, animal manure, and wastewater sludges; hence compost seeding is usually not necessary. Although packages of compost inoculum are commercially available, controlled scientific tests showed no increased benefits over natural sources of organisms (Dindal, 1978). However, some types of agricultural waste, such as rice straw, leaves, and aquatic weeds, which do not readily have these organisms, may require the seeding of nightsoil or sludge at the starting period.

3.4 ENVIRONMENTAL REQUIREMENTS

The effectiveness of a composting process is dependent upon the groups of organisms that inhabit and stabilize the organic wastes. Any process failure may be due to some unbalanced chemical and physical conditions in the compost piles which are favorable for microbial growth. The major environmental parameters needed to be properly controlled in the operation of composting processes are as follows:

3.4.1 Nutrient balance

The most important nutrient parameter is the carbon/nitrogen or C/N ratio. Phosphorus (P) is next in importance, and sulfur (S), calcium (Ca), and trace quantities of several other elements all play a part in cell metabolism.

Alexander (1961) reported that between 20 and 40 percent of C substrate in the organic wastes (compost feed) is eventually assimilated into new microbial cells in composting, the remainder being converted to CO_2 in the energy-producing processes. However, these cells contain approximately 50 percent C and 5 percent N on a dry-weight basis. Accordingly, the requirement of N in the composting feed is 2–4 percent of initial C, i.e. a C/N ratio of about 25 : 1.

The C/N ratios of various wastes are shown in Table 3.1. Except for horse manure and potato tops, the C/N ratios of other wastes should be adjusted to the optimum value of 25 : 1 prior to being composted. In practice, accurate calculations and adjustments of optimum C/N ratios are made difficult by the following factors (Gray *et al.*, 1971):

Table 3.1 C/N ratio of various wastes

Material	Nitrogen (percentage of dry weight)	C/N ratio
Nightsoil	5.5–6.5	6–10
Urine	15–18	0.8
Blood	10–14	3.0
Animal tankage	—	4.1
Cow manure	1.7	18
Poultry manure	6.3	15
Sheep manure	3.8	—
Pig manure	3.8	—
Horse manure	2.3	25
Raw sewage sludge	4–7	11
Digested sewage sludge	2–4	—
Activated sludge	5	6
Grass clippings	3–6	12–15
Nonlegume vegetable wastes	2.5–4	11–12
Mixed grasses	2.4	19
Potato tops	1.5	25
Straw, wheat	0.3–0.5	128–150
Straw, oats	1.1	48
Sawdust	0.1	200–500

From Golueke (1972); reproduced by permission of the JG Press

1. Parts of the C substrate, such as cellulose and lignin, are highly resistant to biological breakdown, and are only decomposed over a long period of time.
2. Some of the nutrients, such as keratin-type proteins, are not easily available to the microbes during much of the decomposing process.
3. Some N fixation can take place through the bacteria *Azotobacter* spp., especially in the presence of adequate phosphatic material.
4. Analysis of C concentration may be difficult and inaccurate.

The following relationship, providing an accuracy of within 2–10 percent is proposed (Gotaas, 1956):

$$\text{Percentage C} = \frac{100 - \text{percentage ash}}{1.8} \tag{3.6}$$

Percentage ash in eq. 3.6 refers to materials remaining after being burnt at 550 °C for 1 h. For some types of wastes that contain a large portion of plastics (which disappear at 550 °C), the use of eq. 3.6 will give a high value of percentage C, which is largely non-biodegradable.

Based on the above reasons, the initial C/N ratios of between 20 and 40 should serve adequately as an optimum range for composting.

If the initial C/N ratio of materials to be composted is greater than the optimum value (such as sawdust and wheat straw), the microorganisms will have growth limitations due to the lack of N. They will have to go through many life cycles, oxidizing the excessive C until a final C/N ratio of about 10 is reached in the composted products. Therefore an extra composting time is needed, and a smaller quantity of final humus is obtained. The relationship between aerobic composting time and C/N ratio observed during studies at the University of California, U.S.A., is as follows (reported in Haug, 1980):

Initial C/N ratio = 20;
composting time about 12 days
Initial C/N ratio = 20–50;
composting time about 14 days
Initial C/N ratio = 78;
composting time about 21 days

With lower than optimum initial C/N ratio (such as the cases of nightsoil and sludge), N will probably be lost as NH_3 gas, especially under conditions of high temperatures, high pH and forced aeration; hence a loss of the valuable nutrient to atmosphere.

An example showing a method of calculation of appropriate mixing ratio of some raw materials to be composted is given below.

Example 3.1 Sludge from a septic tank has the following characteristics:

C/N ratio = 15 : 1 (dry weight basis)
total solids = 10 percent

volatile solids = 90 percent of total solids
bulk density = 1.1 kg/L

Determine the quantity of rice straw needed to be mixed with this septic tank sludge to raise the C/N ratio of the mixture to 30 : 1, suitable for composting. Rice straw has a C/N ratio of 80 : 1, moisture content 50 percent, bulk density = 100 kg/m^3, and N = 2 percent of dry weight.

For the sludge, eq. 3.6 gives

$$\text{percentage C} = \frac{100 - 10}{1.8} = 50 \text{ percent of total solids.}$$

Let x be kg dry weight of rice straw needed to be mixed with 1 kg dry weight of septic tank sludge.

For 1 kg dry weight of septic tank sludge, C content = 1(0.5) kg; N content = 1(0.5)(1/15) kg.

For x kg dry weight of rice straw, N content = x(0.2/100) kg, C content = x(0.2/100)(80/1) kg. Therefore C/N of the mixture is:

$$\frac{1(0.5) + x(0.2/100)(80/1)}{1(0.5)(1/15) + x(0.2/100)} = \frac{30}{1}$$

$$x = 5 \text{ kg}$$

Volume of rice straw required

$$= \frac{5 \text{ kg}}{0.5(100 \text{ kg}/1{,}000 \text{ L})} = 100 \text{ L}$$

1 kg dry weight of septic tank sludge

$$= \frac{1}{0.1(1.1)} = 9.09 \text{ L}$$

Therefore 9.09 L septic tank sludge requires 100 L of rice straw to be mixed to raise the C/N ratio to 30 : 1.

3.4.2 Particle size and structural support of compost pile

The particle size of composting materials should be as small as possible so as to allow for efficient aeration (in case of aerobic composting) and to be easily decomposed by the bacteria, fungi, and actinomycetes. Therefore municipal solid wastes and agricultural residues, such as aquatic weeds and straws, should be shredded into small pieces prior to being composted. Nightsoil, sludge, and animal manure usually contain fine solid particles suitable for microbial decomposition. However, other materials such as organic amendments and/or bulking materials need to be added to these wastes to raise the C/N ratio, provide structural support for the compost pile, and create void spaces in case of aerobic composting. Organic amendments are materials added to the composting feed to increase the quantity of degradable organic C, reduce bulk weight, and increase air voids of the compost mixture; examples of these materials are sawdust, rice straw, peat, rice hulls, and domestic refuse. Bulking materials can be either organic or inorganic, of sufficient size which, when added to sludge, will provide structural support and maintain air space in the compost mixture. Dried water hyacinth (*Eicchornia crassipes*) and rice straw, shredded into small pieces (2–3 cm long), were found to be suitable as both organic amendments and bulking materials for the composting of nightsoil (Polprasert *et al.*, 1980).

3.4.3 Moisture control

An optimum moisture content of the compost mixture is important for the microbial decomposition of the organic waste. As water is essential for nutrient solubilization and cell protoplasm, a moisture content below 20 percent can severely inhibit the biological process. A too-high moisture content will cause leaching of nutrients and pathogens from the compost pile. In aerobic composting, too much water will block the passage of air, causing the compost pile to become anaerobic. A moisture content between 50 and 70 percent (average 60 percent) is most suitable for composting and should be maintained during the periods of active bacterial reactions, i.e. mesophilic and thermophilic growth (Figure 3.2).

As nightsoil, sludge, and animal manure usually have moisture contents higher than the optimum value of 60 percent (see Chapter 2), the addition of organic amendments and bulking materials will help reduce the moisture content to a certain degree. On the other hand, most agricultural residues have moisture contents lower than 60 percent, and some water will have to be added during the composting period of such wastes. For batch operation of composting, the moisture content of the compost mixture can be controlled by adding water to the compost piles once or twice daily. The moisture content should be controlled at the optimum range until the thermophilic period is completed, which is evidenced by a temperature decrease in the compost pile and the appearance of the second- and third-level consumers (Figure 3.3).

For a composting system which is continuously operated, the control of moisture content in the compost mixture can be achieved through a recycling of the composted product, as schematically shown in Figure 3.4 (Haug, 1979).

Assuming organic amendments or bulking materials have been added to the compost feed already, a mass balance on total wet solids is written as:

$$x_c + x_r = x_m \tag{3.7}$$

For dry solids:

$$S_c x_c + S_r x_r = S_m x_m \tag{3.8}$$

$$S_c x_c + S_r x_r = S_m (x_c + x_r) \tag{3.9}$$

where: x_c = total wet weight of compost feed produced per day;
x_p = total wet weight of compost product produced per day;
x_r = total wet weight of compost product recycled per day;

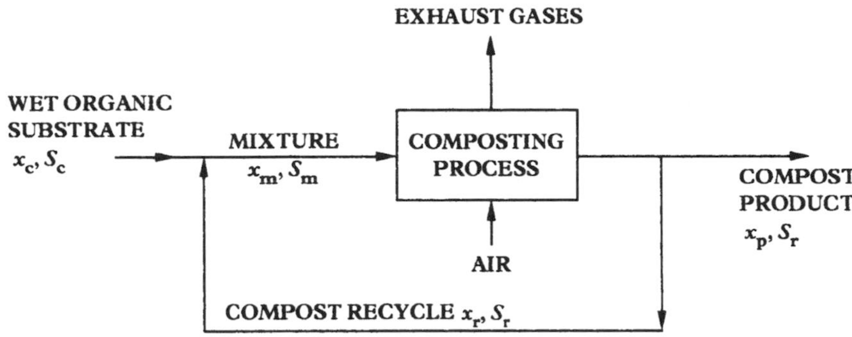

Figure 3.4 Schematic mass balance diagram for sludge composting showing inputs of compost feed and compost recycle, and compost product. (Adapted from Haug, 1979; reproduced by permission of the Water Pollution Control Federation, U.S.A.)

x_m = total wet weight of mixed material entering the compost per day;
S_c = fractional solids content of compost feed;
S_r = fractional solids content of compost product and compost recycle;
S_m = fractional solids content of mixture before composting;
R_w = recycle ratio, based on total wet weight of compost product recycled to total wet weight of compost feed;
R_d = recycle ratio, based on dry weight of compost product recycled to dry weight of compost feed.

$$R_w = \frac{x_r}{x_c} \tag{3.10}$$

Substituting eq. 3.10 into eq. 3.9 and rearranging:

$$R_w = \frac{S_m - S_c}{S_r - S_m} \tag{3.11}$$

Similarly

$$R_d = \frac{S_r x_r}{S_c x_c} \tag{3.12}$$

Substituting eq. 3.12 into eq. 3.9 and rearranging:

$$R_d = \frac{(S_m/S_c) - 1}{1 - (S_m/S_r)} \tag{3.13}$$

Equations 3.11 and 3.13 can be used to calculate the compost recycle ratios based on wet weight and dry weight, respectively. For the desired moisture content of 60 percent in the compost pile ($S_m = 0.4$), these two equations are graphically interpreted in Figures 3.5 and 3.6, respectively.

Example 3.2 From the data of Example 3.1, suppose that the moisture content of the composted product is 20 percent. Determine the quantity of the composted product needed to be recycled to achieve a 40 percent mixture solids content.

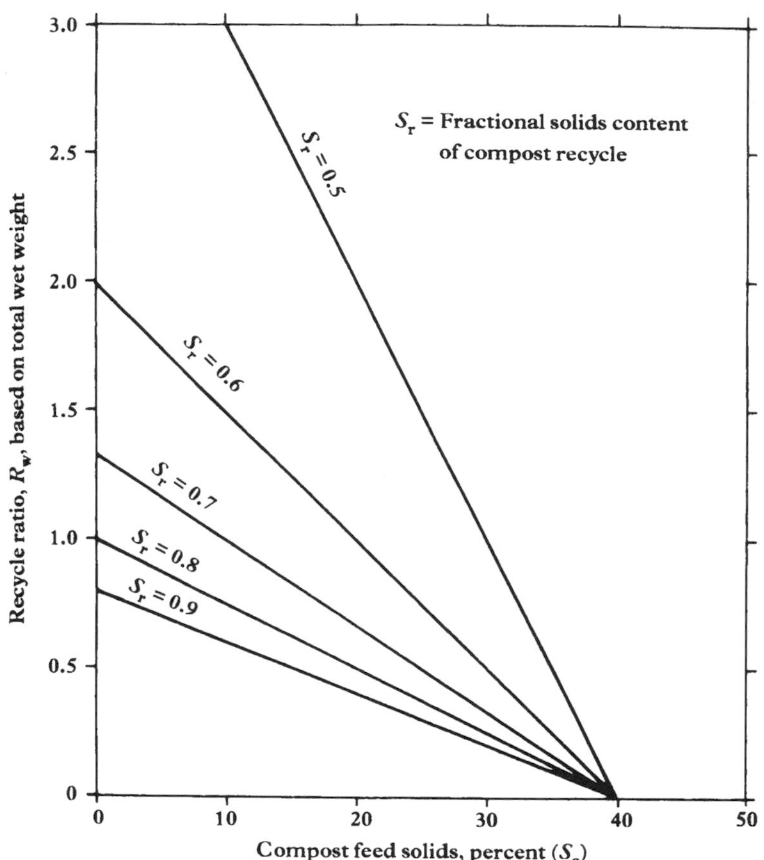

Figure 3.5 Effect of compost feed solids content on the wet weight recycle ratio needed to achieve a 40 percent mixture solids content (Haug, 1979; reproduced by permission of the Water Pollution Control Federation, U.S.A.)

From Example 3.1, moisture content of the mixture between septic tank sludge and rice straw is:

$$\frac{(1/0.1)(90) + (5/0.5)(50)}{(1/0.1) + (5/0.5)} = \frac{14}{20}(100) = 70 \text{ percent}$$

Or the compost feed solids are 30 percent. From eq. 3.13,

$$R_d = \frac{(S_m/S_c) - 1}{1 - (S_m/S_r)} = \frac{(0.4/0.3) - 1}{1 - (0.4/0.8)} = 0.67$$

From eq. 3.11:

$$R_w = \frac{S_m - S_c}{S_r - S_m} = \frac{0.4 - 0.3}{0.8 - 0.4} = 0.25$$

Also from Figure 3.6, at $S_r = 0.8$ and the compost feed solids of 30 percent, the value of $R_d = 0.7$; and from Figure 3.5, $R_w = 0.25$.

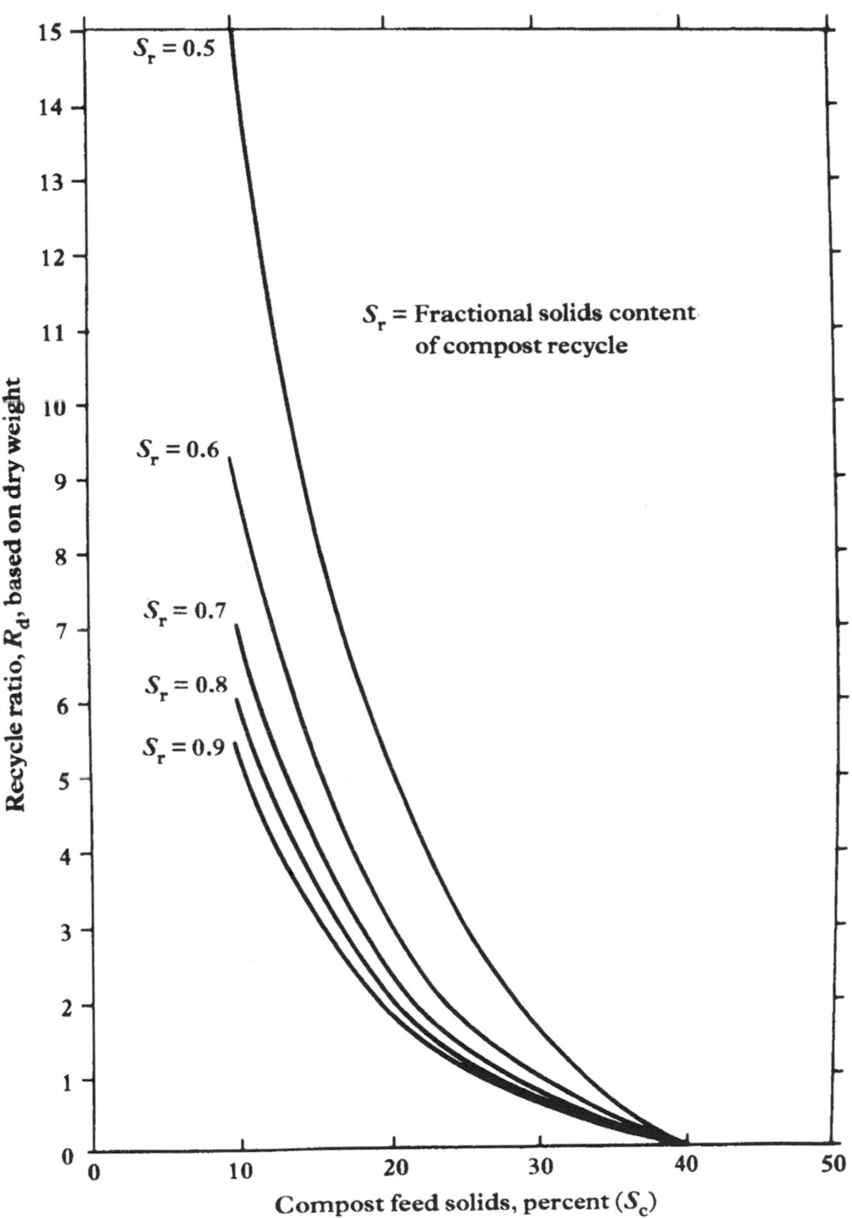

Figure 3.6 Effect of compost feed solids content on the dry-weight recycle ratio needed to achieve a 40 percent mixture solids content (Haug, 1979; reproduced by permission of the Water Pollution Control Federation, U.S.A.)

Therefore, for composting 6 kg dry weight of the mixture of septic tank sludge and rice straw the quantity of compost recycle is:

$$6 \times 0.67 = 4 \text{ kg dry weight or } 20 \times 0.25 = 5 \text{ kg wet weight,}$$

in order to achieve a 40 percent solids content of the compost mixture.

3.4.4 Aeration requirements

Aerobic composting needs proper aeration to provide sufficient oxygen for the aerobic microbes to stabilize the organic wastes. This is accomplished through some non-mechanical means such as periodic turning of the compost piles, insertion of perforated bamboo poles into the compost piles, or the dropping of compost heaps from floor to floor. A more effective, mechanical way is forced-air aeration, in which air is pumped through perforated pipes and orifices into the compost heaps.

Because non-mechanical aeration cannot supply sufficient oxygen to the microorganisms, aerobic conditions prevail only at the outer surface of the compost heaps, while facultative or anaerobic conditions exist inside. The composting rate is accordingly slow and a longer composting period is required. When mechanical aeration is to be employed, the quantity or rate of air flow has to be properly controlled. Too much aeration is wasteful and can cause a loss of heat from the compost piles, while too little aeration would lead to the anaerobic conditions inside the compost piles.

A simple method to determine aeration requirements is based on the stoichiometric reaction of waste oxidation. A knowledge of the chemical compositions of the wastes to be composted or oxidized (Table 3.2) is useful for the calculation. As compost feed is a combination of various organic compounds, for practical purposes, the formulae $C_{10}H_{19}O_3N$ and $C_5H_7O_2N$ have been used to represent its chemical composition. The stoichiometric equations for complete oxidation of these compounds are as follows:

$$C_{10}H_{19}O_3N + 12.5O_2 \rightarrow 10CO_2 + 8H_2O + NH_3 \quad (3.14)$$

$$C_5H_7O_2N + 5O_2 \rightarrow 5CO_2 + 2H_2O + NH_3 \quad (3.15)$$

Parts of NH_3 generated from the above reactions will be lost through volatilization if the pH of the compost piles is above 7. The remaining NH_3 will be nitrified to become NO_3^- during the maturation or curing of the composted products. Example 3.3 shows a method to determine aeration requirements based on the stoichiometric equations.

Table 3.2 General chemical compositions of various organic materials (Haug, 1980)

Waste component	Typical chemical composition
Carbohydrate	$(C_6H_{10}O_5)_x$
Protein	$C_{16}H_{24}O_5N_4$
Fat and oil	$C_{50}H_{90}O_6$
Sludge	
Primary	$C_{22}H_{39}O_{10}N$
Combined	$C_{10}H_{19}O_3N$
Refuse (total organic fraction)	$C_{64}H_{104}O_{37}N$
	$C_{99}H_{148}O_{59}N$
Wood	$C_{295}H_{420}O_{186}N$
Grass	$C_{23}H_{38}O_{17}N$
Garbage	$C_{16}H_{27}O_8N$
Bacteria	$C_5H_7O_2N$
Fungi	$C_{10}H_{17}O_6N$

Example 3.3 Suppose eq. 3.14 represents the reaction occurring during aerobic composting of septic tank sludge. Calculate the amount of air required to completely oxidize this sludge.

$$\underset{(201)}{C_{10}H_{19}O_3N} + \underset{(400)}{12.5O_2} \rightarrow 10CO_2 + 8H_2O + NH_3 \qquad (3.14)$$

$$1 \text{ g of } C_{10}H_{19}O_3N \text{ requires } O_2 = \frac{400}{201} \simeq 2 \text{ g}$$

Assume septic tank sludge contains 75 percent volatile solids, of which 50 percent is biodegradable (as represented by $C_{10}H_{19}O_3N$). Therefore, the amount of O_2 required per 1 g of septic tank sludge is $2(0.75)(0.5) = 0.75$ g.

As air contains 23 percent O_2 (by weight), the amount of air required is $0.75 : 0.23 = 3.26$ g.

The specific weight of air is 1.20 g/L at 25 °C and 1 atmosphere pressure.

The volume of air required to oxidize 1 g (dry weight) of septic tank sludge is $3.26/1.20 = 2.72$ L.

Note: This is a theoretical method of calculating aeration requirement. In practice the amount of air supply should be adjusted according to the biological reactions taking place, e.g. thermophilic reactions will require quantities of oxygen several times higher than those of mesophilic and maturation reactions (see Figure 3.2). Additional amounts of air need to be supplied to compensate for loss to the atmosphere, which can be up to 95–99 percent.

3.4.5 Temperature and pH

The biologically produced heat generated within a composting mass is important for two main reasons:

1. to maximize decomposition rate; and
2. to produce a material which is microbiologically 'safe' for use.

It is generally known that compost temperatures greater than 60–65 °C, above thermophilic range, will significantly reduce the rate of decomposition in compost piles. A recent work using compost samples from a full-scale composting plant showed the optimal temperature for composting, as measured by microbial activity (incorporation of [^{14}C] acetate) was consistently below 55 °C (McKinley *et al.*, 1985). On the other hand, most pathogenic microorganisms are inactivated effectively at temperatures above 50 °C (Figure 3.1). So the key concern is to control temperatures in the compost piles in such a way as to optimize both the breakdown of organic material and pathogen inactivation (approximately 55 °C). Temperature can be controlled by the adjustment of aeration and moisture content, and the utilization of screened compost as insulation cover of the compost piles.

Temperature patterns in compost piles influence the types and species of microorganisms' growth. Mesophilic temperature (25–45 °C) is developed first in composting, followed by thermophilic temperature (50–65 °C). After this phase most organic substrates will have been stabilized, resulting in a temperature decrease to mesophilic and eventually to ambient level (Figure 3.2). In many cases the thermophilic temperature can even reach 55–65 °C and last for a few days, causing an effective inactivation of the pathogens.

Aerobic composting normally proceeds at a pH around neutral, and rarely encounters extreme pH drop or rise. A slight pH drop may occur during the first few days of anaerobic composting due to the production of volatile fatty acids. After this period the pH becomes neutral again when these acids have been converted to methane and CO_2 by the reactions of methane-forming bacteria.

3.5 COMPOSTING MATURITY

There are many criteria to judge the maturity or completion of a composting process. In general, a composted product should contain a low organic content that will not undergo further fermentation when discharged on land, and the pathogens should be inactivated. Some of the approaches to measure the degree of compost stabilization are (Haug, 1980):

1. temperature decline at the end of batch composting;
2. decrease in organic content of the compost as measured by the volatile solid (VS) content, chemical oxygen demand (COD), percentage C or ash content, and C/N ratio;
3. presence of particular constituents such as nitrate, and the absence of others such as ammonia;
4. lack of attraction of insects or development of insect larvae in the final product;

5. absence of obnoxious odor;
6. presence of white or gray color due to the growth of actinomycetes.

In cases where the composted products are to be applied to crops and where public health aspects are of concern, the time required for pathogen die-offs during composting is another important criterion to be considered (see Figure 3.1). More information about pathogen die-off during composting and the public health risks are discussed in section 3.7.

Because compost materials usually contain some biologically resistant compounds, a complete stabilization during composting may not be achieved. The time required for a satisfactory degree of composting would depend on the environmental factors in and around the compost heap, as described in section 3.4. Under suitable conditions, aerobic composting normally takes about 5–20 days, while anaerobic composting can last from 45 to 100 days. The maturing or curing time of compost may be approximately the same as required for organic stabilization. Some manufacturers have produced mechanical composting reactors which claim to yield satisfactory compost within a short period, e.g. 24 h. However, these reactors are both expensive and difficult to operate, and the composted materials will usually need an additional time for curing or nitrification.

It should be noted that, in batch composting, the temperature pattern and biological succession developed in the compost piles are similar to those shown in Figures 3.2 and 3.3, respectively. Because both the stages of waste stabilization and consequently curing occur during batch composting, the composted product is suitable for use in agriculture or horticulture.

The continuous composting process is normally aerobic and has a semi-continuous plug-flow passage of composting materials through a reactor structure, and temperatures of the reactor contents are continuously in the thermophilic range. Depending on the specific design, the material is translocated through the structure as part of the tumbling action in a large revolving drum, by the action of augers force through the mass, or by gravity. These approaches may provide the gas exchange by forcing air through the mass. Residence time in the reactor is 1–10 days, with 5 days being typical. The material displaced from the continuous stage is biologically stabilized (i.e. the reaction of eq. 2.1 is achieved), but mostly not nitrified (i.e. reaction of eq. 3.5 is not achieved). It should be processed further in the curing stage to allow for nitrification to occur, and to make the composted products suitable for agricultural reuse.

An experienced compost operator can, most of the time, recognize the maturity of a composted product. However, it is advisable that the quality of the composted products be regularly checked for the nutrient contents and presence of any pathogenic microorganisms. Chemical fertilizers can be added to a composted product to make it suitable for applying to a particular crop, according to its nutrient requirement (see Table 8.7).

Details about the composting systems of both batch and continuous operation are given in section 3.6.

3.6 COMPOSTING SYSTEMS AND DESIGN CRITERIA

In this section the composting systems are described as the on-site and off-site processes. On-site systems are the ones that compost organic wastes at the place of generation, e.g. at home or in the toilets; the composting process is usually not controlled and occurs naturally. Off-site systems involve the collection and transportation of organic wastes to be composted at central treatment plants; the composting process is usually controlled either manually or mechanically. Nowadays there are many manufacturers producing various composting units for the treatment of nightsoil sludge, or municipal refuse; some of these units are described below.

3.6.1 On-site composting

AEROBIC COMPOSTING TOILETS

These types of composting toilets, often referred to as the 'multrum', were originally invented by Lindstrom and put into commercial production about 20 years ago (Rybczynski *et al.*, 1978). The multrum (Figure 3.7) consists of a watertight container with a sloping bottom. Human excreta (without flushing water) is introduced at the upper end of the container, and mixed with organic kitchen and garden wastes, introduced lower down, to increase the C/N ratio. Air ducts and a vent pipe are provided to promote aeration. The composted material moves toward the lower end, from where it is periodically removed. The decomposition period is long, up to 4 years, and the container is quite large ($3 \times 1 \times 1$ m, length $\times$ width $\times$ height). The air ducts also help to evaporate the humidity and eliminate odors, while the sloping bottom permits continuous use of a single container by separating the fresh and the decomposed materials. Another modification of the multrum is the biopit composting toilet (Figure 3.8), which incorporates a gravel soakage pit to treat and dispose of the liquid waste present in the excreta.

ANAEROBIC COMPOSTING

Pit latrines (or holes in the ground) are the simplest form of anaerobic composting toilets. The excreta deposited into the pit is decomposed anaerobically. When the pit is about two-thirds full, it is filled up with dirt and left there for approximately 2–3 years. After this period the excreta would be well stabilized and the pathogens inactivated, rendering it satisfactory for use as a soil conditioner. To avoid the occurrence of obnoxious odors, ventilated improved pit (VIP) latrines were developed (Figure 3.9) which have yielded encouraging results in many places such as Zimbabwe (Morgan and Mara, 1982) and Botswana (Nostrand and Wilson, 1983).

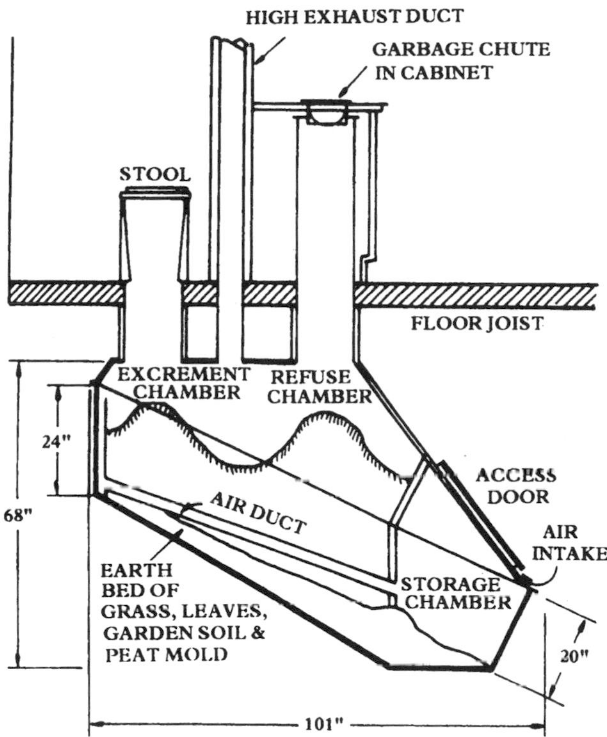

Figure 3.7 The Clivus Multrum (Rybczynski *et al.*, 1978; reproduced by permission of the International Development Research Centre, Canada)

Another version of anaerobic composting is the Vietnamese toilet (Figure 3.10), which was claimed to be the key component of a rural sanitation program for disease prevention and increased food production. It has two watertight tanks serving by turns as receptacles for defecation and composting. A hole is made on the face of each for feces deposition. Kitchen ash is added after each use to reduce odor and increase the C/N ratios. Urine is channeled in a groove into a separate vessel; this method reduces moisture content but does not decrease C/N ratio of feces in the toilet, which is favorable for the composting reactions. Apertures are made in the back wall for the collection of the composted products. The toilets are constructed above ground so as not to be submerged by rainwater.

When the tank is about two-thirds full, it is filled to the brim with dried powdered earth and kept under anaerobic conditions for a few years. During this time another tank is used for defecation, and the cycle is repeated. The anaerobically composted products were reported to be rich in nutrients and hygienically safe for reuse as fertilizers (McMichael, 1976).

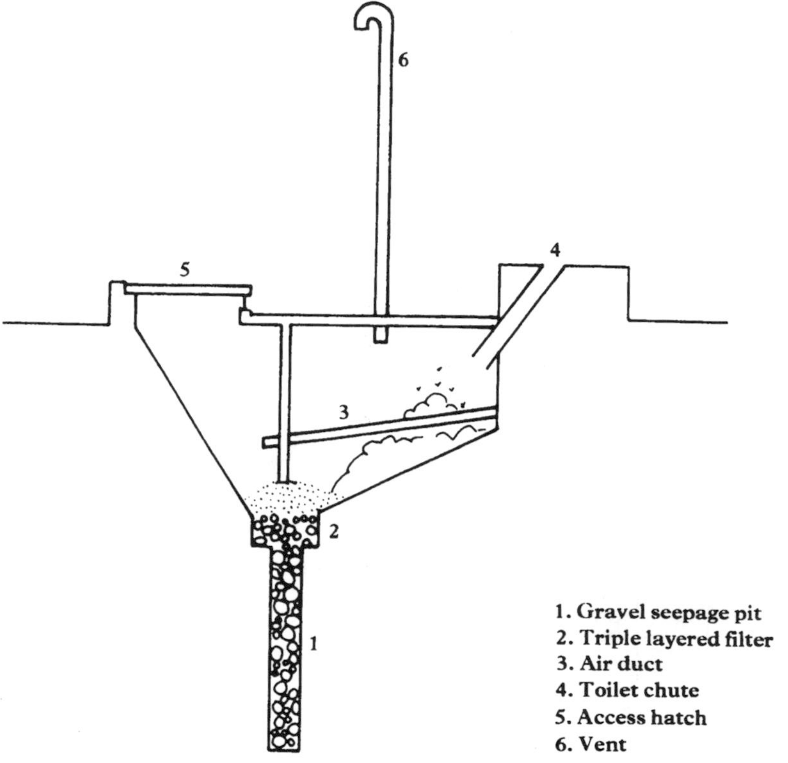

Figure 3.8 The Biopit Composting Toilet (Rybczynski *et al.*, 1978; reproduced by permission of the International Development Research Centre, Canada)

3.6.2 Off-site composting

Because of the large quantities of organic waste to be decomposed, off-site composting should employ aerobic reactions to shorten the composting period and reduce the size of composting plants. Some of the aerobic composting systems currently in operation are listed below.

CHINESE COMPOSTING PILE

As shown in Figure 3.11, the compost feed (a mixture of human or animal manure and vegetable matter) is piled up into a heap of approximately 2 × 2 × 0.5 m (length × width × height). The compost heap is pierced by perforated bamboo poles to facilitate natural aeration and provide a kind of structural support, and no turning of the compost pile is required. To control excessive heat loss the compost

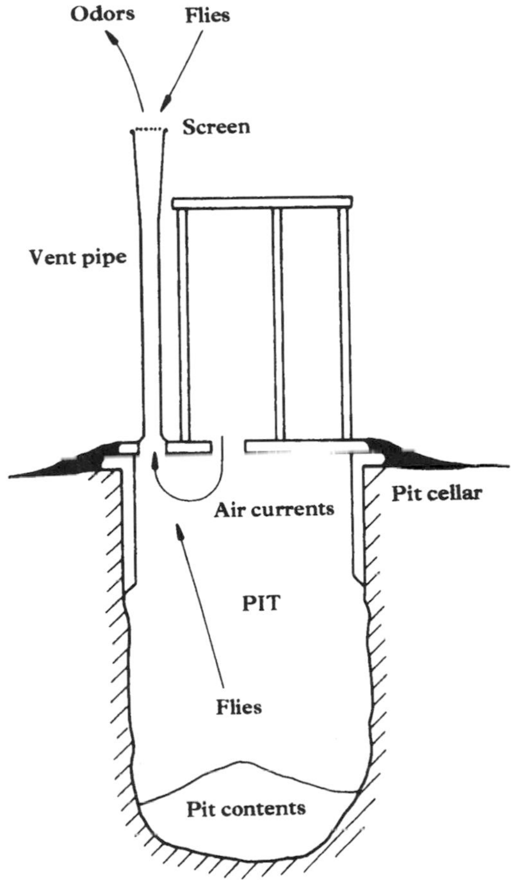

Figure 3.9 Schematic diagram of a ventilated improved pit latrine (Morgan and Mara, 1982; reproduced by permission of the World Bank)

pile is covered with rice straw or a plaster of mud (Figure 3.12). The poles can be removed after 1 or 2 days, when the mud has hardened or the compost pile is structurally stable, but they can remain there for a longer period. An experiment conducted at the Asian Institute of Technology (AIT), Bangkok, Thailand, found the time required for compost stabilization to be about 60 days (Polprasert *et al.*, 1980), and the composted product has become suitable for use as soil conditioner. In China, a greater than 90 percent kill of *Ascaris* ova in the compost was achieved (McGarry and Stainforth, 1978).

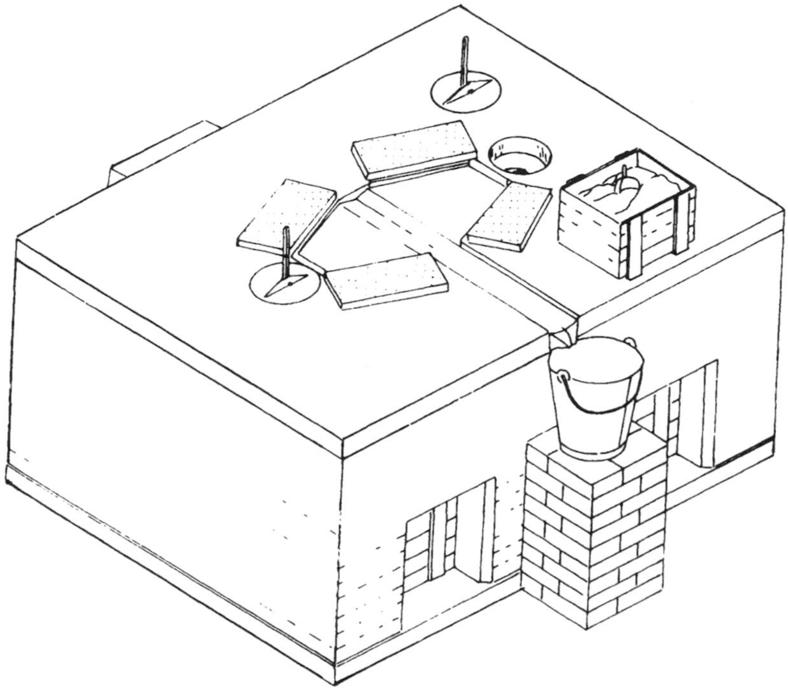

Figure 3.10 The Vietnamese composting toilet (Rybczynski *et al.*, 1978; reproduced by permission of the International Development Research Centre, Canada)

WINDROW COMPOSTING

This system involves periodic turning of the compost piles, manually about once a week, or mechanically once (or more) daily. The purpose of pile turning is to provide aeration and mix the compost materials, hence there is a faster decomposition rate than that of the Chinese composting pile. The approximate size of each pile is 13 × 3 × 1.5 m (length × width × height), but other sizes, subject to convenience, have also been employed. The period required for compost stabilization varies according to the frequency of pile turning, being 20–40 days, and temperatures over 65 °C in the center of the compost pile can be reached.

FORCED-AIR AERATION COMPOSTING

A more efficient composting method which ensures temperatures in the upper thermophilic range and provides an effective inactivation of pathogens is called the Beltsville aerated rapid composting (BARC) method. Developed by Epstein *et al.* (1976), this method involves placing a mixture of sludge and wood chips over a base (consisting of compost and chips) and aeration piping system (Figure 3.13). The approximate size of each pile is 12 × 6 × 2.5 m (length × width × height) and is connected to a centrifugal blower to draw air through the pile (sucking) or to

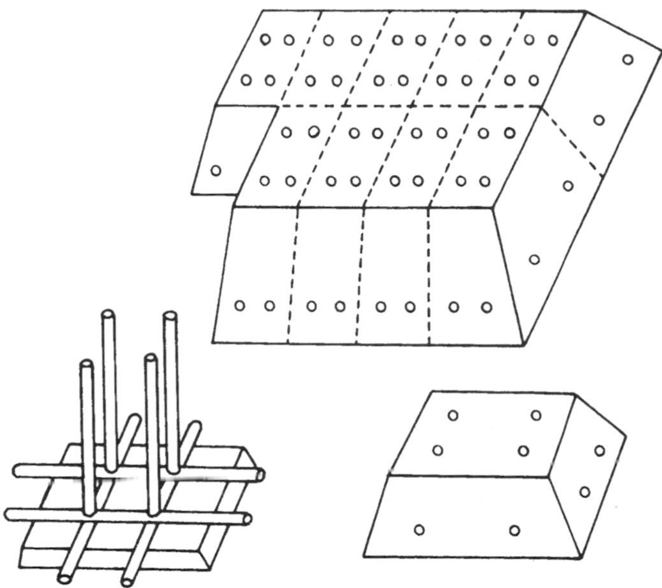

Figure 3.11 Chinese composting pile (McGarry and Stainforth, 1978; reproduced by permission of the International Development Research Centre, Canada)

Figure 3.12 Chinese composting pile covered with rice straw

blow air into the pile (blowing) according to the predetermined aeration requirements. The gases drawn into the pipe are deodorized by passing them into a pile of screened compost (Figure 3.14). The entire pile is covered with a 30 cm layer of the screened compost to minimize odors and to maintain high temperatures in the compost pile.

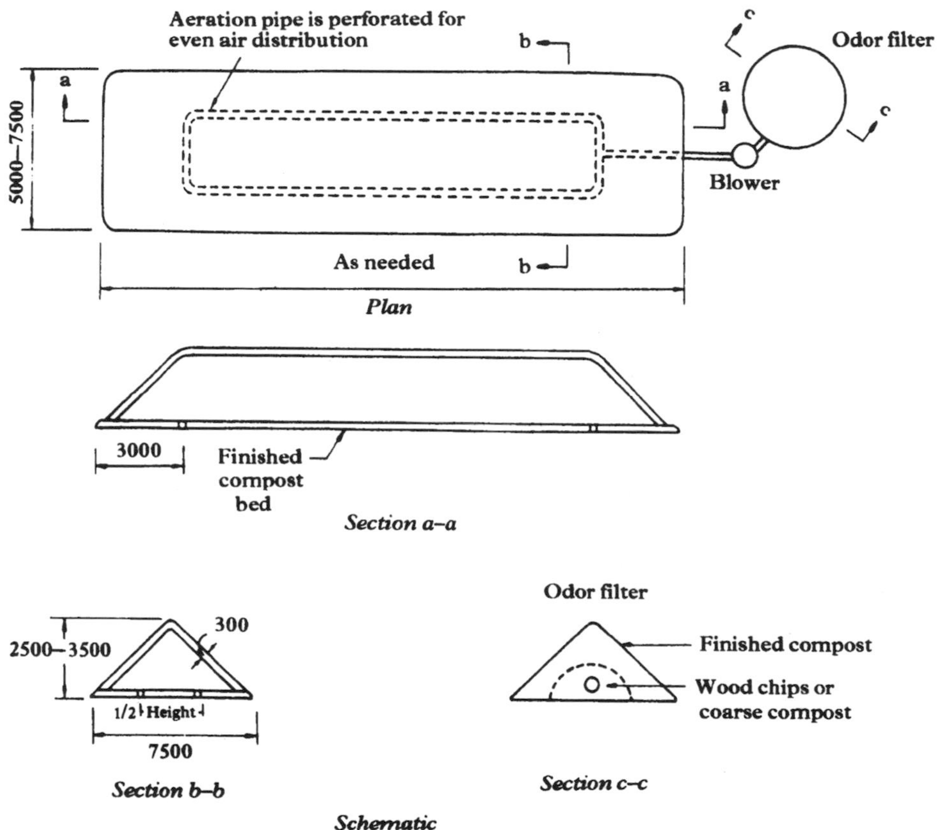

Figure 3.13 Schematic diagram of an aerated pile indicating location of aeration pipe. Loop is perforated for air distribution (units in millimeters) (Shuval *et al.*, 1981; reproduced by permission of the World Bank)

As shown in Figure 3.15, a rapid temperature rise to within 60–80 °C was achieved in 3–5 days of composting, and these temperatures continued for about 10 days. The distributions of temperature in a compost pile for both blowing and sucking types are shown in Figure 3.16 (Stentiford *et al.*, 1985). There appeared to be uneven temperature distribution in the compost pile, with a higher temperature developed at the inner part and lower temperature at the outer portion. This uneven temperature distribution is a typical drawback of the static, non-pile-turning composting method in which pathogens present at the outer portion of the compost pile may not be effectively inactivated by the biologically produced heat.

However, satisfactory reductions of indicator bacteria and viruses have been reported for the BARC system, and the system is essentially unaffected by low ambient temperatures and/or rainfall. It does not require pile turning (which is rather labor-intensive), and should be applicable for nightsoil composting in hot climates (Shuval *et al.*, 1981).

The area required for the BARC system and conventional windrow system has been estimated to be 1 hectare (ha) per 10–12 dry tons per day of sewage sludge.

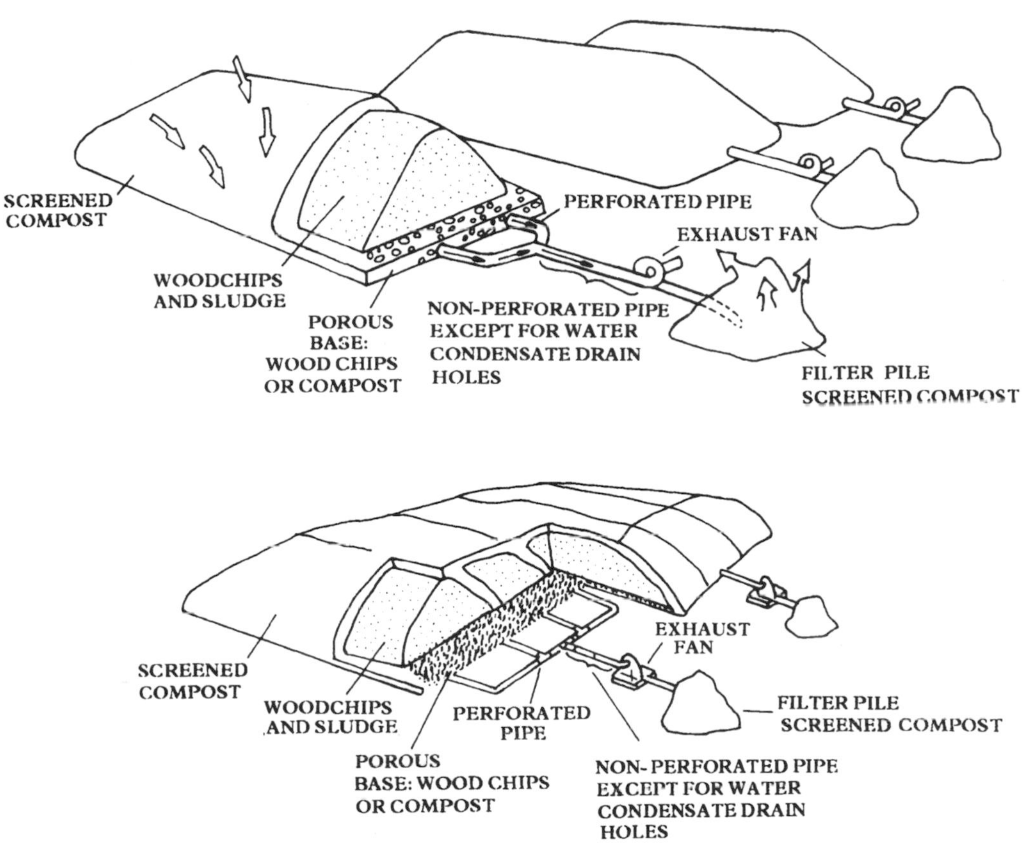

Figure 3.14 Configuration of aerated piles showing construction of pile and the arrangement of aeration pipe—the BARC system (Shuval *et al.*, 1981; reproduced by permission of the World Bank)

This area estimate includes: (i) area for mixing the sludge or nightsoil with bulking materials; (ii) area required for composting; (iii) area required for storage and curing piles and long-term storage before marketing; (iv) area required for screening of the final compost and separation and recycling of wood chips; (v) area required for lagooning or waste stabilization ponds to treat leachate and drainage from the composting area; and (vi) area required for administration, workshops, parking of vehicles, and storage of spare parts (Shuval *et al.*, 1981).

Benedict *et al.* (1986) reported that with the operation capacity of 450 wet tons/day of sludge, the area required for conventional windrow composting of the Los Angeles, U.S.A., facility was about 20 ha. However, this total included areas used for storages of bulking agents, finished compost and equipment, and for a

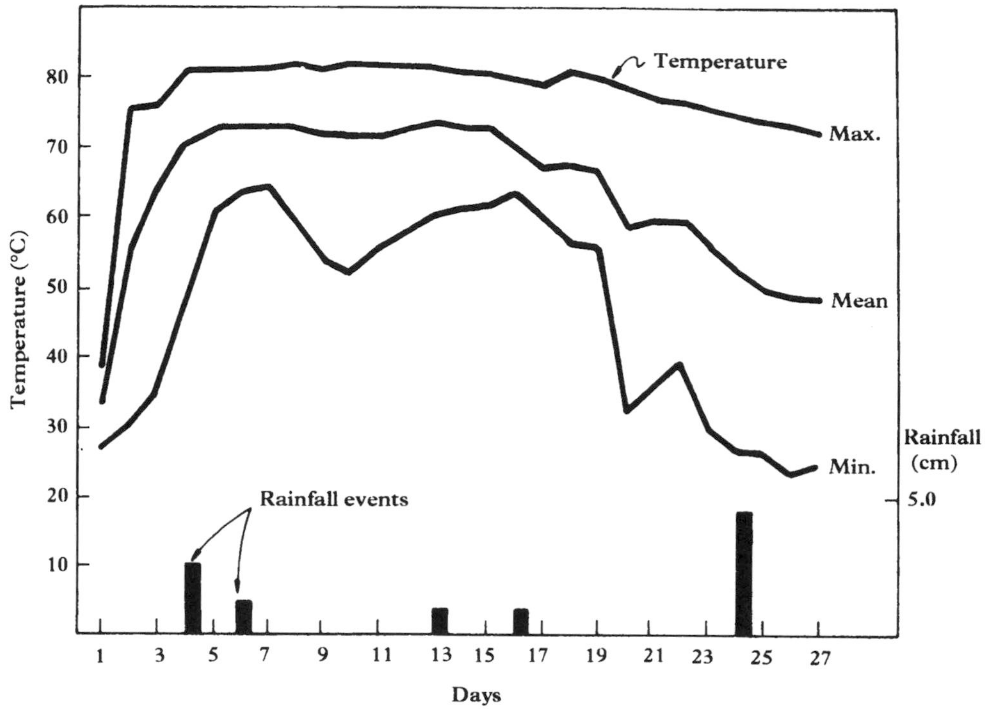

Figure 3.15 Temperature during sludge composting by the BARC system (Shuval *et al.*, 1981; reproduced by permission of the World Bank)

research area; so the operating area for composting was about 15 ha or 0.03 ha for composting of 1 ton/day of wet sludge.

THE DANO SYSTEM

A schematic diagram of a typical DANO plant manufactured for solid waste composting is shown in Figure 3.17. It includes storage hoppers, rotary screens, and magnetic separators to separate out non-compostable materials. Materials to be composted are fed into the 'DANO biostabilizer', a cylindrical chamber tilted slightly from the horizontal, usually about 3–4 m in diameter and varying from 25 to 30 m in length according to the quantity of feeding materials. The cylinder rotates at up to 1 rpm, and air is supplied by fans at a low pressure, through longitudinal ducts, each having several injector nozzles. The above conditions enhance effective aerobic decomposition of the organic matter in the cylinder, in which temperatures of 60 °C and over are developed inside. The steam and waste gases are exhausted through extractor fans. The composting period required for this type of compost reactor ranges from 2.5 to 5 days.

The DANO system, developed in Denmark, is presently being employed for solid waste composting in several countries throughout the world. Although very efficient in composting, the DANO system involves high capital and operation

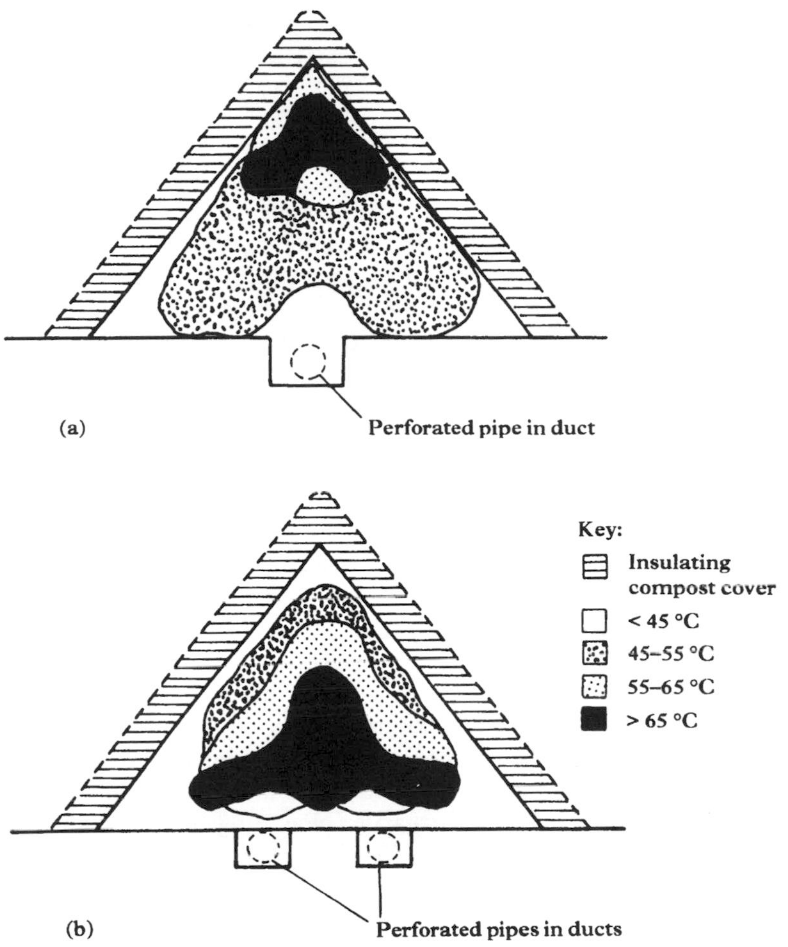

Figure 3.16 Typical temperature cross-section distributions in a pile with (a) positive pressure (blowing) aeration; and (b) negative pressure (sucking) aeration (from Stentiford *et al.*, 1985; by permission of the Institute of Water Pollution Control, U.K.)

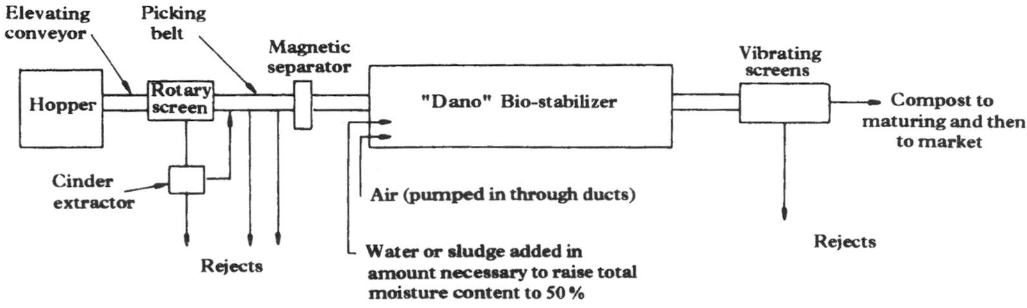

Figure 3.17 Typical DANO plant (after Flintoff and Millard, 1969)

costs, and requiring skillful manpower to maintain the system. Figure 3.18 shows a DANO composting plant treating part of Bangkok city's, Thailand, solid waste.

THE JERSEY SYSTEM

The Jersey composting system, applied mostly to municipal refuse, is an aerobic process in which aeration is accomplished during movement of the composting materials from the top floor to a lower floor. The municipal refuse to be composted has to be pre-processed to remove the inorganic and other non-compostable materials, similar to the DANO system described above. The organic materials are then carried to the fifth floor of the compost plant and stay there for 1 day. These composting materials are then released to the composting basket on the next lower floor (five floors take 5 days). After passing through the composting process for 5 days, the composted products are further cured in an open lot for 30–60 days prior to being used as soil conditioner.

A typical Jersey composting plant, shown in Figure 3.19, is expensive to construct and requires skilled manpower to operate.

From Obeng and Wright (1987) and Haug (1980), composting systems can be classified into non-reactor and reactor (Tables 3.3 and 3.4, respectively). Based on the information given in Table 3.3, it appears that for the non-reactor composting systems (with proper C/N ratio and provided with some types of aeration, natural or forced-air), the time required for the first-phase composting (Figure 3.2) is at least 3–4 weeks, followed by about 4 weeks of maturation or curing. The reactor composting systems (operating under aerobic conditions, Table 3.4) should require at least 3–5 days for the first-phase composting, while the time needed

Figure 3.18 A DANO composting plant in Bangkok

Table 3.3 Non-reactor composting systems

Composting system	Raw material	General description
Aerated static pile	Sewage sludge and wood chips	Raw sewage sludge (22% solids) is mixed with wood chips, and then transferred to a composting pad consisting of wood chips spread over perforated piping. Air is drawn through the pipe into a compost filter. The pile is maintained for 21 days followed by screening and drying. Temperatures of 55 °C are achieved throughout the pile
	Nightsoil, paper, wood chips	The nightsoil is mixed with paper and wood chips on a concrete pad and then transferred to the composting pad, which is a bed of wood chips covering a perforated pipe. Air is drawn through the pipe into a compost filter. The pile is maintained for 21 days at temperatures of 60 °C for most of this time. The compost is then cured for 30 days
	Water hyacinth, nightsoil, rice straw	Experiment. Nightsoil, water hyacinth, and (in some cases) rice straw were mixed and composted in piles for 2–3 months. Aeration through perforated bamboo poles. Temperatures of 43–64 °C were maintained for at least 8 days in the coolest parts of the piles
Bangalore (Indore)	Refuse, nightsoil, earth, straw, etc.	Trench in ground, 2–3 ft deep. Material placed in alternate layers of refuse, nightsoil, earth, straw, etc. No grinding. Turned by hand as often as possible. Detention time of 120–180 days. Used extensively in India
Brikollari (Caspari) (Briquetting)	Refuse	Ground material is compressed into blocks and stacked for 30–40 days. Aeration by natural diffusion and airflow through stacks. Curing follows initial composting. Blocks are later ground. Sludge can be added to mixture up to moisture content of about 53%
Windrow	Sewage sludge and sawdust	The sludge (30% solids) and sawdust are mixed at weight ratios 4 : 1. The windrows are turned for 3 months and then sold as a product 'Grow Rich'. The maximum temperature achieved is 74 °C
	Sewage sludge and wood chips	Digested sewage sludge is mixed with wood chips (1 : 3 volume ratio) in windrows 1.8 m high and 2.1 m wide. The windrows are turned daily for at least 2 weeks, then they are spread out, dried, and cured for 30 days. The wood chips are screened out for reuse
	Raw dewatered sewage sludge and bark	The sludge (25% solids) mixed with bark (1 : 3) and composted in windrows for at least 21 days. In general, temperatures of 50–75 °C are maintained for at least half of this time
	Raw dewatered sewage sludge and straw	(Full-scale experimental plant.) The sewage sludge (25–30% solids) was mixed with straw (1 : 1 volume) and composted in windrows, which were turned eight times for 3 months. At the end of this time the compost was ready for use. Temperatures of 55–60 °C were regularly achieved
	Digested sludge and straw	Experiment. Digested sludge was mixed with straw at a ratio of 1 : 1.25 in windrows for 6 weeks during which temperatures of 65–70 °C were achieved

Table 3.4 Reactor composting systems

Composting system	Raw material	General description
Dano	Municipal solid wastes, sludge	Dispersed flow rotating drum, slightly inclined from the horizontal, 9–12 ft in diameter, up to 150 ft long. Drum kept about half full of refuse. Drum rotation of 0.1–1.0 rpm, 1–5 days digestion followed by windrowing. No grinding. Forced aeration into drum. Temperatures of 60 °C or over are reached
Earp-Thomas		Possibly the oldest reactor system. Silo type with eight decks stacked vertically. Center shaft drives a plow which agitates the compost and moves it downward from deck to deck. Air passes upward through the silo. Digestion of 2–3 days followed by windrowing
Euramca (Roediger, Fermen-technik)	Dewatered sludge and recycled compost	Reactor is of the tower type with a special extraction and agitation mechanism at the bottom. The reactor is batch-operated. Once the reactor is loaded with the mixture of the sludge and recycled compost, material is frequently recycled from the bottom to the top to provide agitation and assure uniform exposure to temperature. Residence time is 4–6 days. Temperatures of over 65 °C are attained in the reactor. After composting, the compost is put in a dryer for a further 4–6 days; it is then pelleted and stored for sale
Fairfield digester reactor	Sewage sludge and shredded paper, municipal refuse	Circular tank. Vertical screws, mounted on two rotating radial arms, keep ground material agitated. Forced aeration through tank bottom and holes in screws. Continuous flow type. Dewatered sewage sludge is mixed with shredded paper and fed into a reactor. The retention time is 5–7 days and temperatures reach 70 °C during this time
Forced aeration through fermenter	Dewatered sewage sludge and recycled compost	The sewage sludge and recycled compost are mixed and ground and fed into the fermenter, where the mixture is aerated and turned. The retention time is 10 days and temperatures of up to 75 °C are attained. The compost is then graded and bagged
Geochemical-Eweson	Municipal refuse, sludge	Cells-in-series type. Unground refuse placed in rotating drum, 11 ft diameter, 110 ft long, slightly inclined from horizontal. Three compartments in drum. Refuse transferred to next compartment every 1–2 days for total digestion time of 3–6 days. Screened output cured in piles
HKS	Sewage sludge and recycled compost	Reactor is of the complete-mix, rotating drum type. The sludge and recycled compost are added to a slowly rotating drum (which is stopped at night). The retention time is 24 h followed by a 2-week maturation period in a windrow. Temperatures of 60–75 °C are attained
Jersey (John Thompson)	Municipal refuse	Structure with five floors, each equipped to dump ground refuse on to the next lower floor. Aeration effected by dropping from floor to floor. Detention time of 5 days; 6–8 weeks of additional curing in unturned piles

Kneer (Later versions known as BAV)	Sludge, mushroom wastes, poultry wastes, organic bulking agents	Reactor consists of a cylindrical tower with no interior floors or other mechanisms. Feed is introduced at the top and flows downward as product is removed by mechanical scraper from the bottom. Oxygen is supplied by forced aeration from the bottom. Residence times of 3–14 days. Temperatures of 60–85 °C are achieved. Then the raw compost is matured for 6–8 weeks in a windrow
Metrowaste	Municipal refuse, animal manure, sludge	Rectangular tanks, 20 ft wide, 10 ft deep, 200–400 ft long. Refuse is ground. Residence time of about 7 days. 'Agiloader' moves on rails mounted on bin walls and provides periodic agitation by turning
Naturizer (International)	Municipal refuse	Five 9-ft wide steel conveyor belts arranged to pass material from belt to belt. Each belt is an insulated cell. Air passes upward through digester. Detection time of 6–8 days
Open baskets	Digested sewage sludge and bark	Dewatered digested sewage sludge was mixed with bark (1 : 3) and composted in large baskets that could be easily stacked. Temperatures of up to 75 °C were attained. The retention time in the baskets was 9–12 weeks followed by 3–4 weeks maturation in piles
Schnorr 'Biocell' System	Dewatered sludge, recycled compost and ground bark	Reactor consists of a vertical tower with 10 floors one above the other. Each floor contains a hydraulically operated valve which allows material to be discharged to the next floor. Oxygen is introduced by forced aeration. Dewatered sludge, recycled compost and ground bark mixed in proportions of 2 : 2 : 1 volume. Composting mixture is about 1 m deep on each floor. Residence time is about 3 days on each, giving a total reactor time of 30 days
T. A. Crane	Municipal refuse	Two cells consisting of three horizontal decks. Horizontal ribbon screws extending the length of each deck recirculate ground refuse from deck to deck. Air introduced in bottom of cells. Three days composting followed by curing for 7 days in a bin
Triga	Municipal refuse, sewage sludge, bark, sawdust	Reactor is a concrete tower called a 'Hygiensator' which is divided into four separate vertical compartments. Residence time of 4–15 days depending on feed and temperatures of 70–80 °C are achieved. Air is pulled out of top of reactor. Screw extractor removes and agitates material from bottom of reactor. Extracted material is recycled three to five times during compost period to avoid compaction at bottom. Curing period of 2–4 months used after reactor, during which temperatures often reach 50 °C
Trough fermenter	Sewage sludge and rice hulls and recycled compost	The digested sewage sludge is mixed with rice hulls and finished compost (1 : 1 : 1 volume) and fed into a trough where it is composted for 2 weeks by forced aeration and turning. This is followed by 1–2 months of maturation. Temperatures reach up to 70 °C
Varro	Municipal refuse	Ground refuse placed in eight-deck digester and moved downward from deck to deck by plows. Each deck pair has its own recirculating air supply to control CO_2 level. Output dried, reground and used as base material for fertilizer, soil conditioner, wall board, etc. Digestion time 40 h

Figure 3.19 A Jersey composting plant in Bangkok

for maturation is about 4 weeks. These composting systems, when properly operated according to the environmental requirements listed in section 3.4, should result in a temperature increase in the compost mixture up to 50 °C or more which is essential in inactivating the pathogens.

3.7 PUBLIC HEALTH ASPECTS OF COMPOSTING

The composting of human or animal manure involves the following disease risks:

1. from the pathogens normally present in the raw wastes (called primary pathogens); and
2. from fungi and actinomycetes that grow during composting (called secondary pathogens).

The primary pathogens such as bacteria, viruses, protozoa, and helminths can initiate an infection in healthy individuals, while secondary pathogens usually infect people with debilitated immune systems, such as those weakened by primary infections or respiratory diseases (U.S. EPA, 1981). Examples of primary and secondary pathogens are listed in Table 3.5; more examples of these pathogens can be found in Chapter 2.

DIE-OFFS OF PRIMARY PATHOGENS

Time and temperature are the two most important parameters responsible for pathogen die-offs during composting. It should be noted that a complete inactivation of pathogens in a compost pile is rarely achieved. This is due to many reasons, such as:

1. The heterogeneous character of the compost materials, which may form clumps with the pathogens and protect them from being fully exposed to thermophilic temperatures.
2. The uneven temperature distribution in the compost piles. Unless completely mixed continuously, the outer surfaces of a compost heap normally have a lower temperature than the inside, causing a lower efficiency of pathogen kill (see Figure 3.16).
3. Partial inactivation of pathogens. Many pathogens—such as spore-forming bacteria, cysts, and helminthic ova—are only partially inactivated during composting. They can re-grow and become infective again if exposed to a favorable environment such as under moistened conditions in crop fields.

A number of workers (Kawata *et al.*, 1977; Cooper and Golueke, 1979) have reported several orders of magnitude reduction of bacteria (total coliforms, fecal coliforms, and fecal streptococci) and viruses (poliovirus and coliphages) during composting. Better inactivation rates were observed with the BARC system, in which these bacteria, including *Salmonella*, were non-detectable after 10 days of composting (Figure 3.20) and for F_2 bacteriophages about 15–20 days (Figure 3.21). The thermal death points of some common pathogens and parasites are

Table 3.5 Examples of pathogens found in, or generated during, composting of sewage sludge, together with human diseases associated with these pathogens (U.S. EPA, 1981)

Group	Example	Disease
Primary pathogens		
Bacteria	*Salmonella enteritidis*	Salmonellosis (food poisoning)
Protozoa	*Entamoeba histolytica*	Amoebic dysentery (bloody diarrhea)
Helminths	*Ascaris lumbricoides*	Ascariasis (worms infecting the intestines)
Viruses	Hepatitis virus	Infectious hepatitis (jaundice)
Secondary pathogens		
Fungi	*Aspergillus fumigatus*	Aspergillosis (growth in lungs and other organs)
Actinomycetes	*Micromonospora* spp.	Farmer's lung (allergic response in lung tissue)

shown in Table 3.6 and the relationships between time and temperature that cause pathogen inactivation, as shown in Figure 3.1, provide useful information when determining the required composting period.

HEALTH RISKS FROM SECONDARY PATHOGENS

The contact or inhalation of air containing a high density of spores of secondary pathogens can cause health hazards to compost workers and users. An epidemiological study was conducted on compost workers at nine sludge composting plants in the U.S.A. to evaluate the associated potential health hazard (Clark *et al.*, 1984). The results, summarized in Table 3.7, indicate a higher health risk for compost workers than for control groups not involved in compost activities. Nose and throat cultures positive for *Aspergillus fumigatus* were more common for these compost workers than for others. As *Aspergillus fumigatus* can cause serious infections to lungs and other human organs, proper care must be taken to avoid the uptake of these spores.

Table 3.6 Thermal death points of some common pathogens and parasites

Organism	Thermal death point
Salmonella typhosa	No growth beyond 46 °C; death within 30 min at 55–60 °C
Salmonella spp.	Death within 1 h at 55 °C; death within 15–20 min at 60 °C
Shigella spp.	Death within 1 h at 55 °C
Escherichia coli	Most die within 1 h at 55 °C, and within 15–20 min at 60 °C
Entamoeba histolytica cysts	Thermal death point is 68 °C
Taenia saginata	Death within 5 min at 71 °C
Trichinella spiralis larvae	Infectivity reduced as a result of 1 h exposure at 50 °C; thermal death point is 62–72 °C
Necator americanus	Death within 50 min at 45 °C
Brucella abortus or *suis*	Death within 3 min at 61 °C
Micrococcus pyogenes var. *aureus*	Death within 10 min at 50 °C
Streptococcus pyogenes	Death within 10 min at 54 °C
Mycobacterium tuberculosis var. *hominis*	Death within 15–20 min at 66 °C, or momentary heating at 67 °C
Corynebacterium diphtheriae	Death within 45 min at 55 °C

From Gotaas (1956); reproduced by permission of the World Health Organization, Geneva

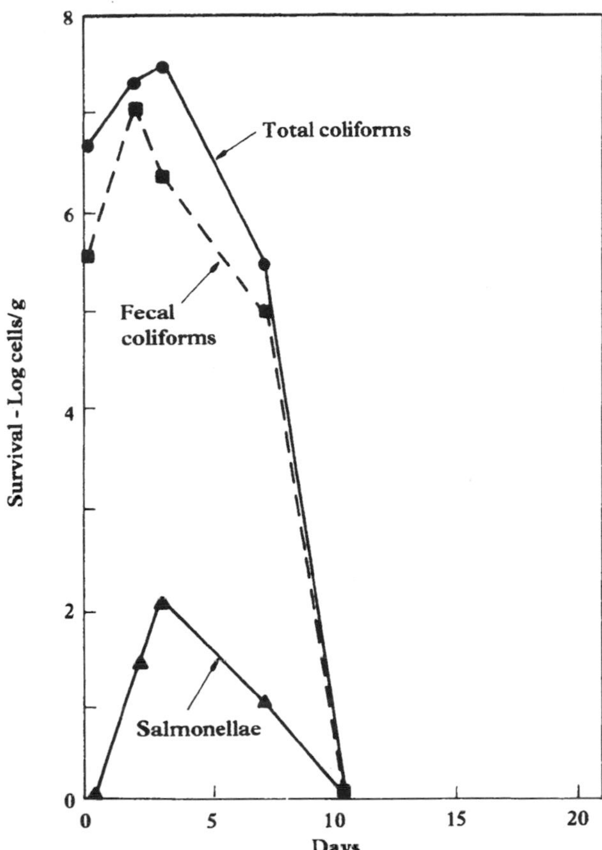

Figure 3.20 Destruction of Salmonellae, fecal coliforms, and total coliforms during composting by the BARC system (U.S. EPA, 1985)

Table 3.7 Epidemiological effects on compost workers

Excess of nasal, ear and skin infections (by physical exams) in compost and intermediate-exposed workers
Symptoms of burning eyes and skin irritation higher in compost and intermediate-exposed workers
Fungal colonies greater in cultures from compost workers
Legionella infections not associated with compost exposure but were higher among all workers exposure categories in Philadelphia–Camden area than in Washington, D.C., area
Evidence of higher white blood cell counts and hemolytic complement in compost workers
Higher antibody levels to compost endotoxin in compost workers

From Clark *et al.* (1984); reproduced by permission of the Water Pollution Control Federation, U.S.A.

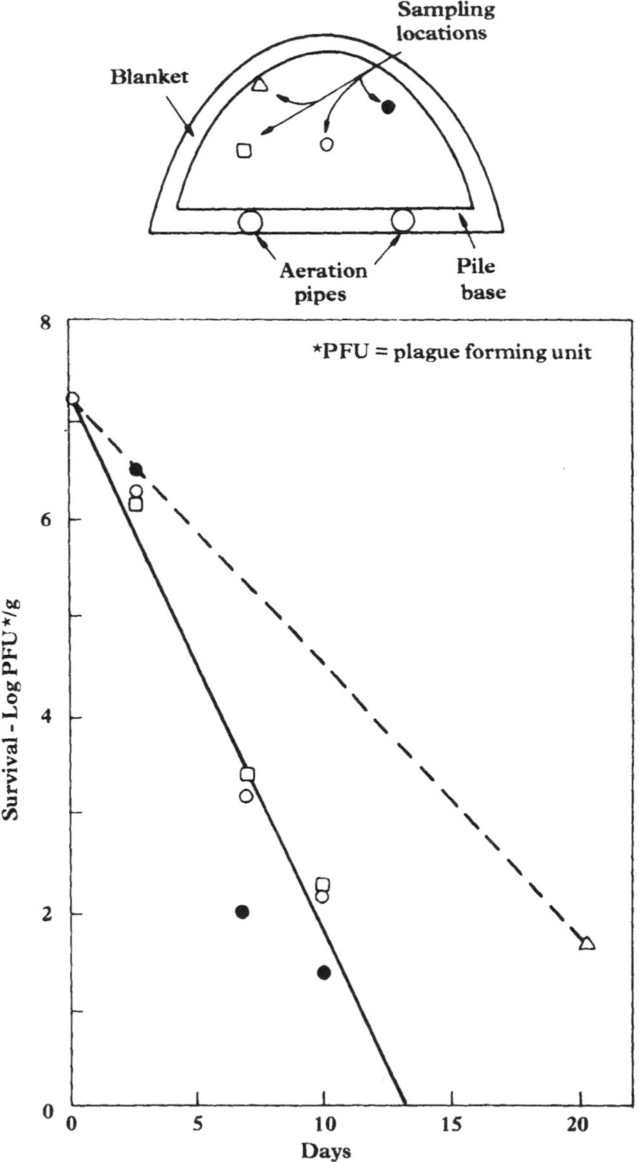

Figure 3.21 Destruction of F_2 bacterial virus during composting by the BARC system (U.S. EPA, 1985)

The growth of *Aspergillus fumigatus* in composting may be controllable through moisture control management practice (Finstein *et al.*, 1980) because fungi tend to thrive in material that is slightly too dry for profuse bacterial growth (Millner *et al.*, 1977). However, routine monitoring for the presence of this type of fungus in compost piles should be carried out, and the compost workers must be informed of the potential infection from *Aspergillus fumigatus*.

Although such an epidemiological study for compost workers in developing countries has not been reported, the health risks originating from the primary and secondary pathogens should be similar to, or even greater than, those reported in the U.S.A. Therefore, skin contact with the composted materials should be avoided and proper measures—such as wearing gloves, boots and masks—should be undertaken to prevent the inhalation of the spores of secondary pathogens, especially during turning of compost piles. The composting processes without pile turning, such as the BARC (Figure 3.14) and the Chinese composting system (Figure 3.12), should pose less health hazards to the compost workers than those with pile turning.

3.8 UTILIZATION OF COMPOSTED PRODUCTS

Compost has been used as: (i) fertilizer; (ii) soil conditioner; (iii) feed for fish in aquaculture; (iv) landfill material; (v) horticultural medium on parkland, ornamental and recreational areas and in highway right-of-ways.

Screening, grinding, or a combination of similar processes should be done to remove plastics, glass, and other materials from the compost that might be objectionable in its use. For some uses, such as landfilling and land reclamation, compost need not be finished or processed further. For general agriculture/aquaculture a coarse grind is satisfactory whereas for horticulture and luxury gardening the compost product must be finer. Compost to be used as fertilizer or soil conditioner is usually mixed with chemical fertilizers to make its nutrient contents suitable for crop growth. A schematic diagram of compost fertilizer production process is given in Figure 3.22.

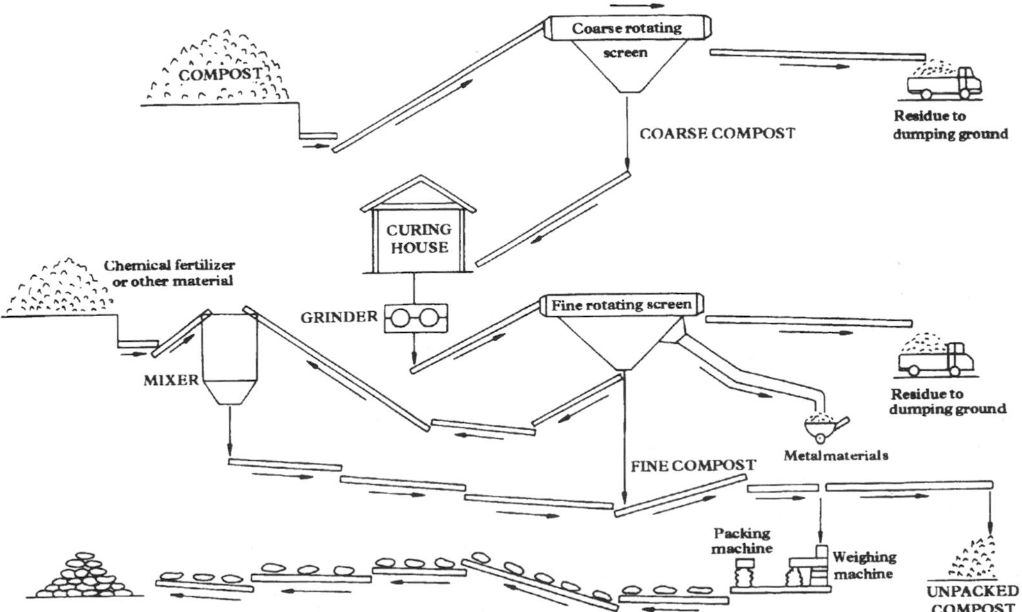

Figure 3.22 Flow chart of compost fertilizer production process

3.8.1 Utilization as fertilizer and soil conditioner

Compost can improve overall soil fertility through the addition of organic matter and plant nutrients, and can modify the soil pH. The long-term effect on soil fertility is also important. The result is that soil errosion can be reduced, water retention capacity raised, soil structure improved, and subsequently, vegetation can be established quickly.

The utilization of composted municipal wastes as fertilizers and soil conditioners will depend on three principal factors:

1. socio-economic considerations;
2. product quality; and
3. soil and plant responses.

SOCIO-ECONOMIC CONSIDERATIONS

Socio-economic factors relate to health regulations, state laws regarding the sale of fertilizers or soil conditioners, public acceptance, and marketing and distribution.

The distribution and utilization of composts produced from sewage sludge or refuse may need regulations that will require stabilization, pathogen reduction, and chemical analysis. The general public is more attuned to acceptance of composted sludge or refuse than to unstabilized and potentially malodorous materials. If the application of chemical fertilizer is cheap, subsidized, or practiced for a long time, then farmers will be reluctant to use compost products as fertilizer. The extent of utilization of composts will depend on the type of market and its proximity to the processing facility. Transportation costs and distance to markets will affect product value and its potential use.

PRODUCT QUALITY

The compost quality is greatly dependent on the chemical and physical characteristics of the raw materials and on the processing system used to produce it. N, P, and K are essential plant micronutrients. The higher the content of these elements, the greater the fertilizer value of the compost. Most of the N in the compost is in an organic form and must be mineralized (such as through curing) to inorganic ammonium or nitrate before it is available to plants. Composted organic wastes function as slow-release N fertilizers when applied to soil. The K content of sludge compost is usually low (0.5 percent) because it is water-soluble and remains in the effluent after wastewater treatment. However, in refuse compost the level of K may exceed 1 percent. The N, P, and K values of compost depend not only on the type of materials composted but also on the initial C/N ratio of that material.

Table 3.8 shows the composition of a compost produced from municipal refuse of Bangkok city. The percentages of N, P, K content were 2.58, 1.67, and 0.58, respectively, i.e. several times lower than those present in chemical fertilizers. This compost thus needs to be mixed with chemical fertilizers prior to being used as fertilizer.

Table 3.8 Composition of compost produced from municipal refuse

Items	Properties of composition (based on dry solids)
Total N	2.58%
Total P (P_2O_5)	1.67%
Total K (K_2O)	0.58%
Magnesium (Mg)	0.49%
Calcium (Ca)	6.20%
Sulfur (S)	0.37%
Iron (Fe)	4.389%
Manganese (Mn)	0.158%
Copper (Cu)	0.094%
Zinc (Zn)	0.30%
Boron (B)	0.0513%
Molybdenum (Mo)	0.001%
Chlorine (Cl)	0.66%
Moisture content	23.86%
pH	7.2

Source: Office of Bangkok Municipal Fertilizer (personal communication)

Table 3.9 Nutrient contents of nightsoil, water hyacinth and leaves (dry-weight basis)

Raw materials	Carbon (C) (%)	Nitrogen (N) (%)	Phosphorus (P) (%)	C/N ratio
Nightsoil	54.1	3.6	0.9	15.2
Water hyacinth	45.0	2.9	0.5	15.8
Leaves	49.9	0.8	0.2	60.8

From Polprasert (1984); reproduced by permission of Pergamon Books Ltd

An experiment was carried out at AIT by composting nightsoil, water hyacinth and leaves (compositions are shown in Table 3.9) with different initial C/N ratios (Table 3.10). The method of composting employed was similar to that shown in Figures 3.11 and 3.12. The percentages of N, P, and K of the composted products were found to be highest when the C/N ratio of the mixture was 20 (Table 3.11) (Polprasert, 1984). It is apparent that the magnitude of increase in N, P, and K percentages in the composted products was related to the initial characteristics of the composting materials and the percentage loss of volatile solids after composting.

Organic chemicals such as pesticides, polychlorinated biphenyls (PCBs), and heavy metals may be present in some wastes in sufficiently high concentrations to make the compost undesirable for land application.

Table 3.10 Percentage of raw materials required in preparing compost piles (dry-weight basis)

Raw materials	Initial C/N ratio		
	20	30	40
Nightsoil (%)	14.4	10.9	7.3
Water hyacinth (%)	57.4	25.4	10.9
Leaves (%)	28.2	63.7	81.8
Water hyacinth : nightsoil ratio	80 : 20	70 : 30	60 : 40

From Polprasert (1984); reproduced by permission of Pergamon Books Ltd

SOIL AND PLANT RESPONSE

Composts can be considered as fertilizers rich in organic matter. The organic matter is an excellent soil conditioner because it has been stabilized, decomposes slowly, and thus remains effective over a longer period of time. Composts improve the physical properties of soils, as evidenced by increased water content and water retention; enhanced aggregation; increased soil aeration, soil permeability and water infiltration; and decreased surface crusting. The greatest improvement in soil physical properties occurs in sandy and clay soils.

There are two principal objectives in applying sludge compost as a source of N for crops: to supplement the amount of potentially available N, and to ensure that the level of N applied as compost does not greatly exceed that necessary for attainable crop yields. Based on the N requirements of different crops (see Table 8.7), the application rates will be different and the rates should be such as to produce maximum growth and yield.

3.8.2 Utilization as feed for fish

In waste recycling through aquaculture, herbivorous fish species (e.g. tilapia) feeding on phytoplankton are commonly used. The addition of compost to fish ponds could increase the growth of phytoplankton, and will increase fish growth and yield.

A preliminary experiment was carried out at AIT on tilapia ponds fed with the composts whose characteristics are given in Table 3.11. The fish ponds were earthen, each having a working dimension of 20 × 10 × 1 m, length × width × depth. Three experimental runs, each lasting 6 months, were carried out on four fish ponds. In experiment 1, ponds 2 and 4 served as control without compost feeding. Although the compost feeding rate for ponds 1 and 3 was 50 kg VS or COD/(ha day), the compost materials fed were obtained from the compost piles where initial C/N ratios were 30 : 1 and 20 : 1, respectively. Consequently, because of the higher nutrients content of the compost pile with the initial C/N ratio of 20 : 1 (Table 3.11), the fish yield in pond 3 was almost double of that in pond 1, while ponds 2 and 4 without compost feeding had fish yields several times lower

Table 3.11 Average nutrient contents (dry-weight basis) of compost piles at the starting period and during the 30–60 days of composting

	Initial C/N ratio					
	20		30		40	
Nutrients	Starting period	30–60 days of composting	Starting period	30–60 days of composting	Starting period	30–60 days of composting
N, %	2.0	2.2	1.3	1.9	1.3	1.9
P, %	0.3	0.6	0.2	0.4	0.3	0.3
K, %	1.0	1.0	0.6	0.6	0.6	0.6

From Polprasert (1984); reproduced by permission of Pergamon Books Ltd

than those of the other two ponds (Table 3.12). Based on these results, the compost materials used in experiments 2 and 3 were obtained from the compost piles with the initial C/N ratio of 20 : 1.

The data from Table 3.12 showed that the compost of nightsoil mixed with water hyacinth and vegetable leaves added to fish ponds increases the fish yield considerably, and the amount of fish yields was almost proportional to the rate of compost feeding. Also these experimental results strongly indicated the technical feasibility of using composted nightsoil as feed for tilapia growth. More details of the waste-fed fish ponds are given in Chapter 6.

Table 3.12 Fish yields from compost-fed ponds (Edwards *et al.*, 1983)

Experiment number	Pond number	Compost feeding rate, kg COD/(ha-yr)	Fish yield, kg/pond per 6 months	Extrapolated fish yield, kg/(ha-yr)
1	1	50[a]	26.50	2,845
	2	0	7.60	816
	3	50[b]	51.90	5,572
	4	0	8.15	875
2	1	25[b]	15.69	1,507
	2	100[b]	32.20	3,093
	3	0	6.60	634
	4	50[b]	30.17	2,898
3	1	25[b]	36.14	3,527
	2	50[b]	35.86	3,500
	3	100[b]	57.75	5,636
	4	0	11.25	1,098

[a] Compost with initial C/N ratio 30
[b] Compost with initial C/N ratio 20
Reproduced by permission of Elsevier Science Publishers B.V.

The public health hazard resulting from this practice is considered to be much lower than from the direct feeding of septic tank sludge or nightsoil to fish ponds. This is because most of the enteric microorganisms had been inactivated by heat during composting, and that remaining in the compost, when applied to fish ponds, would be diluted and eventually, subject to natural die-off. However, in areas where certain helminthic diseases are endemic, care should be taken to control the transmission of these helminths, whose life cycles include pond fauna such as snails and/or fish as their intermediate hosts.

REFERENCES

Alexander, M. (1961). *Introduction to Soil Microbiology.* Wiley, New York.

Benedict, A. H., Epstein, E., and English, J. N. (1986). Municipal sludge composting technology evaluation. *J. Water Pollut. Control Fed.*, **58**, 279–89.

Clark, C. S., Bjornson, H. S., Schwartz-Fulton, J., Holland, J. W., and Gartside, P. S. (1984). Biological health risks associated with the composting of wastewater treatment plant sludge. *J. Water Pollut. Control Fed.*, **56**, 1269–76.

Cooper, R. C. and Golueke, C. G. (1979). Survival of enteric bacteria and viruses in compost and its leachate. *Compost Science/Land Utilization*, **20**, 29–35.

Dindal, D. (1978). Soil organisms and stabilizing wastes. *Compost Science/Land Utilization*, **19**, 8–11.

Edwards, P., Polprasert, C., Pacharaprakiti, C., Rajput, V. S., and Suthirawut, S. (1983). Compost as fish food, a practical application of detritivory for cultivation of Tilapia. *Aquaculture*, **32**, 409–13.

Epstein, E., Wilson, G. B., Burge, W. D., Mullen, D. C., and Enkiri, N. K. (1976). A forced aeration system for composting wastewater sludge. *J. Water Pollut. Control Fed.*, **48**, 688–94.

Feachem, R. G., Bradley, D. J., Garelick, H., and Mara, D. D. (1983). *Sanitation and Disease: Health Aspects of Excreta and Wastewater Management.* Wiley, Chichester.

Finstein, M. S., Morris, M. L., and Strom, P. F. (1980). Microbial ecosystem responsible for anaerobic digestion and composting. *J. Water Pollut. Control Fed.*, **52**, 2675–85.

Flintoff, F. and Millard, R. (1969). *Public Cleansing*. Maclaren and Sons, London.

Golueke, C. G. (1972). *Composting: A Study of the Process and its Principles*. Rodale Press, Emmaus.

Golueke, C. G. (1982). Composting: a review of rationale, principles and public health. In *Composting: Theory and Practise for City, Industry and Farm*, pp. 19–25. JG Press, Emmaus.

Gotaas, H. B. (1956). *Composting: Sanitary Disposal and Reclamation of Organic Wastes.* World Health Organization Monograph Series No. 31, Geneva.

Gray, K. R., Sherman, K., and Biddlestone, A. J. (1971). A review of composting—Part 1. *Process Biochem.*, **6**, 32–6.

Haug, R. T. (1979). Engineering principles of sludge composting. *J. Water Pollut. Control Fed.*, **51**, 2189–206.

Haug, R. T. (1980). *Compost Engineering: Principles and Practice.* Ann Arbor Science, Michigan.

Kawata, K., Cramer, W. W., and Burge, W. D. (1977). Composting destroys pathogens in sewage solids. *Water and Sewage Works*, **124**, 76–9.

McGarry, M. G. and Stainforth, J. (1978). *Compost, Fertilizer, and Biogas Production from Human and Farm Wastes in the People's Republic of China.* IDRC-T58e. International Development Research Centre, Ottawa.

McKinley, V. L., Vestal, J. R., and Eralp, A. E. (1985). Microbial activity in composting. *Biocycle*, **26**, 47–50.

McMichael, J. K. (1976). *Health in the Third World—Studies from Vietnam*. Spokesman Books, London.
Metcalf and Eddy, Inc. (1991). *Wastewater Engineering: Treatment, Disposal and Reuse*, 3rd edition. McGraw-Hill, New York.
Millner, P. D., Marsh, P. B., Snowden, R. B., and Parr, J. F. (1977). Occurrence of *Aspergillus fumigatus* during composting of sludge. *Appl. Environ. Microbiol.*, **34**, 765–72.
Morgan, P. R. and Mara, D. D. (1982). *Ventilated Improved Pit Latrines: Recent Developments in Zimbabwe*. World Bank Technical Paper No. 3. Washington, D.C.
Nostrand, J. V. and Wilson, J. G. (1983). *The Ventilated Improved Double-pit Latrine: a Construction Manual for Botswana*. TAG Technical Note No. 3. World Bank, Washington, D.C.
Obeng, L. A. and Wright, F. W. (1987). *The Co-composting of Domestic Solid and Human Wastes*. World Bank Technical Paper No. 57. The World Bank, Washington, D.C.
Pereira-Neta, J. T. (1987). On the treatment of municipal refuse and sewage sludge using aerated static pile composting—a low-cost technology approach. Ph.D. dissertation, Leeds University, U.K.
Polprasert, C. (1984). Utilization of composted night soil in fish production. *Conservation and Recycling*, **7**, 199–206.
Polprasert, C., Wangsuphachart, S., and Muttamara, S. (1980). Composting nightsoil and water hyacinth in the tropics. *Compost Science/Land Utilization*, **21**, 25–7.
Rybczynski, W., Polprasert, C., and McGarry, M. G. (1978). *Low-cost Technology Options for Sanitation*. IDRC-102e. International Development Research Centre, Ottawa.
Shuval, H. I., Gunnerson, C. G., and Julius, D. S. (1981). *Nightsoil Composting*. World Bank, Washington, D.C.
Stentiford, E. I., Taylor, P. L., Leton, T. G., and Mara, D. D. (1985). Forced aeration composting of domestic refuse and sewage sludge. *Water Pollut. Control*, **84**, 23–32.
Strom, P. F. (1985). Identification of thermophilic bacteria in solid-waste composting. *Appl. Environ. Microbiol.*, **50**, 906–13.
U.S. EPA (1981). *Composting Processes to Stabilize and Disinfect Municipal Sewage Sludge*. U.S. Environmental Protection Agency Report No. 430/9–81–011, Cincinnati.
U.S. EPA (1985). *Composting of Municipal Wastewater Sludges*. U.S. Environmental Protection Agency Report No. EPA/625/4–85/014, Cincinnati.

EXERCISES

3.1 You are to conduct a batch composting experiment at home using raw materials generated in the house. Then, prepare a report describing the composition and characteristics of the organic wastes used as raw materials, types of composting (aerobic, anaerobic, aeration, or pile turning, etc.), temperature profiles, composting period, and changes (physical, chemical, biological characteristics) observed in the compost materials.

3.2 You are to conduct a survey of fertilizer usage in your country. Then estimate the amount of compost fertilizer that can be produced from the organic wastes, and calculate economic savings that could be achieved from usage of the compost fertilizer.

3.3 You are to conduct a public opinion survey to assess social acceptability of the usage of compost fertilizer in your community. The survey is to be done through a questionnaire in which you (or a group of colleagues) need to set up suitable questions. The main aim of the survey is to find out means to increase public acceptability of the compost fertilizer.

3.4 A farm near Bangkok has 10 beef cattle. The manure of the cattle is to be composted together with rice straw, and the composted product is to be used as fertilizer. The composting is carried out one batch each month. From the data in Tables 2.7 and 3.1, estimate the quantity of rice straw needed to mix with the cattle manure to obtain a C/N ratio of 25 : 1. Other data necessary for the estimation are as follows:

Manure:
  Total solids = 10 percent
Rice straw:
  C/N ratio = 80 : 1
  Moisture content = 30 percent
  Bulk density = 150 kg/m^3
  N content = 0.2 percent of dry weight

3.5 In Exercise 3.4, calculate the initial moisture content of the manure–straw mixture. What would you suggest if the moisture content is greater than 70 percent.

3.6 A sludge contains 70 percent of volatile portion (dry-weight basis) and 60 percent of this volatile portion is biodegradable. The chemical composition of the volatile portion is $C_{10}H_{19}O_2N$. Calculate the weight and volume of air required to oxidize 1 kg of the sludge. Assume that the N is first converted to ammonia (NH_3) and then to nitrate (NO_3^-). Air contains 23 percent O_2 by weight with a specific weight of 1.2 kg/m^3 at 25 °C and 1 atmospheric pressure.

3.7 Compare the advantages and disadvantages of aerobic composting and anaerobic composting?

3.8 What measures should be taken in using the composted products as soil conditioner or fertilizer in order to minimize the health risk?

4

Biogas production

The energy crisis in the 1980s caused economic problems for many countries, especially poor countries that depend on imported oil and gases. Biogas (also called 'marsh gas'), a by-product of anaerobic decomposition of organic matters (eq. 2.3), has been considered as an alternative source of energy. The biogas can be used in small family units for cooking, heating, and lighting, and in larger institutions for heating or power generation.

The common raw materials used for biogas generation are often defined as 'waste materials', e.g. human excreta, animal manure, sewage sludge, and vegetable crop residues, all of which are rich in nutrients suitable for the growth of anaerobic bacteria. Although some of these materials can be used directly as fuels and fertilizers, they could be used for biogas production, as illustrated in Figure 4.1, to gain some additional heat value (from the biogas) while the other benefits are still retained. Depending on factors such as the composition of the raw materials, organic loading applied to the digesters, and the time and temperature of anaerobic decomposition, some variations in the composition of biogas can be noticed, but it approximately conforms to the following:

Methane	(CH_4)	55–65 percent
Carbon dioxide	(CO_2)	35–45 percent
Nitrogen	(N_2)	0–3 percent
Hydrogen	(H_2)	0–1 percent
Hydrogen sulphide	(H_2S)	0–1 percent

Of the different gases produced, CH_4 is the most desirable gas, because it has a high calorific value ($\approx 9{,}000$ kcal/m^3). The approximate heat value of the biogas is 4,500–6,300 kcal/m^3, depending on the contents of the other gases besides CH_4.

4.1 OBJECTIVES, BENEFITS, AND LIMITATIONS OF BIOGAS TECHNOLOGY

The main purposes and benefits of biogas technology are classified as follows:

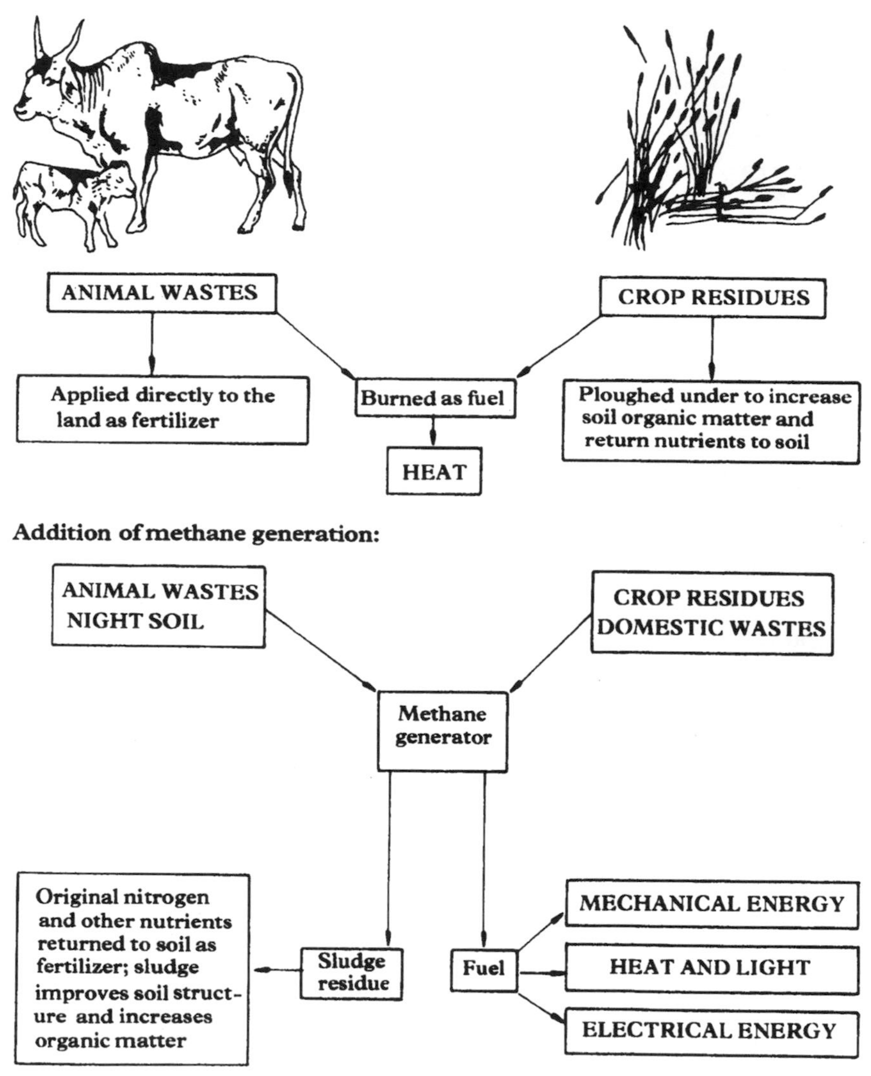

Figure 4.1 Impact of anaerobic fermentation on use of organic wastes (NAS, 1977)

4.1.1 Production of an energy source

The production of an energy resource (biogas) from anaerobic digestion of organic wastes is the most tangible benefit of biogas technology. Practicing biogas production in rural areas can have several advantages, such as relieving demand on electricity, coal, oil, firewood, and problems associated with the administrative/distribution network. The organic matter required for biogas production is abundant and readily available. Decreasing the demand for firewood spares the forest and furthers afforestation efforts.

4.1.2 *Waste stabilization*

The biological reactions occurring during anaerobic digestion in the biogas digester reduce the organic content of waste materials by 30–60 percent and produce a stabilized sludge which can be used as a fertilizer or soil conditioner.

4.1.3 *Nutrient reclamation*

The nutrients (N, P, K) present in the waste are usually in complex organic forms, difficult to be taken up by the crops. After digestion at least 50 percent of the N present is in the form of dissolved ammonia, which can be nitrified to become nitrate, for application to crops (eq. 3.5) so as to be readily available for uptake. Thus digestion increases the availability of N in organic wastes to above its usual range of about 30–60 percent. The phosphate and potash contents are not decreased, and their availability of about 50 and 80 percent, respectively, is not changed during digestion. Anaerobic digestion does not remove or destroy any of the nutrients from domestic and farm wastes, but makes them more available to plants. In addition to being used as a fertilizer, the biogas digester slurry also acts as a soil conditioner and helps to improve the physical properties of the soil. The application of digester slurry to unproductive soils would eventually improve the soil quality, or useless land could be reclaimed.

4.1.4 *Pathogen inactivation*

During the digestion process the waste is kept without oxygen for a long period of time (15–50 days), at about 35 °C. These conditions are sufficient to inactivate some of the pathogenic bacteria, viruses, protozoa, and helminth ova.

However, biogas technology has some drawbacks. When compared with other alternatives, such as composting, biogas production can be construed as the only major advantage of this technology. Other benefits, e.g. waste stabilization and pathogen inactivation, are fulfilled better by composting. A comparative analysis between biogas technology and composting is summarized in Table 4.1. Other limitations include high capital cost, seasonal variations in gas production, as well as operation and maintenance problems.

Because pathogen inactivation in anaerobic digesters is generally incomplete, and the digested slurry is in liquid form, special care has to be taken in the handling and reuse of the digested slurry. Perhaps this is a reason for the negative attitude towards the use of human nightsoil for biogas production.

Table 4.2 summarizes some of the advantages and disadvantages of the use of biogas technology as a method of processing biodegradable organic wastes.

Table 4.1 Comparative analysis of biogas technology and composting (adapted from Tam and Thanh, 1982)

Operating conditions	Composting (aerobic/anaerobic)	Biogas technology
Materials added to nightsoil or animal manure for C/N ratio and moisture adjustments	Vegetation	Water + vegetation
Temperatures	50–70 °C	Ambient
Period of operation	6–8 weeks (including maturation and curing)	4–8 weeks
Nitrogen loss	Low to high	Low to medium
Space required	Same	Same
Modes of operation	Range from traditional to complicated	Complicated
Odor	Problem for open system	Less problems due to closed system
End-product	**Composted materials**	**Digested slurry (and CH_4 gas)**
Weight	Reduced due to water loss	Increased in density due to biomass production
Water content	40–50%	88–92%
Humus content	Abundant	Less than composted product
Pathogen destruction	Good	Moderate
Transport	Easier (solid matter)	More difficult (liquid matter)
Further handling	Not necessary	Drying usually needed
Storage	Easy, little loss of nitrogen	Difficult, with possible loss of nitrogen

4.2 BIOCHEMICAL REACTIONS AND MICROBIOLOGY

The anaerobic digestion of organic material is biochemically a very complicated process, involving hundreds of possible intermediate compounds and reactions, each of which is catalyzed by specific enzymes or catalysts. However, the overall chemical reaction is often simplified to:

$$\text{Organic matter} \xrightarrow[\text{digestion}]{\text{anaerobic}} CH_4 + CO_2 + H_2 + NH_3 + H_2S \tag{4.1}$$

In general, anaerobic digestion is considered to occur in the following stages:

1. liquefaction or polymer breakdown;
2. acid formation; and
3. methane formation.

Table 4.2 Advantages and disadvantages of biogas technology (NAS, 1977)

Advantages	Disadvantages
Produces large amount of methane gas; methane can be stored at ambient temperature	Possibility of explosion
Produces free-flowing, thick, sludge	High capital cost (however, if operated and maintained properly, the system may pay for itself)
Sludges are almost odorless, with the odor not being disagreeable	May develop a volume of waste material much larger than the original material, as water is added to the substrate (this may not be a disadvantage in the rural areas of developing countries where farm fields are located close to the village, thus permitting the liquid sludge to be applied directly to the land, serving both for irrigation and fertilization)
Sludge has good fertilizer value and can be used as a soil conditioner	Liquid sludge presents a potential water pollution problem if handled incorrectly
Reduces organic content of waste materials by 30–50 percent and produces a stabilized sludge for ultimate disposal	Maintenance and control are required
Weed seeds are destroyed and pathogens are either destroyed or greatly reduced in number	Certain chemicals in the waste, if excessive, have the potential to interfere with digester performance (however, these chemicals are encountered only in sludges from industrial wastewaters and therefore not likely to be a problem in a rural village system)
Rodents and flies are not attracted to the end-product of the process; access of pests and vermin to wastes is limited	Proper operating conditions must be maintained in the digester for maximum gas production
Provides a sanitary way for disposal of human and animal wastes	Most efficient use of methane as a fuel requires removal of impurities such as CO_2 and H_2S, particularly when the gas is to be used in internal-combustion engines
Helps conserve scarce local energy resources such as wood	

Figure 4.2 shows the main intermediate compounds formed during anaerobic decomposition of protein, carbohydrate, and fat. Descriptions of the reactions occurring in each of the three stages are as follows:

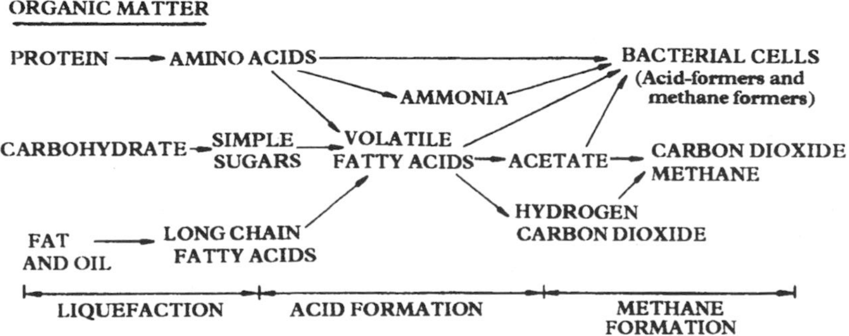

Figure 4.2 Anaerobic digestion of organic compounds

Stage 1: Liquefaction

Many organic wastes consist of complex organic polymers such as proteins, fats, carbohydrates, cellulose, lignin, etc., some of which are in the form of insoluble solids. In this stage these organic polymers are broken down by extracellular enzymes produced by hydrolytic bacteria, and dissolved in water. The simple, soluble, organic components (or monomers) which are formed are easily available to any acid-producing bacteria. It is difficult to distinguish this stage from what is known as stage 2 (acid-formation stage), because some molecules will be absorbed without further breakdown and can be degraded internally.

As shown in Figure 4.2, the hydrolysis reactions in this stage will convert protein into amino acids, carbohydrate into simple sugars and fat into long-chain fatty acids. However, the liquefaction of cellulose and other complex compounds to simple monomers can be the rate-limiting step in anaerobic digestion, as this bacterial action is much slower in stage 1 than in either stage 2 or 3 (NAS, 1977). The hydrolysis rate is dependent on substrate and bacterial concentrations, as well as on environmental factors such as pH and temperature.

Stage 2: Acid formation

The monomeric components released by the hydrolytic breakdown due to bacterial action in stage 1 are further converted to acetic acid (acetates), H_2 and CO_2 by the acetogenic bacteria. Volatile fatty acids are produced as the end-products of bacterial metabolism of protein, fat, and carbohydrate; in which acetic, propionic, and lactic acids are the major products. Carbon dioxide and hydrogen gas are also liberated during carbohydrate catabolism, with methanol, and other simple alcohols, being other possible by-products of carbohydrate breakdown. The proportion of these different substrates produced depends on the flora present, as well as on the environmental conditions.

Stage 3: Methane formation

The products of stage 2 are finally converted to CH_4 and other end-products by a group of bacteria called methanogens. Methanogenic bacteria are obligate anaerobes whose growth rate is generally slower than the bacteria in stages 1 and 2.

The methanogenic bacteria use acetic acid, methanol, or carbon dioxide and hydrogen gas to produce methane. Acetic acid or acetate is the single most important substrate for methane formation, with approximately 70 percent of the methane produced from acetic acid. The remaining methane comes from carbon dioxide and hydrogen. A few other substrates can also be used, such as formic acid, but these are not important, as they are not usually present in anaerobic fermentation. The methanogenic bacteria are also dependent on the stage 1 and stage 2 bacteria to provide nutrients in a usable form. For example, organic N compounds must be reduced to ammonia to ensure efficient utilization by the methanogenic bacteria.

The methane formation reaction in stage 3 is very important in anaerobic digestion. Besides producing CH_4 gas, the methanogens also regulate and neutralize the pH of the digester slurry by converting the volatile fatty acids into CH_4 and other gases. The conversion of H_2 into CH_4 by the methanogens helps reduce the partial pressure of H_2 in the digester slurry, which is beneficial to the activity of the acetogenic bacteria. If the methanogenic bacteria fail to function effectively there will be little or no CH_4 production from the digester, and so waste stabilization is not achieved because the organic compounds will be converted only to volatile fatty acids, which can cause further pollution if discharged into a water course or on land. As the methanogenic bacteria are obligate anaerobes their growth is inhibited even by small amounts of oxygen, and it is essential that a highly reducing environment be maintained to promote their growth. The methane bacteria are also sensitive to other environmental factors, which are discussed in detail in section 4.3.

The current understanding of the microbiology of anaerobic digestion is illustrated in Figure 4.3. There are four main groups of bacteria involved in the process, namely (Brown and Tata, 1985):

I. acid-forming (hydrolytic and fermentative) bacteria,

II. acetogenic (acetate and H_2-producing) bacteria,

III. acetoclastic (methane-forming) bacteria, and

IV. hydrogen-utilizing methane bacteria.

Acid-forming and acetogenic bacteria are collectively called non-methanogenic bacteria in which the major bacterial species are given in Table 4.3. The acetoclastic and hydrogen-utilizing methane bacteria are collectively called methanogenic bacteria as shown in Table 4.4.

The acid-forming bacteria are involved in the hydrolysis and breakdown of complex organic compounds into simple products such as CO_2, H_2, and other volatile fatty acids via two main pathways (Gunnerson and Stuckey, 1986):

$$\text{Substrate} \rightarrow CO_2 + H_2 + \text{acetate} \tag{4.2}$$

$$\text{Substrate} \rightarrow \text{propionate} + \text{butyrate} + \text{ethanol} \tag{4.3}$$

The products from eq. 4.2 can be utilized directly by the acetoclastic bacteria (eq. 4.4) and the hydrogen-utilizing methane bacteria (eq. 4.5) to produce CH_4.

$$\underset{\text{(acetate)}}{CH_3COO^-} + H_2O \rightarrow CH_4 + HCO_3^- + \text{energy} \tag{4.4}$$

$$4H_2 + HCO_3^- + H^+ \rightarrow CH_4 + 3H_2O + \text{energy} \tag{4.5}$$

McInerney and Bryant (1981) reported that the reaction in eq. 4.2 will predominate in a digester having a low H_2 partial pressure. At high H_2 partial pressure eq. 4.3 will be favored with the formation of volatile fatty acids having more than two C atoms (e.g. propionate and butyrate) and ethanol. These products are converted further to methanogenic substrate such as acetate, H_2, and CO_2 by the acetogenic bacteria through the acetogenic dehydrogenation reaction. Some acetogenic bacteria can also convert H_2 and CO_2 to acetate through the acetogenic hydrogenation (Figure 4.3).

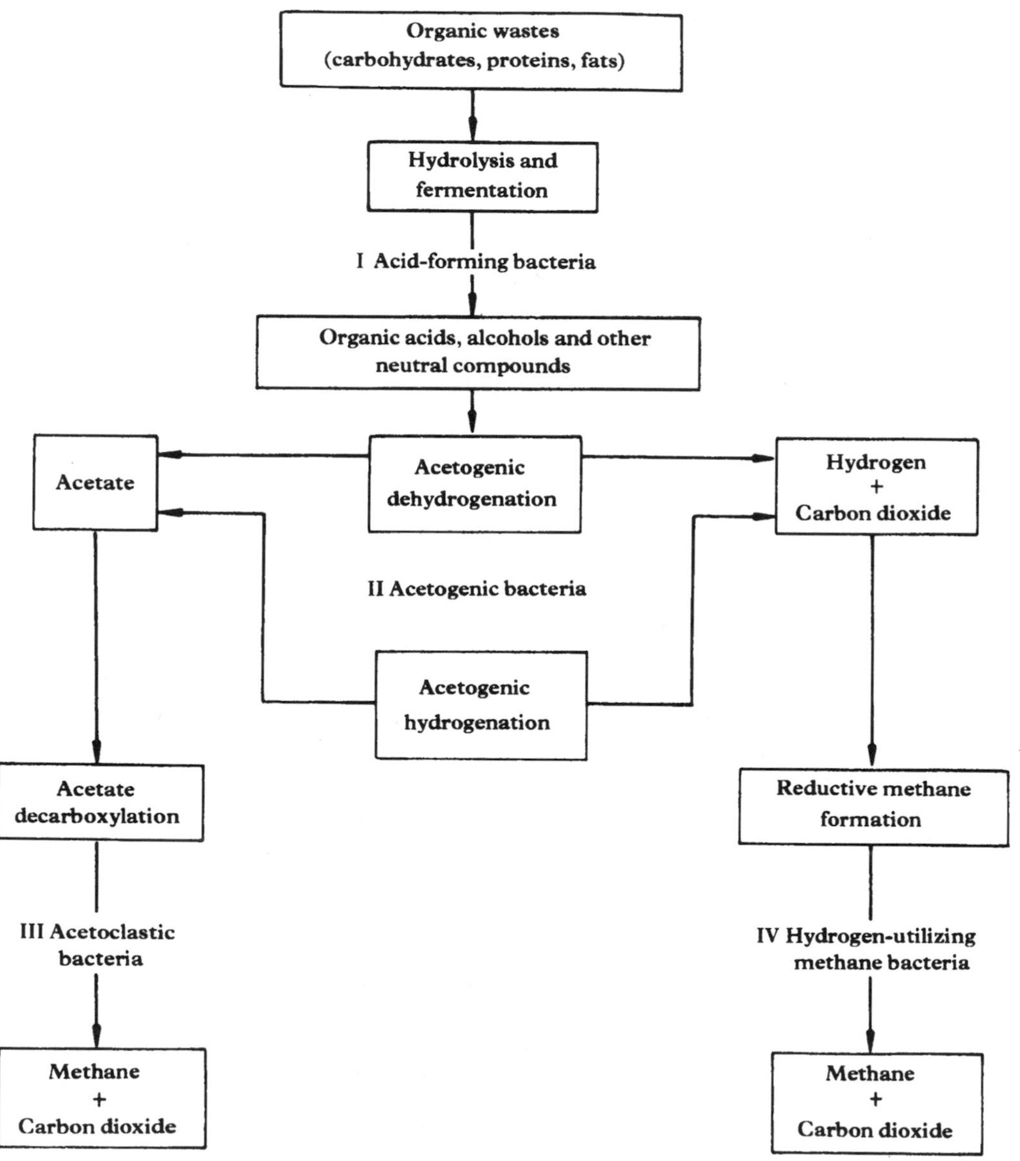

Figure 4.3 Biomethanation of organic wastes (Brown and Tata, 1985)

Brown and Tata (1985) reported that the acetoclastic bacteria have a longer generation time than the acid-forming bacteria (i.e. 2–3 days vs 2–3 h at 35 °C, under optimum conditions). Thus, anaerobic digesters should not receive too high organic loadings because the acid-forming bacteria will produce volatile fatty acids faster than the rate at which the acetoclastic bacteria can utilize them.

It is currently known that the growth of acetogenic bacteria is sensitive to the H_2 partial pressure in the anaerobic digestion system (McInerney and Bryant, 1981).

Table 4.3 Non-methanogenic bacteria

Bacterium	Substrate	Product
1. Cellulose-splitting *Acetivibrio cellulolyticus* *Bacteroides fibrisolvens* *B. succinogenes* *Clostridium cellulovorans* *Cl. dissolvens* *Cl. omelianskii* *Cl. papyrosolvens* *Cl. populeti* *Cl. thermocellum* *Neocallimastix frontalis* *Ruminococcus flavefaciens*	Cellulose	Smaller compounds of low molecular weight, e.g. formate, acetate, succinate, lactate, H_2, CO_2, ethanol
2. Semi-cellulose-splitting *Bacteroides fibrisolvens* *B. ruminicola*	Semi-cellulose	Xylose, arabinose, galactose, mannose, formate, butyrate, lactate, H_2, CO_2, propionate, acetate, succinate
3. Starch-splitting *Bacterium butylicum* *Bacteroides* spp. *Clostridium acetobutylicum* *Lactobacillus* spp.	Starch	Acetobutanol, butyrate, acetate, H_2, glucose, maltose, oligosaccharides
4. Protein-splitting *Bacteroides amylophilus* *B. ruminicola* *Clostridium* spp.	Protein	Amino acids, organic acids, NH_3, H_2S
5. Fat-splitting *Alcaligenes* spp. *Bacillus* spp. *Micrococcus* spp. *Pseudomonas* spp. *Streptomyces* spp.	Fat and oil	Long-chain fatty acids
6. Acetogenic Species of the Enterbacteriaceae and Bacillaceae families	e.g. alcohol, butyric acid, aromatic acids, long-chain fatty acids	Acetate, H_2, CO_2

Sources: Chynoweth and Isaacson (1987); United Nations (1984); Smith *et al.* (1988); ESCAP (1975)

If the H_2 partial pressure is above 0.0001 atmospheres or 0.01 percent, the eq. 4.3 reaction will take place, and the production of acetate will be minimized. As about 70 percent of CH_4 is formed by the reaction in eq. 4.4 (as previously stated), the rate of biogas production will be decreased.

It appears that the reaction in eq. 4.5 is important to the anaerobic digestion process because it removes H_2 gas from the system and helps maintain the low H_2 partial pressure. Mosey (1982) suggested that it may be more useful to monitor the H_2 partial pressure in a digester with a simple electronic instrument, to control the anaerobic digestion process. However, a malfunctioning digester might also be caused by other environmental factors such as pH decrease and overloading; these factors are described in section 4.3.

Table 4.4 Type species of methanogenic bacteria (Ferguson and Mah, 1987), reproduced by permission of Chapman & Hall

Bacterium	Substrate
Methanobacterium alcaliphilum WeN4	H_2–CO_2
Methanobacterium bryantii M.O.H.	H_2–CO_2
Methanobacterium formicicum DSM 863	H_2–CO_2, formate
Methanobacterium thermoautotrophicum ΔH	H_2–CO_2
Methanobacterium uliginosum P2St	H_2–CO_2
Methanobacterium wolfei DSM 2970	H_2–CO_2
Methanobrevibacter arboriphilicus DH1	H_2–CO_2
Methanobrevibacter ruminantium M1	H_2–CO_2, formate
Methanobacterium smithii PS	H_2–CO_2, formate
Methanococcoides methylutens TMA-10	Methanol, trimethylamine[a]
Methanococcus halophilus INMI Z-7982	Methanol, trimethylamine[a]
Methanococcus jannaschii JAL-1	H_2–CO_2
Methanococcus maripaludis JJ	H_2–CO_2, formate
Methanococcus thermolithotrophicus DSM 2095	H_2–CO_2, formate
Methanococcus vanniellii DSM 1224	H_2–CO_2, formate
Methanococcus voltae PS	H_2–CO_2, formate
Methanogenium aggregans MSt	H_2–CO_2, formate
Mathanogenium bourgense MS2	H_2–CO_2, formate
Mathanogenium cariaci JR1	H_2–CO_2, formate
Mathanogenium marisnigri JR1	H_2–CO_2, formate
Mathanogenium thermophilicum	H_2–CO_2, formate
Methanolobus tindarius Tindari 3	Methanol, trimethylamine[a]
Methanomicrobium mibile BP	H_2–CO_2, formate
Methanomicrobium paynteri G-2000	H_2–CO_2
Methanoplanus limicola DSM 2279	H_2–CO_2, formate
Methanosarcina acetivorans C2A	H_2–CO_2, methanol, trimethylamine,[a] acetate
Methanosarcina barkeri MS	H_2–CO_2, methanol, trimethylamine,[a] acetate
Methanosarcina mazei S-6	Methanol, trimethylamine,[a] acetate
Methanosarcina thermophila TM-1	Methanol, trimethylamine,[a] acetate
Methanosphaera stadmanae MCB-3	Methanol plus H_2[b]
Methanospirillum hungatei JF1	H_2–CO_2, formate
Methanothermus fervidus DSM 2088	H_2–CO_2
Methanothrix concilii GP6	Acetate
Methanothrix soehngenii Opfikon	Acetate

[a]May also use mono- and dimethylamine
[b]Requires the combination of methanol and H_2

4.3 ENVIRONMENTAL REQUIREMENTS

Anaerobic reactions in a digester can start quickly with the presence of a good inoculum or seed, such as digested sludge. During start-up or acclimation the seed material should be added to the influent feed material in sufficient quantity, e.g. at least 50 percent. The seed volume can then be progressively reduced while increasing the proportion of the influent feed over a 3–4 week period. At the end of this period the influent feed can be fed alone to the digester to support the growth of anaerobic bacteria. For treatment of animal wastes, nightsoil or crop residues, solid content of the feed material should be about 5–10 percent, the remainder being water.

Like any other biological processes, anaerobic digestion is a multi-parameter controlled process, each individual parameter having control over the process either through its own effect on the system or through interaction with other parameters. These parameters are described below:

4.3.1 Temperature

Temperature, and its daily and seasonal variation, has a pronounced effect on the rate of gas production. Generally two ranges of temperature are considered in methane production. These are mesophilic (25–40 °C) and thermophilic (50–65 °C), similar to those described in Chapter 3. The rate of methane production increases as the temperature increases, but there is a distinct break in the rise at about 45 °C, as this temperature favors neither the mesophilic nor the thermophilic bacteria (Figure 4.4). However, no definite relation other than increasing rate of gas production (within certain limits) can be established. Below 10 °C gas production decreases drastically; therefore operation below this level is not recommended due to the limited amount of gas production (among other technical problems). Above 30–35 °C operation of the digester depends upon a substantial energy input for digester heating, and this in turn will make the operation economically impractical. This suggests that the mesophilic range provides the optimal operational range of temperature, although pathogen inactivation will be less than that to be achieved in the thermophilic range (Figure 3.1).

During the winter period heating of biogas digesters may be necessary, so that growth of anaerobic bacteria, especially the methanogens, is possible. The heating of a digester can be accomplished by heating the influent feeding materials (e.g. with the biogas produced) and feeding it to the digester or by recirculating hot water through pipe coils installed inside the digester. Other means of heating a digester include (Brown and Tata, 1985):

1. housing digester in an enclosure lined with a thick transparent plastic film—the heat within the enclosure can be 5–10 °C higher than the ambient temperature;
2. designing the digester in such a way that water can be held on the roof of the digester and heated by solar radiation;
3. insulating the digester with suitable materials available locally, or placing compostable material such as leaves in an annular space built around the digester.

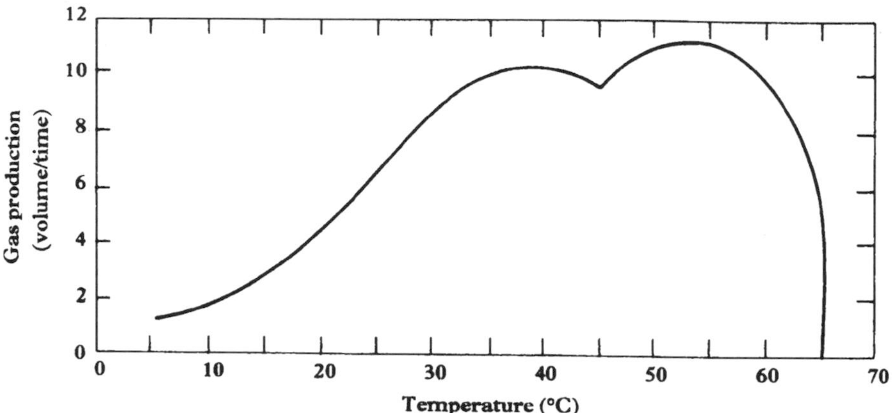

Figure 4.4 Effect of temperature on gas production (Price and Cheremisinoff, 1981)

4.3.2 pH and alkalinity

The operational range of pH in anaerobic digesters should be between 6.6 and 7.6, with the optimum range being 7–7.2. Although acid-forming bacteria can tolerate a pH as low as 5.5, the methanogenic bacteria are inhibited at such low pH values. The pH of a digester may drop to below 6.6 if there is an excessive accumulation of volatile fatty acids. Such an accumulation may occur when the organic loading rates are excessively high and/or when toxic materials are present in the digester, all producing inhibitory effects to the methanogenic bacteria.

Appropriate measures should be taken promptly when there is a lowering of pH in an anaerobic digester, due to accumulation of volatile fatty acids or increase in H_2 partial pressure, and the rate of CH_4 production decreases. In general, the feeding of the digester should be stopped to allow the methanogens to utilize the accumulated volatile fatty acids and H_2 at their own pace. When the optimal gas production rates are re-established the normal loading of the digester can be resumed. In addition, the pH of the digester needs to be adjusted to neutrality by the addition of lime or other basic materials. If the alkalinity of the digester slurry is maintained within the range 2,500–5,000 mg/L, a good buffering capacity is normally obtained in the digester.

4.3.3 Nutrient concentration

Most of the information available in this area has been obtained from studies of rumen bacteria. The energy for growth of most rumen bacteria is obtained through anaerobic fermentation of carbohydrates. N is utilized for cell structure. To guarantee normal biogas production it is important to mix the raw materials in accordance with a proper C/N ratio. Bacteria use up C 25–30 times faster than they use N. Therefore, at this ratio of C/N (25–30/1) the digester is expected to operate at the optimal level of gas production, similar to that required for composting (Chapter 3). The importance of other elements such as P, Na, K, and Ca in gas production is also indicated. However, C/N ratio is considered to be the essential factor.

The C/N ratios of various organic wastes are given in Table 3.1. As human nightsoil, animal manures, and sewage sludge have C/N ratios lower than the optimum values, they may be mixed with other agricultural residues that have high C/N ratios. Examples of these residues are wheat straw, rice straw, water hyacinth, and duckweeds, all of which are usually biodegradable, and can be made more so by physically reducing their size (e.g. shredding) or by precomposting. However, problems can arise with these agricultural residues because they float to the top, thereby forming a hard layer of scum on the slurry surface inside the digester (Polprasert *et al.*, 1982).

When a combination of organic wastes is considered for anaerobic digestion, Table 3.1 should be used as a reference, to select the proper combination and appropriate C/N ratio according to the method shown in Example 3.1.

4.3.4 Loadings

This term can be expressed as organic loading (kg COD or volatile solids (VS)/m^3-day) and hydraulic loading or retention time (HRT). A high organic loading will normally result in excessive volatile fatty acid production in the digester (sour condition) with a consequent decrease in pH, and will adversely affect the methanogenic bacteria. A low organic loading will not provide a sufficient quantity of biogas for other uses, and will make the digester unnecessarily large. Because organic materials fed to anaerobic digesters are in semi-solid form, organic loading to a digester can be conveniently interpreted in terms of VS.

Section 4.4 describes the two main types of anaerobic digesters, i.e. dispersed-growth digesters (those employing dispersed-growth bacteria) and attached-growth digesters (those employing attached-growth bacteria). Optimum organic loadings to dispersed-growth digesters have been reported to be 1–4 kg VS/(m^3-day) and 1–6 kg COD/(m^3-day); for attached-growth digesters they are 1–15 and 5–30 kg COD/(m^3-day) for anaerobic filters and up-flow sludge blanket digesters, respectively (Barnet *et al.*, 1978; Brown and Tata, 1985).

HRT has an equally significant effect on digester performance. Too short an HRT will not allow sufficient time for anaerobic bacteria, especially the methane-forming bacteria, to metabolize the wastes. Too long an HRT could result in an excessive accumulation of digested materials in the digester, and construction of a digester which is too large. Similar to the organic loadings, an optimum HRT depends on the characteristics of influent feed materials and environmental conditions in the digesters. For dispersed-growth digesters the optimum HRT falls within the range 10–60 days; while for the attached-growth digesters, optimum HRT values are 1–10 and 0.5–6 days for anaerobic filters and up-flow sludge blanket digesters, respectively (Brown and Tata, 1985).

It appears from the above information that the attached-growth digesters can be operated at higher organic loadings or shorter HRT than the dispersed-growth digesters. This advantage is attributed to the nature of attached-growth bacteria that attach to the media and/or stay in the digester for a long time. They are thus present in the digesters in high concentrations, not easily washed out or overflowed in the digester slurry, and are well acclimatized to incoming wastes. To increase the process performance or achieve higher organic loading rates, the dispersed-growth digesters can have part of their slurry recycled back to the digesters in order to retain more active biomass and increase the solids retention time.

4.3.5 *Presence of toxic compounds*

For anaerobic digestion of organic wastes such as human excreta, animal manure, and other agricultural residues, accumulation of volatile fatty acids, H_2, and undissociated ammonia is commonly associated with digester failure. The presence of molecular oxygen is also inhibitory to the methanogenic bacteria.

A list of common inhibitors to anaerobic digestion is given in Table 4.5. Organic wastes containing those inhibitors as listed in Table 4.5 should be pretreated or diluted so that the inhibitor concentrations are below the inhibiting concentrations prior to being fed to anaerobic digesters.

Table 4.5 Inhibitors of biomethanation (U.S. EPA, 1979)

Parameter	Inhibiting concentration (mg/L)
Volatile acids	> 2,000 (as acetic acid)[a]
Ammonia nitrogen	1,500–3,000 (at pH > 7.6)
Sulfide (soluble)[b]	> 200; > 3,000 toxic
Calcium	2,500–4,500; 8,000 strongly inhibitory
Magnesium	1,000–1,500; 3,000 strongly inhibitory
Potassium	2,500–4,500; 12,000 strongly inhibitory
Sodium	3,500–5,500; 8,000 strongly inhibitory
Copper	0.5 (soluble metal)
Cadmium	150[c]
Iron	1,710[c]
$Chromium^{+6}$	3
$Chromium^{+3}$	500
Nickel[d]	2

[a] Within the pH range 6.6–7.4, and with adequate buffering capacity, volatile acids concentrations of 6,000–8,000 mg/L may be tolerated
[b] Off-gas concentration of 6 percent is toxic
[c] Millimol of metal per kg of dry solids
[d] Nickel promotes methane formation at low concentrations. It is required by methanogens

4.3.6 Mixing

Mixing digester slurry is important to provide better contact between the anaerobic bacteria and the incoming organic wastes, so that biogas production is enhanced. It reduces the settling of solids or accumulation of digested solids at the bottom of the digester and helps to prevent and break up scum formation at the slurry surface. For small-scale digesters mixing of the digester's slurry can be accomplished manually, as described in section 4.4. In large-scale digesters mixing can be done mechanically by stirring and recirculating the gas and/or the digested slurry.

4.4 MODES OF OPERATION, TYPES OF BIOGAS DIGESTERS AND TROUBLE SHOOTING

There are different types of anaerobic digesters for experimental purposes, pilot plant investigations and actual field use. Their design, materials, system performance, price, etc. naturally vary a great deal. Operationally, it is required that air is excluded from the content of the digester, and sufficient volume is provided within the digester for the biological reactions to take place.

4.4.1 Modes of operation

Major modes of digester operation can be classified into three groups as follows:

BATCH OPERATION

In this mode of operation the digester is filled completely with organic matter and seed inoculum, sealed, and the process of decomposition is allowed to proceed for a long time until gas production is decreased to a low rate (duration of process varies based on regional variation of temperature, type of substrate, etc.). Then it is unloaded, leaving 10–20 percent as seed, then reloaded and the operation continues. In this type of operation the gas production is expected to be unsteady and the production rate is expected to vary from high to low. Digestion failures due to shock load are not uncommon. This mode of operation, however, is suitable for handling large quantities of organic matter in remote areas. It may need separate gas-holders if a steady supply of gas is desired.

SEMI-CONTINUOUS OPERATION

This involves feeding the digester on a more regular basis. Feeding is done usually once or twice a day. The digested organic matter is also removed at the same time intervals. This type of operation is suitable when there is a steady flow of organic matter. The digester volume will have to be large enough to serve both as a reactor and a gas storage tank. Total gas production per unit organic weight of organic matter loaded is usually high. Most of the operational biogas plants in the field are of this type.

CONTINUOUS OPERATION

In this mode of operation the feeding and removal of organic matter take place continuously. The amount of material to be digested is kept constant in the digester by overflow or pumping. The process has been used in treatment of liquid wastes or organic wastes with low solid concentrations. Continuous operation relies heavily on external energy inputs for pumping and mixing, and therefore has limited application in areas where energy resources are limited.

It should be noted, however, that seed inoculum is added at the beginning of the process of anaerobic digestion (start-up). Actual operation starts as soon as the microbial population establishes itself, and gas production, including percentage of methane gas in the total gas production, stabilizes. In field implementation of anaerobic digestion the animal manure itself acts as seed inoculum, and the process may reach stability within 20–30 days of operation (depending upon temperature, digester size and type of substrate).

4.4.2 Types of digesters

Various designs of biogas digesters for actual field operation range from a simple design to a sophisticated one. It is generally known that an increased level of sophistication in design results in a higher demand for manpower with appropriate skills, which are normally in short supply. Furthermore, more advanced designs increase the cost associated with construction and operation, with little apparent increase in the level of gas production.

Double-stage digesters (i.e. first stage for acid formation and second stage for CH_4 formation) are mostly designed for experimental purposes to provide more insight into and understanding of the complex nature of the anaerobic process. Single-stage digesters, on the other hand, are of a more practical nature.

In general, types of digesters can be divided into two main groups: those utilizing dispersed-growth bacteria and those utilizing attached-growth bacteria. Because most digesters are operated as flow-through without sludge recycling, the dispersed bacteria overflow into the digester slurry, making the HRT equal to the mean cell residence time (θ_c) in the digester. To provide longer θ_c in the digesters, attached-growth bacteria have been employed; in this case the anaerobic bacteria, attached to artificial media or settled as a blanket in the digesters, decompose the organic wastes. Descriptions of dispersed-growth digesters and attached-growth digesters are given below.

DISPERSED-GROWTH DIGESTERS

Combined digester and gas-holder—fixed dome (Chinese)

In this type of digester (Figure 4.5) the gas storage volume is directly above the digesting contents of the reactor, with the volume of the digester equalling the volume of slurry and gas combined. The small-capacity digesters of this type (6–

12 m^3) are suitable for a single family or a group of families. The larger sizes (50 m^3) are designed for community gas requirements. The roof, walls, and bottom of the reactors are constructed either of bricks *in situ* or of precast concrete. The inlet opening and displacement tank are made of lime clay. The top and the bottom are hemispherical, and are joined together by straight sides. If the digester is constructed from bricks the inside surface is sealed by several thin layers of mortar to make it watertight and gastight.

This digester is completely buried underground to ensure uniform temperature distribution, to save space, and to make use of soil support. The inlet pipe is straight and ends at mid-level in the digester. The outlet is at the same level, and consists of a fairly large storage tank. A manhole is provided at the top to gain access to the digester during cleaning. When the digester is in operation this manhole is covered and sealed with clay.

The gas produced during digestion is stored under the dome, and displaces some of the digester contents into the effluent chamber; this creates gas pressure in the dome of between 1 and 1.5 m of water. The high pressure exerted on the construction is partially balanced by the weight of soil which covers the digester's top.

This type of digester is common in developing countries. Approximately seven million of this type have been constructed in China, but, due to various reasons, more than 50 percent of them are reportedly not functioning well. Most of the operational parameters of this digester are semi-continuous, although some digesters are batch-operated. Figure 4.6 shows the construction of a 50 m^3 fixed-dome digester in a rural village in Thailand.

Typical feed to these digesters is usually a mixture of animal manure, nightsoil, water hyacinth, and agricultural residues, depending on their availability and C/N ratios. The rate of gas production is reported to be 0.15–0.2 m^3/day per m^3 of digester capacity, but in tropical areas it can be as much as 0.3–0.4 m^3/day per m^3 of digester capacity.

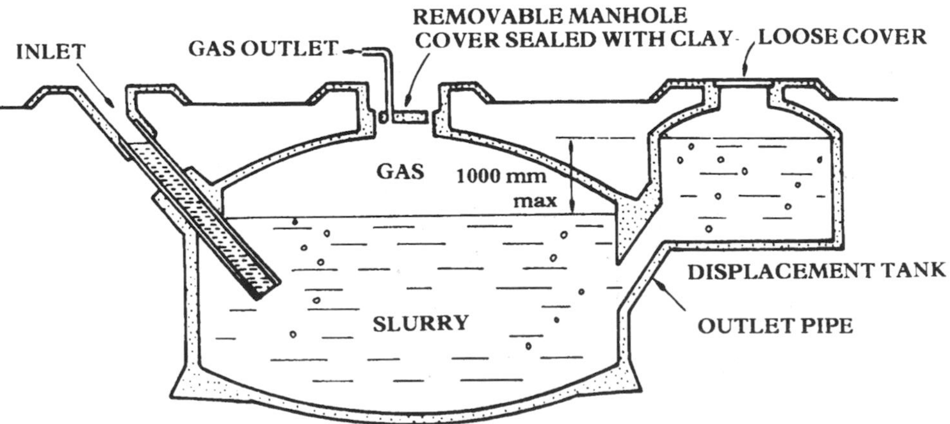

Figure 4.5 A fixed-dome digester

Figure 4.6 Construction of a 50 m^3 fixed-dome digester

Mixing of the digester contents is accomplished through feeding of the influence materials and withdrawal of effluent slurry. An improvement of the mixing efficiency of the digester contents is shown in Figure 4.7. In this digester design, mixing is accomplished by pulling the plastic rope back and forth at the inlet and outlet, causing the plastic blades to agitate the slurry to provide better contact between the organic matter and the anaerobic bacteria. This method of mixing prevents both the blocking of the inlet and outlet pipes and accumulation of the digested residue in the digestion chamber.

Another modification of the Chinese-type fixed-dome digester is given in Figure 4.8. The mixing device comprises four steel plates attached to a steel bar, and mixing is done manually by moving the steel plates up and down a few times daily.

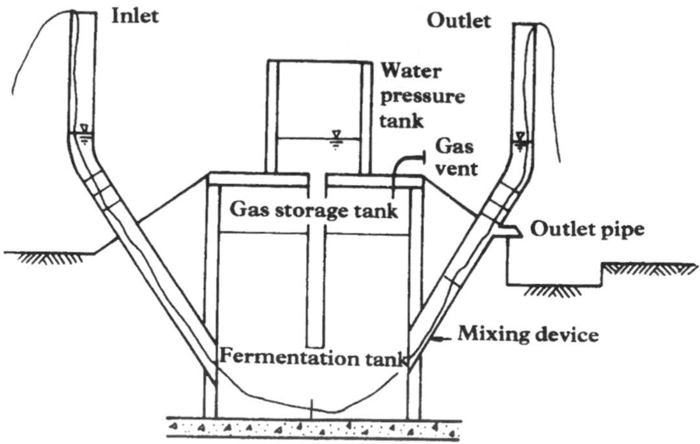

Figure 4.7 A low-cost biogas digester (note a unique mixing device)

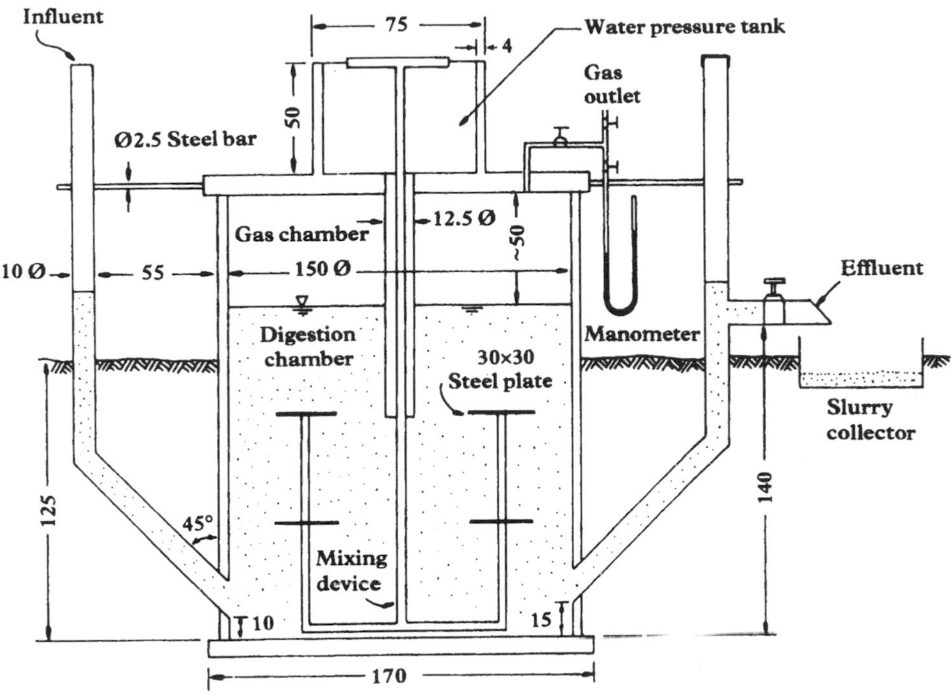

Figure 4.8 A pilot small-scale digester (adapted from Polprasert *et al.*, 1982) (dimensions in cm)

In a study conducted at the Asian Institute of Technology (AIT), Bangkok, Thailand, four pilot small-scale digesters (size 3.5 m^3), as shown in Figure 4.8, were fed with heterogeneous substrate comprising nightsoil, chopped and sun-dried water hyacinth, and rice straw (Polprasert *et al.*, 1982). These digesters were operated under ambient conditions (mean temperature 30 °C). The characteristics of these feeding materials are given in Table 4.6; the proportions of these materials mixed to obtain the desired C/N ratio of 25 are shown in Table 4.7. Among the three hydraulic retention times (HRT) studied, biogas production was highest from the digesters operated at an HRT of 30 days (Figure 4.9). A comparison using volumetric methane production rate per unit volume of digester (Table 4.8) showed the digesters operated at the 30-day HRT to give the highest value of 0.33 $m^3/(m^3\text{-day})$; this parameter is important for operational design because it relates the size (or capital cost) of digester to gas production.

Table 4.6 Characteristics of raw materials used to prepare influent mixture

Material	Total organic carbon (%)	Total Kjeldahl nitrogen (%)	Total phosphorus (%)	Total volatile solids (%)	Carbon/ nitrogen ratio	Moisture content (%)
Nightsoil	47.6	4.5	0.7	85.6	10.7	82.3
Rice straw	42.7	0.9	0.1	77.3	50.3	14.1
Water hyacinth	37.8	1.4	0.4	68.0	27.0	43.7

Table 4.7 Proportions of raw materials used to obtain the desired C/N ratio of 25, and characteristics of influent mixture

			Approximate proportion of raw materials (nightsoil/rice straw/water hyacinth)		Influent mixture				
Hydraulic retention time (days)	Carbon to nitrogen ratio (C/N ratio)	Total solids to water ratio	Dry weight basis	Wet weight basis	Total solids (g/L)	Total volatile solids (g/L)	Chemical oxygen demand (g/L)	Total Kjeldahl nitrogen (mg/L)	pH
30	25	8 : 92	1.0 : 5.6 : 3.0 – 1.0 : 8.0 : 8.0	3.1 : 3.2 : 4.3 – 3.0 : 9.5 : 11.5	46.9	36.1	46.0	887.8	7.4
50	25	8 : 92	1.0 : 3.4 : 1.0 – 1.0 : 0.9 : 1.0	5.5 : 3.8 : 1.3 – 8.3 : 1.4 : 2.1	49.0	37.7	48.7	899.2	7.4
70	25	8 : 92	1.0 : 2.0 : 1.0 – 1.0 : 2.8 : 1.0	6.1 : 2.2 : 1.3 – 5.6 : 3.1 : 1.3	48.1	37.2	49.6	880.1	7.4

Table 4.8 Summary of biogas production from the pilot-scale digesters

Hydraulic retention time (days)	Organic loading rates (kg TVS m^{-3} day^{-1})	Number of digesters	Total biogas production rate [m^3 (kg TVS)$^{-1}$ day^{-1}]	Volumetric biogas production rate (m^3 m^{-3} day^{-1})	Methane production rate [m^3 (kg TVS)$^{-1}$ day^{-1}]	Volumetric methane production rate (m^3 m^{-3} day^{-1})
30	1.20	4	0.28	0.33	0.18	0.22
50	0.75	4	0.32	0.24	0.21	0.16
70	0.53	3	0.25	0.13	0.16	0.09

Polprasert *et al.* (1982) reported that there was floating scum formed in the digesters fed with a mixture of nightsoil, water hyacinth, and rice straw, causing operational problems with mixing, and feeding and withdrawal of slurry. Table 4.9 shows the characteristics of the digested slurry which still contained high concentrations of organic matter (based on total volatile solids and chemical oxygen demand) and N, and thus need to be treated further prior to disposal. On the other hand, this slurry is suitable for reuse as fertilizer or soil conditioner (Figure 4.1).

Floating gas holder digester (Indian)

This type of digester, designed by the Khadi and Village Industries Commission (KVIC) in India, consists of a cylindrical well, most commonly made from bricks, although chicken-wire mesh reinforced concrete has been used. Pressure in the digester remains constant because the gas produced is trapped under a floating cover on the surface of the digester which rises and falls on a central guide (Figure 4.10a,b). The cover is usually constructed of mild steel, although, due to corrosion problems, other materials such as ferrocement, bamboo cement, different kinds of plastics, and fiberglass have been used. A major cause for the loss of heat from this kind of digester is its cover. The digester may be buried under the ground to prevent heat loss, leaving the gas holder more or less above the ground.

This digester is fed semi-continuously through a straight inlet pipe, and displaces an equal amount of slurry through an outlet pipe. When the ratio of height to diameter is high a central partition wall is built inside the digester to prevent any short-circuiting of the substrate.

This design is extensively used in India, and fed with cattle dung only. If agricultural residues are being used, mixed with animal wastes, the residues should be chopped into small pieces to prevent blockage. This design is simple to build and maintain, and does not require an experienced builder. It is used throughout the world, and together with the Chinese-type fixed-dome digester is the most common type of digester used for treating organic waste.

Table 4.10 briefly compares the various aspects of the Chinese and Indian biogas digester systems.

Table 4.9 Characteristics of effluents from biogas digesters

HRT (days)	Number of digesters	Total solids		Total volatile solids		Nonfiltered chemical oxygen demand		Total Kjeldahl nitrogen		pH
		g/L	Reduction (%)	g/L	Reduction (%)	g/L	Reduction (%)	mg/L	Reduction (%)	
30	4	24.7	47	19.1	47.3	24.8	46.7	707.5	20.5	7.4
50	4	18.4	62	13.7	62.7	17.7	63.4	635.8	29.1	7.3
70	3	16.5	68	11.2	70.0	14.9	69.3	619.5	31.4	7.2

Table 4.10 Comparison between Chinese and Indian designs of biogas digesters (Tam and Thanh, 1982)

	Chinese design	Indian design
Construction material	Several different materials Usually locally available	Masonry May be brought from outside village
Construction	Closed, underground masonry or concrete pit with adjacent inlet and outlet Requires skill to build dome, careful lining to prevent gas leaks Usually self-help	Simply above-ground tank Easy to build but hard to install when drum cannot be made or easily carried Self-help possible, but gas holder has to be produced in workshop
Gas storage	In dome combined with digestion chamber Manometer indicates gas volume For use throughout digester lifespan with occasional linings Keeping it gastight is a problem; needs efficient lining	In floating metal drum Height of drum indicates gas volume Drum needs regular painting to prevent corrosion No problem in keeping gastight
Gas pressure	High: up to 1000 mm water column Varying according to gas use Automatic release of excessive gas through manometer	Low: 70–150 mm water column Steady, due to floating drum Automatic release of excessive gas through gas drum
Efficiency	Low, due to gas escape through large inlets and outlets: 0.15–0.30 m^3 gas produced per m^3 digester per day Stable through seasons, due to good insulation of underground construction	Higher, gas escape insignificant: 0.30–0.60 m^3 gas produced per m^3 digester per day Subject to seasonal variations, loss heat through metal drum
Feeding materials	Mostly mixtures of animal wastes, human excreta, household refuse, agricultural residues	Virtually only cow dung, but agricultural residues can be added
Operation	Mostly semi-continuous loading, can be batch operation Effluent removal by pump or bucket Labor-intensive for batch loading, emptying pit and removing effluent	Virtually semi-continuous loading Effluent removal by gravity flow Virtually no attention beyond mix and feeding influent
Maintenance	Wall lining	Drum painting
Cost	Low, because no metal part	High, due to metal drum

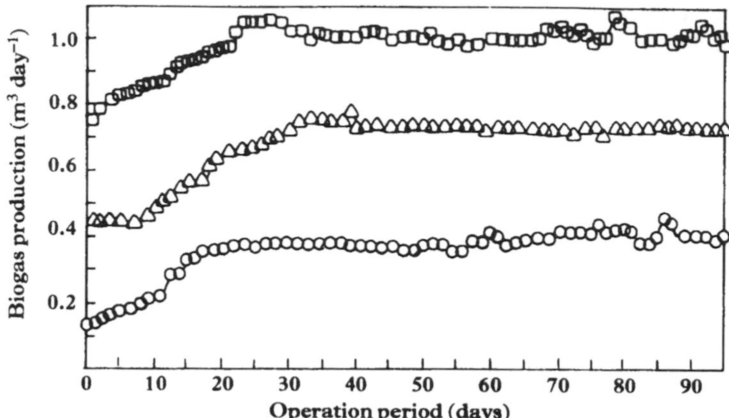

Figure 4.9 Mean daily biogas production at hydraulic retention times of 30 (□), 50 (△), and 70 (○) days from pilot small-scale digesters. Each point is the mean of four digesters

Plug-flow digester (horizontal displacement digesters)

This type of digester consists of a long trench cut into the ground and lined either with concrete or an impermeable membrane. The digester is covered either with a flexible cover anchored to the ground, which acts as a gas holder, or with a concrete or galvanized-iron top. In the latter type a gas storage tank is required. The main difficulty in plug-flow digestion is to ensure true plug-flow conditions, and to achieve this, the length has to be considerably greater than the width and depth (Figure 4.11). The digester is fed semi-continuously through the inlet which is located at one end of the digester. This type of digester has limited use in developing countries, and only a few of them are operating, e.g. in Mexico. It can sustain relatively high loading rates of organic matter with less frequent failure due to shock loads.

Similar to the plug-flow digester is the 'bag digester' which is commonly used in Taiwan and Korea. Early digesters of this type were made of neoprene-coated nylon, which was expensive. The present ones are made from red mud plastic (RMP), which is a residue of aluminum refineries, costing about US\$50/m^3 of digester volume (Yang and Nagano, 1985). There are also bag digesters made of PVC in Central America. This is a very lightweight digester, easy to install and durable (estimated at 20 years). The Chinese have also started producing this kind of digester. A rapid expansion in the use of bag digesters is expected in China, due to their simplicity, durability, and low cost—US\$8/m^3 of digester volume—(Figure 4.12). This digester is fed semi-continuously. The inlet and outlet to the digester are at opposite ends. The biogas is accumulated at the top of the digester and collected through gas pipes. The digester is easily installed by excavating a shallow trench slightly deeper than the radius of the digester. This digester is designed to receive mostly pig manure, but other agricultural residues can also be added. The reported value of gas production is very much temperature-dependent. For pig manure values are 0.14 m^3/day per m^3 of digester capacity at 8 °C to

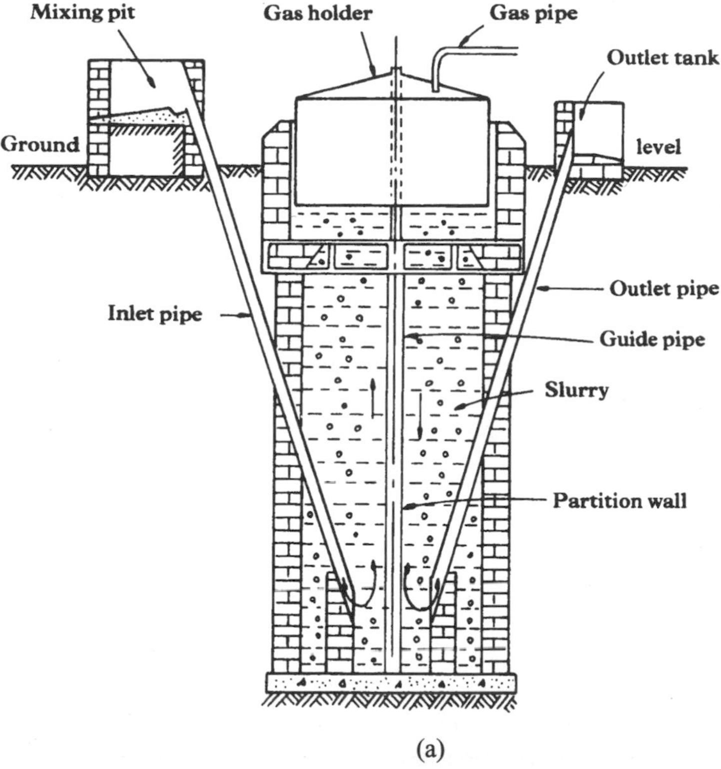

(a)

(b)

Figure 4.10 (a) Schematic of a floating gas holder digester. (b) Operation of a floating gas holder digester (6 m^3 in size)

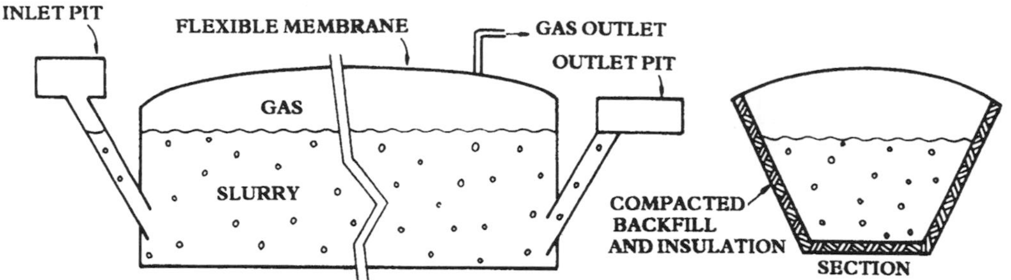

Figure 4.11 A plug-flow digester

0.7 m^3/day per m^3 of digester capacity at 32 °C (Hao, 1981). A higher gas production rate of 1.53 m^3/day per m^3 of digester volume was reported by Yang and Nagano (1985) for the RMP digester operating with slurry recycle at the ratio of 0.25 of the influent flow rate.

Separate gas holder digester

This kind of system could include any of the digesters explained so far. The difference is that a separate tank is added for gas collection (Figure 4.13). In addition, the gas holding tank can be used as a common reservoir for a few digesters and therefore the system as a whole can serve larger communities or industrial plants. The chief advantage of this kind of system is that it provides a continuous supply of biogas (even in the case of a batch system). Furthermore, the gas pressure can be regulated; therefore, appliances and engines may be designed and used at their optimum working conditions. However, this type of digester has not been introduced extensively in developing countries.

Conventional digesters

This type of digester is used in conventional sewage treatment plants to treat sludge. The gas produced is used to augment the energy demand of the treatment

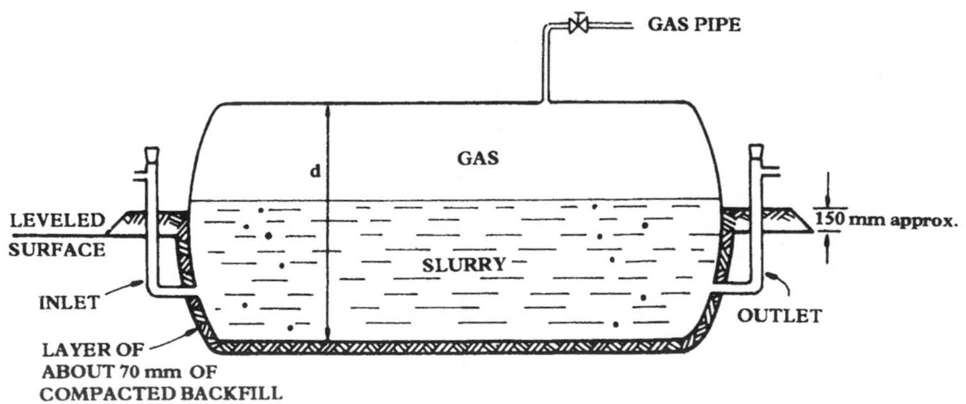

Figure 4.12 A bag digester

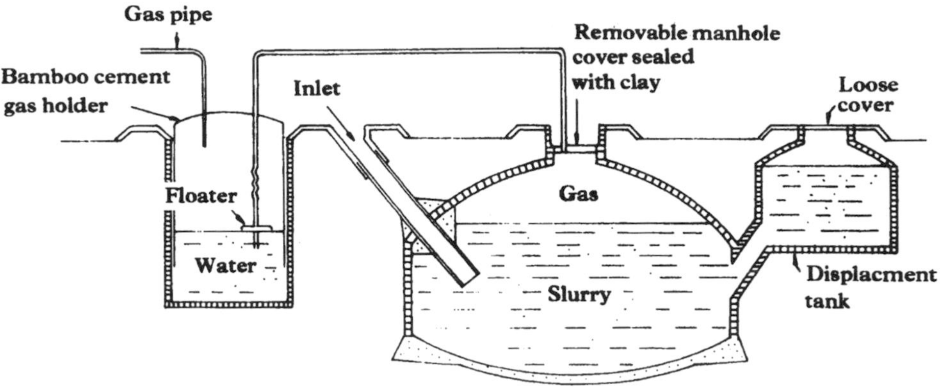

Figure 4.13 A fixed-dome digester with a separate gas holder

plant or to heat the digester. Figure 4.14 shows a typical design of a cylindrical shape digester. Major components of the digester are for mixing and recirculation of the digester contents, elimination of scum, gas collection, and digested sludge withdrawal. The operation of these digesters requires skilled labor and regular monitoring. Variations of designs are in shape and in mixing method. Elimination of scum is easier in a dome-shaped digester because of the low surface area in the top portion of the digester, however its construction requires special techniques. The size of these conventional digesters ranges from 250 to 12,000 m^3 or more.

Besides the above types of digesters, there are numerous modifications of biogas digesters made to suit particular needs and local conditions. Schematic drawings of these modifications can be found in the book *Biomethanation* by Brown and Tata (1985). A summary of design criteria and performance data of dispersed-growth biogas digesters is given in Table 4.11.

ATTACHED-GROWTH DIGESTERS

Anaerobic filter (Young and McCarty, 1969)

This is essentially a filter column packed with stationary media such as rocks, gravels, or different types of proprietary plastic materials (Figure 4.15). Columns with media having a larger specific surface area (surface area per unit volume of the medium) will have more fixed-film bacteria attached to the media and some entrapped within the void spaces of the media. In general, the packing media to be installed in anaerobic filters should have a high specific surface area (surface area to volume ratio) to provide a large surface for the growth of attached biofilms, while maintaining a sufficient void volume to prevent the reactor from plugging either from particulate solids entering with influent waste stream or bacterial floc growing within the reactor (Vigneswaran *et al.*, 1986). Commercial media available for use in anaerobic filters include loose fill media such as Pall rings and modular block media formed from corrugated plastic sheets, in which the

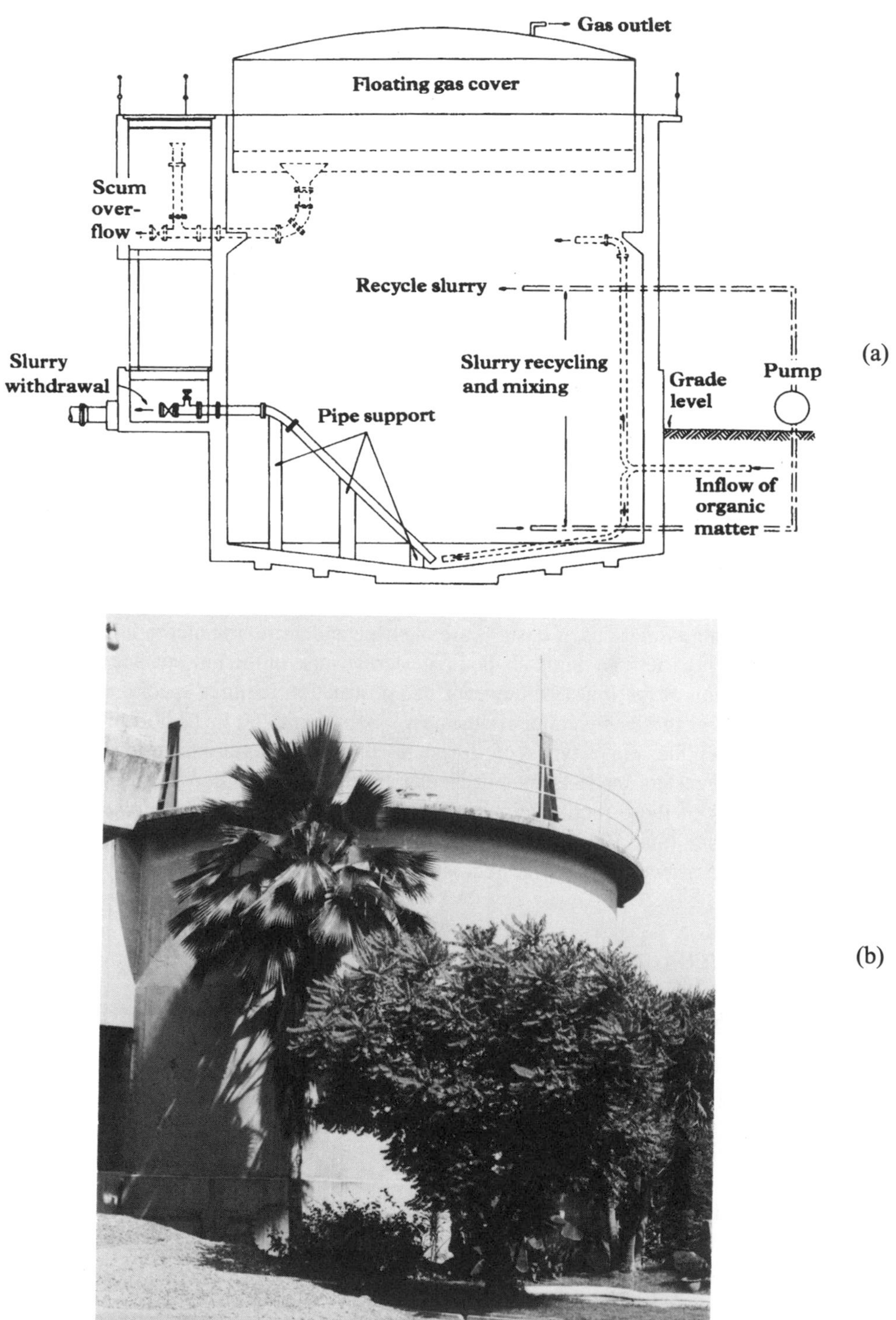

Figure 4.14 (a) Schematic diagram of a conventional digester at Huay Kwang, Bangkok, Thailand. (b) Digester at Huay Kwang, Bangkok, Thailand.

Table 4.11 Design criteria and performance data of dispersed growth biogas digesters

Temperature inside digester (°C)	25–40 (optimum 35, mesophilic)
pH	6.6–7.6 (optimum 7.0–7.2)
Alkalinity (mg $CaCO_3$/L)	2,500–5,000
C/N ratio	25–30
Loading rate	
(kg VS/m^3-day)	1–4
(kg COD/m^3-day)	1–6
Feed solid content (%)	5–10
HRT (days)	10–60 (no sludge recycle)
Gas production (m^3/day per m^3 of digester capacity)	
Fixed-dome digester	0.15–0.40 (part of digester for gas storage)
Floating gas holder digester	0.30–0.60
Volume for gas storage (% of 1-day gas production)	50–100
Seeding material at start-up	> 50% of influent feed
Air tightness	Essential
Water tightness	Strongly required
COD removal (%)	30–70
VS removal (%)	40–70
N reduction (%)	20–35

channels in modular media may be tubular or crossflow (Figure 4.16). The specific surface area of media used in full scale anaerobic filters averages about 100 m^2/m^3. Table 4.12 lists size, specific surface area and porosity of some packing media used for anaerobic filters. The waste flows upward or downward through the anaerobic filter column, contacting the media on which anaerobic bacteria grow and are retained. Because the bacteria are retained on the media and not easily washed off in the effluent, the mean cell residence times (θ_c) in the order of 100 days can be achieved with short HRT (Metcalf and Eddy, 1991). Table 4.13 shows the advantages and disadvantages of the anaerobic filter process, while Table 4.14 reports operating and performance data of some full-scale anaerobic filters in the U.S.A. and Canada. Depending on the flow regime, the enrichment of

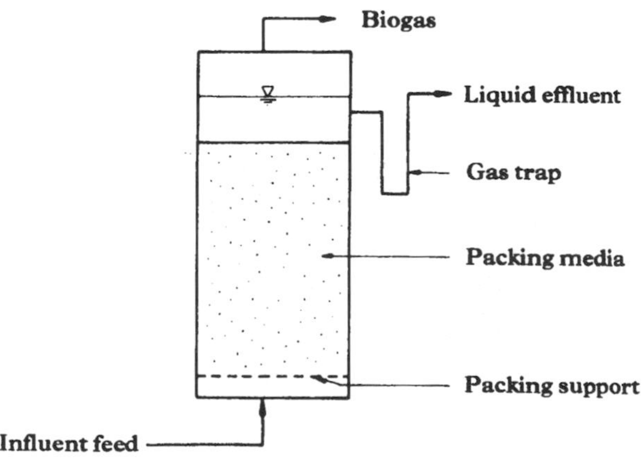

Figure 4.15 Up-flow anaerobic filter

Table 4.12 Packing media for anaerobic filters (Henze and Harremoes, 1983)

Material	Characteristics	Specific surface area (m^2/m^3)	Empty bed porosity
Quartz stone	20–40 mm diam.		0.42
Granite stone	25–38 mm diam.		0.53
Lime stone	20–40 mm diam.		0.49
Gravel	6–38 mm diam.		0.4
Oyster shells	60–100 mm diam.		0.82
Active carbon	1.5 mm diam.		0.60
Red clay blocs	28 · 28 mm pore size	157	0.7
Corrugated blocs		98–132	0.95
Granite chips	12–25 mm		0.40
Pall rings	90 · 90 mm	102	0.95
Norton plastic rings	90 mm	114	0.95
Raschig rings	10 · 16 mm	45–49	0.76–0.78
Polypropylene spheres	90 mm diam.	89	> 0.95

the acid- and methane-forming bacteria takes place at different zones in the filter. For an up-flow column there will be more acid-formers and methane-formers at the bottom and top regions, respectively.

Because of the physical configuration of the filter, only soluble waste, or wastes with low solid contents including domestic wastewaters, should be treated to avoid frequent clogging of the filter.

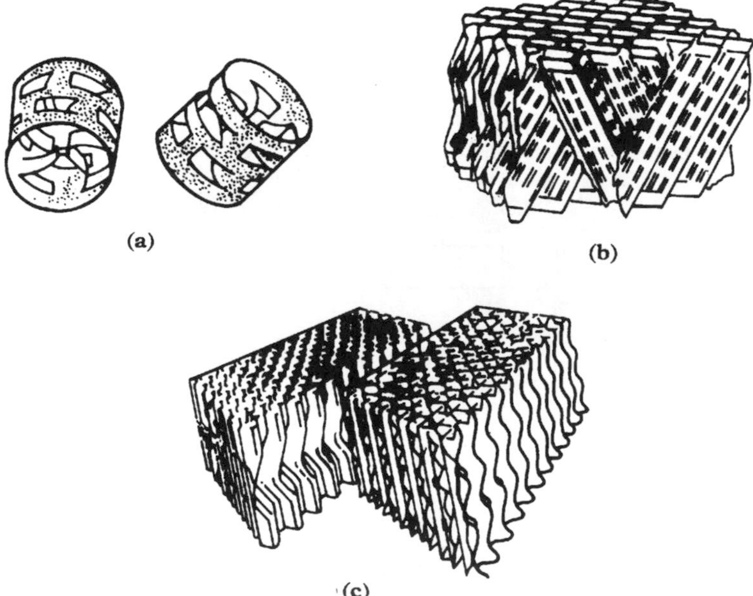

Figure 4.16 Packing media for anaerobic filters: (a) pall rings; (b) crossflow media; (c) tubular media

Table 4.13 Characteristics of anaerobic filter process

Advantages	Disadvantages
High organic removal capacity	Difficult to start up
Short HRT	Risk for clogging
Good adaptation to different wastewater	Restricted to wastewater with low TSS
Application to dilute and high strength wastewater	High TSS and NH_4–N contents in the effluent
No mechanical mixing required	May require periodic biomass removal
Insensitive against load fluctuations	Limited access to reactor interior for monitoring and inspection of biomass accumulation
Fast re-start after shut-downs	High costs for packing media and support systems
Low area demand	

Up-flow anaerobic sludge blanket (UASB) reactor

This type of reactor, developed by Lettinga *et al.* (1983) in the Netherlands, is suitable for the treatment of high-strength organic waste that is low in solid content (e.g. the agro-industrial wastes mentioned in Chapter 2) with or without sludge recycling. The digester has three distinct zones: (i) a densely packed sludge layer at the bottom; (ii) a sludge blanket in the center; and (iii) a liquid layer at the top (Figure 4.17).

Wastewater to be treated enters at the bottom of the reactor and passes upward through the sludge blanket composed of biologically formed granules. Treatment occurs as the wastewater comes in contact with the granules. The gases produced under anaerobic conditions (principally CH_4 and CO_2) cause internal sludge circulation, which helps in the formation and maintenance of the biological granules. Some gas bubbles produced within the sludge blanket become attached to the biological granules and bring them to the top of the reactor. The granules that reach the surface hit the bottom of the inverted pan-like gas/solids separator (degassing baffles), which causes the attached gas bubbles to be released. The degassed granules typically settle to the sludge blanket zone, thus creating a long mean cell residence time (θ_c) and a high solid concentration in the system. About

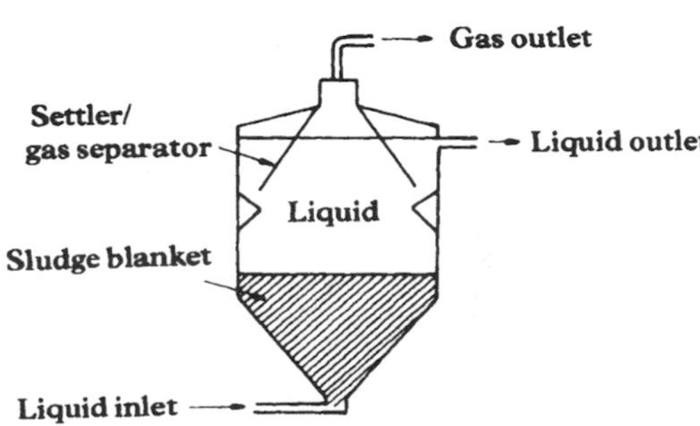

Figure 4.17 Upflow anaerobic sludge blanket digester

Table 4.14 Operating and performance data of some full-scale anaerobic filters in the U.S.A. and Canada (adapted from Young and Yang, 1989)

Type of waste (major constituent)	Feed COD/BOD (mg/L)	COD loading (kg/m^3-day)	Temp. (°C)	HRT	Percent removal	Reactor data[a]
Wheat starch (carbohydrates)	8,000 COD 6,500 BOD	4.4	32	44 h	75–80 COD	Upflow, V = 760 m^3, D = 9 m, H = 6 m; M = 12–50-mm stone
Chemical processing (alcohols)	12,000 COD	8–12	37	24–36 h	75–85 COD	Upflow, V = 6,400 m^3, D = 26 m, H = 12.2 m; M = 90-mm Pall rings
Rum distillery (carbohydrate)	85,000 COD 40,000 BOD	6–7	38	12–14 days	65–75 COD 70–80 BOD	Downflow, V = 12,500 m^3, D = 36 m, H = 12 m; M = tubular modules
Landfill leachate (organic acids)	11,000 COD 8,650 BOD	0.2–0.7	37	30–40 days	90–96 BOD	Upflow, V = 2,800 m^3, D = 18.3 m, H = 11 m; M = tubular modules
Food canning (carbohydrate)	4,000 COD 2,500 BOD	1.2–2.5	35	48–72 h	89 TCOD 97 TBOD	Upflow, V = 3,600 m^3, D = 33 m, H = 4 + m; M = crossflow modules

[a] V = volume; D = diameter; H = height; M = media

Table 4.15 Characteristics of the UASB process (adapted from Weiland and Rozzi, 1991)

Advantages	Disadvantages
High organic removal capacity	Granulation process difficult to control
Short HRT	Granulation depends on wastewater properties
Low energy demand	Start-up eventually needs granular sludge
No need of packing media	Sensitive to organic and hydraulic shock loads
Long experience in practice	Restricted to nearly solid free wastewater
	Ca^{2+} and NH_4^+ inhibit granular formation
	Re-start can result in granular flotation

80–90 percent of the decomposition of the organic matter takes place in the sludge blanket zone, which occupies about 30 percent of the total volume of the reactor. To keep the sludge blanket in suspension, upflow velocities in the range of 0.6–0.9 m/h have been used (Metcalf and Eddy, 1991).

Table 4.15 summarizes the advantages and disadvantages of the UASB process, while the design and performance data of some UASB reactors in the U.S.A. and the Netherlands are given in Table 4.16. To minimize the overflow of granules in the effluent, the design guidelines of the gas-solids separator device for UASB reactors, listed in Table 4.17, should be followed.

It should be noted from Table 4.16 that the volumetric gas production rates from the UASB reactors, ranging from 3.7 to 7.5 m^3/m^3 of reactor volume, are over 10 times higher than those of the conventional biogas digesters (see Table 4.8). This is probably due to three reasons: (i) the UASB reactors were fed with wastes which contain high concentrations of soluble COD and are readily biodegradable; (ii) better mixing and less shorting-circuiting in the UASB reactors; and (iii) due to the internal sludge circulation in the UASB reactors, more active microorganisms in the form of granules are available for anaerobic biodegradation. However, it is much more expensive to build a UASB reactor than a conventional biogas digester.

Besides the anaerobic filter and up-flow anaerobic sludge blanket reactor, other anaerobic fixed-film reactors have been developed such as fluidized and expanded bed reactors. As the fixed-film bacteria stay in the reactor longer, or have longer θ_c than the dispersed bacteria, they are able to withstand shock loadings or withstand higher organic loadings better than the dispersed bacteria. Application of anaerobic fixed-film reactors in waste treatment and biogas production is currently growing.

4.4.3 Trouble-shooting

All biogas digesters will have problems sooner or later with their structure components and in the operation and maintenance. Tables 4.18 and 4.19 deal with problems associated with the dispersed-growth digesters, their causes and remedies; while Table 4.20 deals with those of the attached-growth digesters, i.e. anaerobic filters and UASB.

Table 4.16 Design and performance data of UASB reactors in the U.S. and Netherlands (adapted from Biljetina, 1987)

Reactor description			Operating data				Gas production	
Type of waste	Location	Capacity	HRT (h)	Feed COD (mg/L)	COD loading (kg/m^3-day)	COD removal (%)	Gas prod. (m^3/h)	% CH_4
Brewery	LaCrosse, Wisc.	961.4 m^3/h	4.9	2,500	16.34	80	35,300	75
Potato starch	Caribou, Maine	37.85 m^3/h	47	22,000	11.05	85	7,800	77
Alcohol	ZNSF, Nether.	10,977 kg COD/day	8	5,330	16.02	90	NA[a]	NA[a]
Sugar beet	CSM Breda, Nether.	15,568 kg COD/day	4.8	2,400	12.02	75	200	82

[a] NA = not available

Table 4.17 Guidelines for the design of the gas-solids separator device for UASB reactors (Lettinga and Hulshoff, 1992). Reproduced by permission of Technomic Publishing Co., Inc.

1. The slope of the settler bottom, i.e., the inclined wall of the gas collector, should be between 45 and 60°
2. The surface area of the apertures between the gas collectors should not be smaller than 15–20% of the total reactor surface area
3. The height of the gas collector should be between 1.5 and 2 m at reactor heights of 5–7 m
4. A liquid gas interface should be maintained in the gas collector in order to facilitate the release and collection of gas bubbles and to combat scum layer formation
5. The overlap of the baffles installed beneath the apertures should be 10–20 cm in order to avoid upward flowing gas bubbles entering the settler compartment
6. Generally scum layer baffles should be installed in front of the effluent weirs
7. The diameter of the gas exhaust pipes should be sufficient to guarantee the easy removal of the biogas from the gas collection cap, particularly also in the case where foaming occurs
8. In the upper part of the gas cap anti-foam spray nozzles should be installed in the case where the treatment of the wastewater is accompanied by heavy foaming

4.5 BIOGAS PRODUCTION

The rate of biogas yield per unit weight of organic wastes can vary widely depending on the characteristics of influent feed and environmental conditions in the digesters, as stated in section 4.3. Data on biogas production from various types of wastes can be obtained from the literature, or can be theoretically estimated from chemical stoichiometry and kinetic reactions. Further information is given below.

A compilation of data of biogas yield from various organic waste materials is presented in Table 4.21. The range of biogas production is 0.20–1.11 m^3/kg of dry solids, with CH_4 content being 57–69 percent. Theoretically, CH_4 production can be determined according to the method outlined in Metcalf and Eddy (1991):

During anaerobic digestion biodegradable organic matter (BOD_L) is converted to mainly:

1. CH_4, CO_2, NH_3, H_2 and trace quantities of other gases, and
2. biological cells.

Organic matter used for CH_4 production

$$= \text{Organic matter stabilized} - \text{Organic matter used for cell production} \quad (4.6)$$

$$= BOD_L \text{ stabilized} - BOD_L \text{ for cell production} \quad (4.7)$$

The relationship between BOD_L and CH_4 production is described as follows:

$$C_6H_{12}O_6 \rightarrow 3CH_4 + 3CO_2 \quad (4.8)$$

$$3CH_4 + 6O_2 \rightarrow 3CO_2 + 6H_2O \quad (4.9)$$

Table 4.18 Problems with the digester, gas holder, and pipes (adapted from Tata Energy Research Institute, 1982)

	Problem	Possible cause	Remedy
Digester	1. Cracking of the digester wall due to hydraulic pressure from inside or from outside in horizontal digester plants	Lack of consolidation of soil on the outside of the digester wall, resulting in cavities in soil surrounding the masonry wall, and leading to the development of cracks in the digester wall	Outlet part of the digester should be packed in layers of soil of 30 cm thickness each, with sufficient water, after the construction of the digester is completed. This would increase the life of the structure and also reduce seepage loss to some extent
	2. Rising of the digester floor with the rise of ground water table	(a) Faulty compaction of the bottom (b) Hydraulic pressure of ground water	Pressure is neutralized when the digester is full. So where the water table is high the digester should always remain full
	3. Bursting of the digester with excessive gas pressure	(a) Gas holder at the top becomes jammed in the digester due to drying of the scum in between the gas holder and digester and restricting movement of the holder	(a) Scum should not be allowed to dry and should be forced down with a rod 2–3 times a week
		(b) Guide frame gets loosened from its support and the holder cannot move freely	(b) Guide pipe should be rewelded and rivetted and guideframe, etc., should be checked before installing it in the digester
Gas holder	1. Corrosion of the gas holder	Gas holders, commonly made of mild steel, remain in contact with digester slurry and with the gas containing methane and other gases, including H_2S which is highly corrosive	Painting of the gas holder with black paint or even coal tar each year. Alternative materials like PVC, ferrocement, galvanized iron, fiberglass, etc. may be used for manufacturing the holder
Inlet and outlet and central guide pipes	1. Clogging of the inlet/outlet pipes	1. Accumulation of feed or scum	1. Should be washed or flushed regularly with clean water or should be cleaned with a pole moving up and down
	2. Breakdown of central guide pipe	2. Rusting	2. Replacement (as it is mostly beyond repair)
Gas pipe carrying gas to the kitchen	1. Collection of condensate	Presence of water vapors in the gas	Condensate should be blown out 2–3 times a week. The pipe may be kept at a slope to ease blowing off

Table 4.19 Problems with the operation of the biogas digester (adapted from Tata Energy Research Institute, 1982)

Problem	Cause	Remedy
1. Gas holder (of floating type) does not rise or bag (of flexible bag type) does not inflate or pressure (of fixed dome type) does not rise at the start of the operation	(a) Very few bacteria (b) Lack of time (c) Feeding of slurry while waiting for gas holder to fill (d) No water in water outlet device (e) Leak in gas holder or gas pipe (f) Gas tap or cock or condensate tap open	(a) Seed with slurry from a working digester before adding the new slurry to the digester; rotate the gas holder daily (b) In cold weather it takes about 3 weeks to fill the gas holder for the first time (c) No slurry should be fed in until the third gas holder of burnable gas has been produced (d) About 0.25 L of water should be poured into the dipper pipe and the excess removed (e) Should be located and repaired (f) All open taps should be closed
2. Gas holder or bag (or pressure in fixed-dome-type digester) goes down very quickly once main gas valve is opened	(a) Condensate or water outlet tap open (if fitted) or gas tap for burner open or gas cock for lamp open (b) No water in water outlet device or syphon (c) Major leak in the pipe-work	(a) Close the taps (b) Pour water into water outlet device to remove excess water (c) Leaks should be located and repaired
3. Gas holder or bag or pressure (in fixed-dome digester) rises very slowly	(a) Temperature too low (b) Thick scum on top of slurry (c) Too much slurry put in daily (d) Slurry mixture suddenly changed a lot (e) Chemicals, oil, soap or detergent put into slurry (f) Gas leak (g) Slurry mixture too thick or too thin (h) Washing out mixing tank with extra water and allowing it to go into the digester	(a) Temperature may be raised by, e.g. solar heating, heating with biogas and composting around the digester wall (b) Remove the scum layer by agitating the drum daily and seeing to it that no straw or grass, etc. get into the digester (c) Correct amount of slurry should be added daily (d) Slurry mixture should not be altered too much at a time (e) Daily feed with dung and water only, to be continued (f) Leak should be located and repaired (g) Slurry to be made of the right consistency (h) No extra water should get into the digester
4. Too little gas, although pressure is correct	Gas jet is blocked	Gas jet to be cleaned with split bamboo or needle
5. No gas at the burner	(a) Main valve is closed (b) Gas tap is blocked (c) Burner or pipe is completely blocked by condensate (d) Leakages in the main/feeding pipes	(a) Open the valve (b) Clean the tap (c) Remove the condensate

(continued)

Table 4.19 Problems with the operation of the biogas digester (adapted from Tata Energy Research Institute, 1982) (*continued*)

Problem	Cause	Remedy
6. First gas produced will not burn	(a) Wrong kind of gas (b) Air in the gas pipe	(a) First gas should not be burned as air may be mixed with it and could explode. Also, frequently the first gas produced in winter has a high percentage of CO_2 (b) Air should be allowed to escape until there is a definite smell of gas
7. Formation of scum	(a) Presence of undigested vegetable matter (e.g. straw, water hyacinth or coarse dung such as elephant dung), animal bedding (e.g. straw, sawdust), animal clothing (e.g. pig hair, chicken feathers) and not properly broken dung (b) Slurry not properly mixed	(a) Mixing the slurry to disperse the floating material (b) Agitation or semi-circular rotation of the drum may help mix the slurry. It may also be necessary to introduce special stirrers or mixing devices for this purpose
8. pH too acidic ($pH < 6$)	(a) Adding raw material too fast (b) Wide temperature fluctuation (c) Build-up of scum	(a) Reduce feeding rate (b) Stabilize temperature (c) Remove scum
9. pH too alkaline ($pH > 7$)	Initial raw material too alkaline	Have patience. Never put acid into the digester
10. Slow rate of gas production or no gas production	(a) Increase in toxicity with retention time (b) Increase in solid content (c) pH too acidic or too alkaline (d) Low temperature (e) Addition of significant quantity of different types of feed material to a working digester, for example, addition of an equal mass of vegetable matter to the digester utilizing pig waste (f) Digester filled with exhausted dung heaps (g) Digester filled up with raw dung followed by water to make the slurry	(a) Dilution or low loading makes ammonia toxicity less critical (b) Stirring, dilution or low loading reduces viscosity (c) As discussed earlier in Problems 8 and 9 (d) Raise the temperature by heating (e) Do not change the slurry mixture of a working digester (f) Do not use exhausted dung heaps (g) Make the slurry as recommended

Table 4.20 Problems with anaerobic filters and UASB reactors (Lettings *et al.*, 1983; Henze and Harremoes, 1983; Vigneswaran *et al.*, 1986)

	Problem	Cause	Remedy
Anaerobic filter	Start-up difficulty	Improper seeding or improper initial loading	Inoculation with the right microbial culture Low initial loading during start-up (approx. 0.1 kg COD/kg VSS/day)
	Channeling and short circuiting	Gas bubbles adhere to flocs/bed particles and cause these to rise in the reactor	Backwashing to remove excess solid build-up in the voids
	High levels of SS in reactor effluent	Clogging due to too much solids retained in the filter	Remove the solids from packing by draining and backwashing the media
UASB	Start-up difficulty	Improper seeding or improper initial sludge loading	Amount of seed sludge: 10–15 kg VSS/m^3. Initial organic load: 0.05–0.1 kg COD/kg VSS/day. No increase of the organic load unless all VFAs are more than 80% degraded. Permit the washout of poorly settling sludge. Retain the heavy part of sludge
	Buoying sludge pushed through the aperture and overflow with the liquid effluent	Ineffective separation of the entrapped or attached gas from the sludge flocs at the gas–liquid interface	Keep the gas–liquid interface well stirred to allow entrapped gas to readily escape
		Shock loading	Avoid shock loading
	Excessive foaming	Poor treatment efficiency due to overloading or nutrient deficiency	Add small quantity of anti-foaming agent to the feed solution

Table 4.21 Yield of biogas from various waste materials[a]

Raw materials	Biogas production per unit weight of dry solids		Temperature		CH_4 content in gas (%)	Fermentation time (days)
	ft^3/lb	m^3/kg	°F	°C		
Cow dung	5.3	0.33	—	—	—	—
Cattle manure	5	0.31	—	—	—	—
Cattle manure (India)	3.6–8.0	0.23–0.50	52–88	11.1–31.1	—	—
Cattle manure (Germany)	3.1–4.7	0.20–0.29	60–63	15.5–17.3	—	—
Beef manure	13.7[b]	0.86[b]	95	35	58	10
Beef manure	17.7	1.11	95	35	57	10
Chicken manure	5.0[c]	0.31[c]	99	37.2	60	30
Poultry manure	7.3–8.6[f]	0.46–0.54[f]	90.5	32.5	58	10–15
Poultry manure	8.9[c]	0.56[c]	123	50.6	69	9
Pig manure[d,e]	11.1–12.2	0.69–0.76	90.5	32.5	58–60	10–15
Pig manure[d,e]	7.9	0.49	91	32.8	61	10
Pig manure	16.3	1.02	95	35	68	20
Sheep manure[d]	5.9–9.7	0.37–0.61	—	—	64	20
Forage leaves	8	0.5	—	—	—	29
Sugar beet leaves	8	0.5	—	—	—	14
Algae	5.1	0.32	113–122	45–50	—	11–20
Nightsoil	6	0.38	68–79	20–26.1	—	21

[a] Compiled by NAS (1977)
[b] Based on total solids
[c] Based on volatile solids fed
[d] Includes both feces and urine
[e] Animals on growing and finishing rations
[f] Based on volatile solids destroyed. On the basis of conversion efficiencies, these results may be expressed as 4.0–4.7 ft^3/lb dry solids added or 0.26–0.30 m^3/kg

180 kg glucose produces 48 kg CH_4
180 kg glucose is equivalent to 192 kg BOD_L
i.e. 1 kg BOD_L produces

$$
\begin{aligned}
&= (48/180) \times (180/192) \text{ kg } CH_4 \\
&= 0.25 \text{ kg } CH_4 \\
&= 0.25 \text{ kg} \times (10^3 \text{ mol}/16 \text{ kg}) \\
&\quad \times (22.4 \text{ L}/1 \text{ mol}) \times (1 \text{ m}^3/10^3 \text{ L}) \\
&= 0.35 \text{ m}^3 \ CH_4 \text{ at STP (standard temperature and pressure)}
\end{aligned}
$$

Volume of CH_4 produced, m^3/day

$$= 0.35 \ (BOD_L \text{ stabilized} - BOD_L \text{ for cell production}) \tag{4.10}$$

To solve eq. 4.10, the terms 'BOD_L stabilized' and 'BOD_L for cell production' need to be determined.

$$BOD_L \text{ stabilized} = EQS_0 \ (10^3 \text{ g/kg})^{-1} \text{ kg/day} \tag{4.11}$$

where E = efficiency of waste stabilization fraction
Q = influent flow rate, m^3/day, and
S_0 = ultimate influent BOD, g/m^3.

From eq. 2.5:

$$\frac{\text{kg } O_2}{\text{kg cells}} = \frac{160}{113} = 1.42$$

$$\text{or } BOD_L = 1.42 \ P_x \tag{4.12}$$

where P_x = net mass of cells produced.

From the Monod equation, and considering mass balance in a digester without cell recycle, the term 'P_x' in kg/day can be expressed as follows:

$$P_x = \frac{QYES_0}{1 + k_d\theta_c} \times 10^{-3} \tag{4.13}$$

where Y = yield coefficient of anaerobic bacteria (normally = 0.05), and
k_d = decay coefficient of anaerobic bacteria (normally 0.01–0.03 day^{-1})
θ_c = mean cell residence time, day.

Substituting eqs 4.11, 4.12 and 4.13 into eq. 4.10, the following is obtained:

$$\text{Volume of } CH_4 \text{ produced, m}^3\text{/day} = 0.35EQS_0(10^{-3})\left(1 - \frac{1.42Y}{1 + k_d\theta_c}\right) \tag{4.14}$$

If the kinetic coefficients Y and k_d for cell production and the efficiency of waste stabilization (E) in a digester are known, then CH_4 production can be predicted from eq. 4.14. An experimental method to determine values of the kinetic coefficients for a particular type of mixture of influent feed is outlined in Metcalf and Eddy (1991), or these values can be obtained from literature.

It should be noted that eq. 4.14 is applicable only to dispersed-growth digesters without sludge recycling. Although kinetic models for methane production and waste stabilization for attached-growth digesters have been developed, their applications to field-scale digesters are not yet widely adopted.

An example showing the application of eq. 4.14 is given in Example 4.1. Example 4.2 is a simplified method to design a biogas digester using data obtained from literature.

Example 4.1 A tapioca factory produces wastewater with the characteristics as shown below.

$$\begin{aligned} \text{Flow rate} &= 100\ \text{m}^3/\text{day} \\ \text{COD} &= 20{,}000\ \text{mg/L} \\ \text{N} &= 400\ \text{mg/L} \end{aligned}$$

Determine the quantity of CH_4 that can be produced from a dispersed-growth digester treating this tapioca wastewater and the digester size. Assume that, at the θ_c value of 10 days, there is a waste stabilization efficiency of 65 percent and the values of Y and k_d were experimentally found to be 0.05 and 0.02 day^{-1}, respectively.

Solution From eq. 4.14:

$$\begin{aligned} CH_4 \text{ produced} &= 0.35(0.65)(100)(20{,}000)(10^{-3}) \times \left\{1 - \frac{1.42(0.05)}{1 + 0.02(10)}\right\} \\ &= 428.1\ \text{m}^3/\text{day} \end{aligned}$$

Note: if COD represents total C in the wastewater, the ratio of COD/N = 50 : 1, which is more than the optimum value of 25 : 1. This tapioca wastewater requires the addition of some nitrogenous compounds (such as urea) to adjust the C/N ratio prior to feeding to the digester.

$$\begin{aligned} \text{The volume of this digester is} &= 100\ \text{m}^3/\text{day} \times 10\ \text{days} \\ &= 1{,}000\ \text{m}^3 \end{aligned}$$

Construct 2 conventional digesters, each with a working volume of 500 m^3.

Example 4.2 A family of five persons needs about 10 m^3 of methane for daily family use. Determine the size of biogas digester required and other operating procedures necessary to maximize gas production. Raw materials are nightsoil and rice straw.

Characteristics of raw materials:

	Nightsoil	*Rice straw*
Organic carbon (C), % total solids	48	43
Total kjeldahl nitrogen (N), % total solids	4.5	0.9
Total volatile solids (TVS), % total solids	86	77
Percentage moisture	82	14

Let the dry weights of nightsoil and rice straw required to be fed to the digester be m_1 and m_2 kg/day, respectively. An optimal condition for anaerobic digestion should have a C/N = 25 : 1 in the influent feed:

$$\frac{0.48\ m_1 + 0.43\ m_2}{0.045\ m_1 + 0.009\ m_2} = 25$$

$$m_2/m_1 = 0.645/0.205 = 3.15 \qquad (4.15)$$

From Tables 4.8 and 4.21, choose a methane production rate:

$$= 0.3\ \text{m}^3/\text{kg TVS added.}$$

$$\text{TVS required/day} = 10/0.3 \approx 33.33\ \text{kg, or} \qquad (4.16)$$

$$0.86\,m_1 + 0.77\,m_2 = 33.33$$

From eqs 4.15 and 4.16

$m_1 = 10.16$ kg/day

$m_2 = 32$ kg/day

Wet weight of nightsoil required = 10.16/0.18 = 56.44 kg/day

Wet weight of rice straw required = 32/0.86 = 37.22 kg/day

$$\text{Volume of digester} = \frac{\text{Volatile solids added/day}}{\text{Volatile solids loading}}$$

The normal range of volatile solids loading

= 1–4 kg VS/(m^3-day) (Barnett *et al.*, 1978)

Choose a volatile solids loading rate
= 2 kg VS/(m^3-day)

The volume of digester required = 33.33/2 ≈ 17 m^3

To provide additional space for gas storage the digester volume should be approximately 22–25 m^3.

The normal hydraulic retention time = 10–60 days (Barnett *et al.*, 1978 and section 4.3).

Choose a hydraulic retention time of 30 days
Waste volume to be added/day

= 17/30 = 0.57 m^3/day = 570 L/day

Assume bulk densities of 1.1 and 0.1 kg/L for nightsoil and rice straw, respectively.

Volume of raw wastes to be added each day

$$=(56.44/1.1)+(37.22/0.1)$$

$$=423.5 \text{ L}$$

Volume of dilution water required to be added to the influent mixture

$$=570-423.5$$

$$=146.5 \text{ L/day}$$

The design and operation of a biogas digester can follow those outlined in the above example. In practice the process of digester size selection may be as follows. First, the digester size is determined. This requires an estimate of the daily requirements of biogas (see Table 4.22), coupled with the availability of the organic materials, which is given in Chapter 2. In calculating the required volume for the digester, the amount of dilution water needed and additional space for gas storage should be considered.

Table 4.22 Quantities of biogas required for a specific application[a]

Use	Specification	Quantity of gas required	
		ft^3/h	m^3/h
Cooking	2-inch burner	11.5	0.33
	4-inch burner	16.5	0.47
	6-inch burner	22.5	0.64
	2–4-inch burner	8–16	0.23–0.45
	per person/day	12–15	0.34–0.42
Gas lighting	per lamp of 100 candle power	4.5	0.13
	per mantle	2.5	0.07
	per mantle	2.5–3.0	0.07–0.08
	two-mantle lamp	5	0.14
	three-mantle lamp	6	0.17
Gasoline or diesel engine[b]	converted to biogas, per horsepower	16–18	0.45–0.51
Refrigerator	per ft^3 capacity	1	0.028
	per ft^3 capacity	1.2	0.034
Incubator	per ft^3 capacity	0.45–0.6	0.013–0.017
	per ft^3 capacity	0.5–0.7	0.014–0.020
Gasoline	1 L	47–66[c]	1.33–1.87[c]
Diesel fuel	1 L	53–73[c]	1.50–2.07[c]
Boiling water	1 L	3.9[d]	0.11[d]

[a] Compiled by NAS (1977)
[b] Based on 25 percent efficiency
[c] Absolute volume of biogas needed to provide energy equivalent of 1 L of fuel
[d] Absolute volume of biogas needed to boil off 1 L of water

4.6 END-USE OF BIOGAS AND DIGESTED SLURRY

4.6.1 Biogas

Based on the heat value of the biogas (4,500–6,300 kcal/m^3), Hesse (1982) estimated that on complete combustion 1 m^3 of biogas is sufficient to:

1. run a 1 horsepower engine for 2 h,
2. provide 1.25 kW-h of electricity,
3. provide heat for cooking three meals a day for five people,
4. provide 6 h of light equivalent to a 60-W bulb,
5. run a refrigerator of 1 m^3 capacity for 1 h,
6. run an incubator of 1 m^3 capacity for 0.5 h.

Therefore 1 m^3 of biogas is equivalent to 0.4 kg of diesel oil, 0.6 kg petrol, or 0.8 kg of coal. Quantities of biogas required for specific application are presented in Table 4.22. Figure 4.18 shows examples of stoves and lamps that use biogas as fuel for their operation.

In China some 5.2 percent of the rural population (about 30 million Commune members) and in India 0.8 percent of rural households are being served by biogas (Barnett *et al.*, 1978). About 95 percent of all the biogas plants in Asia are of the family-size type, and therefore the principal uses of their output are cooking and lighting. The remaining 5 percent of biogas plants are being used for other purposes such as refrigeration, electricity generation, and running irrigation pumps. For these uses it becomes necessary to compress and store the gas in portable containers and carry it to the place of application. Flexible bags made of a variety of materials such as PVC, rubber and polyethylene can be used for gas storage, and are commercially available in several countries in Asia.

Engines are designed to run either on pure methane gas or on digester gas. Either type of gas is a suitable fuel for petrol and diesel engines. Diesel engines that run on dual fuel (biogas/diesel oil) or diesel fuel only are now manufactured in India. Kerosene- and gas-operated engines can also be modified to use biogas. Stationary engines located near a large biogas plant can be an economical and practical proposition. It is more efficient to use biogas to generate electricity rather than to use direct lighting. However, the high cost of engine and generator might be prohibitive for the farmer or biogas owner.

The biogas from digestion of animal manure and vegetable matter normally does not contain sufficient H_2S to require purification before use. For cooking and lighting the biogas does not need to be purified. However, if the gas is to be stored or transported, then any H_2S should be removed to prevent corrosion of storage bags. CO_2 should also be removed as there is no advantage in compressing it. Biogas purification is not normally practiced for small-scale digesters. For large-scale or institutional digesters there might be economic reasons favoring biogas purification. Some practical methods of biogas purification are described below.

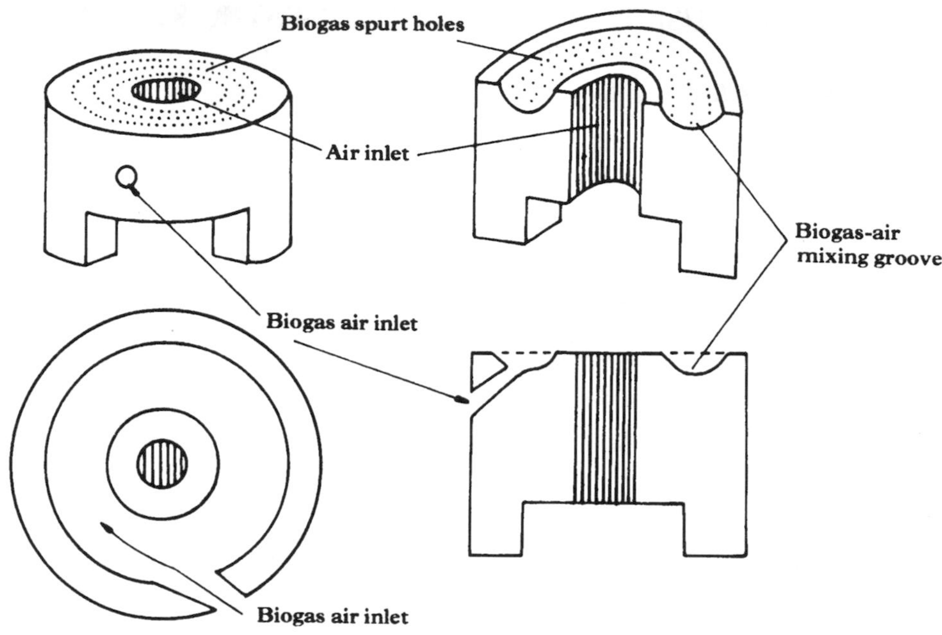

(a) Earthen stove with mixing groove

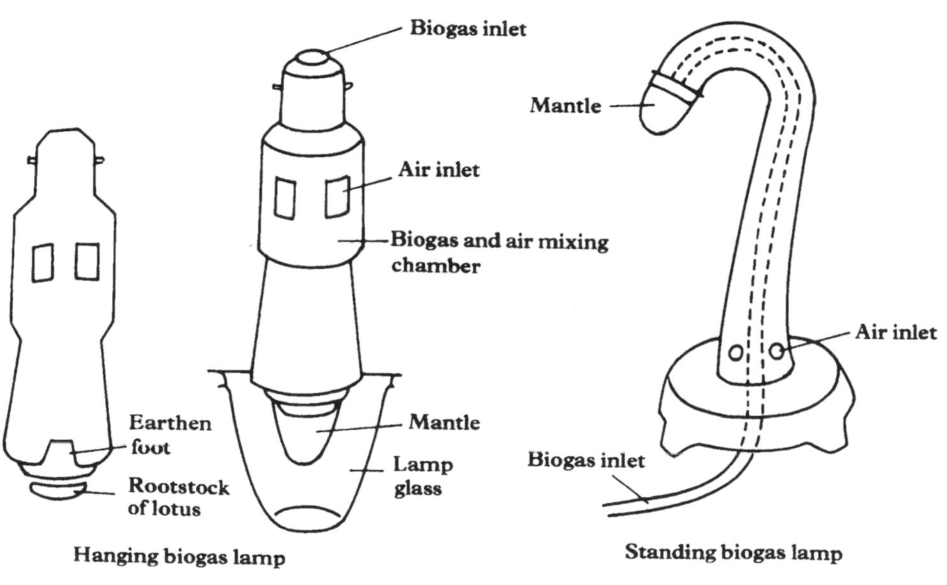

(b) Biogas lamps

Figure 4.18 (a) Biogas stove and (b) biogas lamps (after McGarry and Stainforth, 1978; reproduced by permission of the International Development Research Centre, Canada)

CO_2 REMOVAL

As CO_2 is fairly soluble in water, water scrubbing is perhaps the simplest method of CO_2 removal from biogas. However, this method requires a large quantity of water as can be estimated from Table 4.23. Assuming that a biogas has a 35 percent CO_2 content and a CO_2 density of 1.84 kg/m^3 at 1 atmosphere pressure, and 20 °C, the amount of water required is 429 L to scrub 1 m^3 of this biogas.

CO_2, being an acid gas, can be absorbed in alkaline solution. The three common alkaline reagents are NaOH, $Ca(OH)_2$ and KOH. Two consecutive reactions of CO_2 removal in NaOH solution are:

$$2NaOH + CO_2 \rightarrow Na_2CO_3 + H_2O \tag{4.17}$$

$$Na_2CO_3 + CO_2 + H_2O \rightleftarrows 2NaHCO_3 \downarrow \tag{4.18}$$

$NaHCO_3$ is the precipitate formed which can be removed from the solution.

Lime or $Ca(OH)_2$ is readily available in most areas and is inexpensive. The reaction of CO_2 removal in lime solution is

$$Ca(OH)_2 + CO_2 \rightarrow CaCO_3 \downarrow + H_2O \tag{4.19}$$

$CaCO_3$ is the precipitate which must be removed from the solution. Based on the stoichiometry in eq. 4.19, a mixture of 1 kg burnt lime in 1 m^3 of water is sufficient to remove about 300 L of CO_2 or 860 L of biogas (assuming a CO_2 content of 35 percent).

The availability of KOH is not as extensive as $Ca(OH)_2$ and its application in CO_2 scrubbing is also limited.

Hesse (1982) showed a model of a lime water scrubber for scrubbing CO_2 from biogas (Figure 4.19) which was developed by Ram Bux Singh in India. A stirring paddle creates agitation which aids in the diffusion of gas molecules into the alkaline solution and extends contact time between the liquid and the gas.

Table 4.23 Approximate solubility[a] of CO_2 in water[b]

Pressure		Temperature °F (°C)				
atm.	kg/cm^2	32 (0)	50 (10)	68 (20)	86 (30)	104 (40)
1	1.03	0.40	0.25	0.15	0.10	0.10
10	10.3	3.15	2.15	1.30	0.90	0.75
50	51.7	7.70	6.95	6.00	4.80	3.90
100	103	8.00	7.20	6.60	6.00	5.40
200	207	—	7.95	7.20	6.55	6.05

[a] Solubility is expressed as lb CO_2 per 100 lb H_2O (kg CO_2 per 100 kg H_2O)
[b] Adapted from Nonhebel (1972); reproduced by permission of Butterworth & Co., London

H_2S REMOVAL

Na_2CO_3 formed in eq. 4.17 can also be used to remove H_2S from the biogas, provided that the contact time is sufficiently long for the reaction (eq. 4.20) to occur in the scrubber:

$$H_2S + Na_2CO_3 \rightarrow NaHS + NaHCO_3 \downarrow \tag{4.20}$$

A simpler and more economical method of H_2S removal, when other constituents need not be removed, is to pass the biogas over iron filings or ferric oxide (Fe_2O_3) mixed with wood shavings (NAS, 1977). This method, called 'dry gas scrubbing', can be effected as shown in Figure 4.20 (this model was developed by Ram Bux Singh, as reported by Hesse, 1980). The reaction occurring during H_2S scrubbing is:

$$Fe_2O_3 + 3H_2S \rightarrow Fe_2S_3 + 3H_2O \tag{4.21}$$

The Fe_2O_3 can be regenerated by exposing or heating Fe_2S_3 in air (or oxygen) according to the following reaction:

$$2Fe_2S_3 + 3O_2 \rightarrow 2Fe_2O_3 + 3S_2 \tag{4.22}$$

Other processes of biogas purification are available (Nonhebel, 1972), but they are technically sophisticated and expensive, and unsuitable for the biogas systems described in this chapter.

4.6.2 Digested slurry

Although there is a considerable degree of reduction of organic, solid and nitrogenous matters in anaerobic digesters, digester slurry still contains high concentrations of the above materials (see Table 4.9) and so needs further treatment prior to disposal. The digester slurry also contains various types of

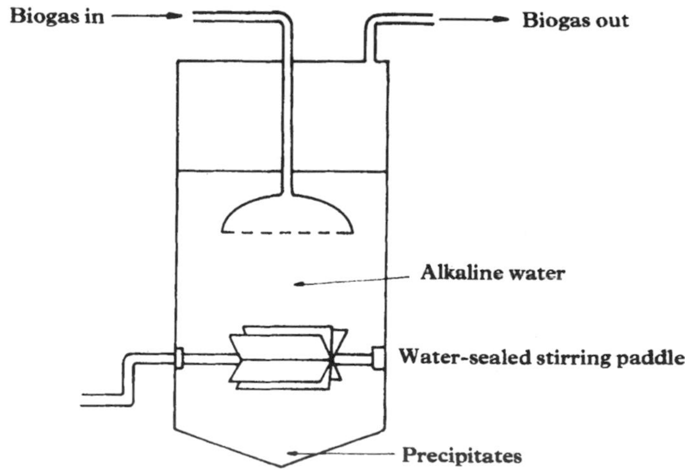

Figure 4.19 Model of CO_2 and H_2S scrubber in alkaline water

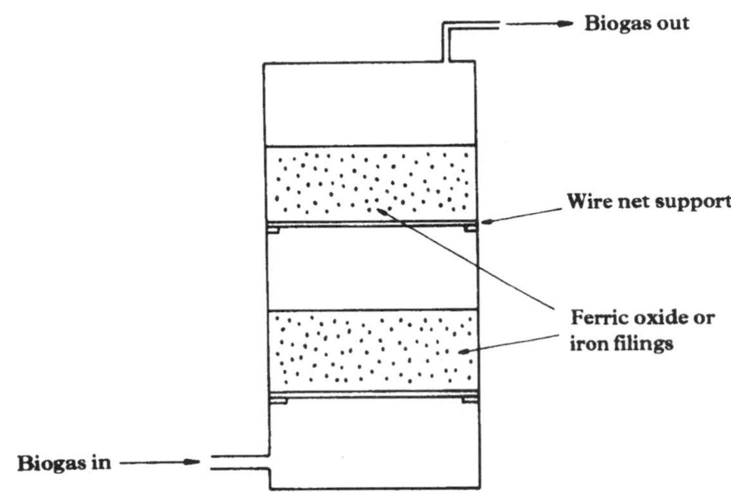

Figure 4.20 Model of dry gas scrubber for H_2S removal

pathogens (as stated in section 4.1), requiring great care in handling and disposal. On the other hand, high nutrient contents in the slurry make it suitable for reuse as compost fertilizer (Chapter 3) and in application to fish ponds (Chapter 6) or on land (Chapter 9).

REFERENCES

Barnett, A., Pyle, L., and Subramanian, S. K. (1978). *Biogas Technology in the Third World: A Multidisciplinary Review.* IDRC-103e. International Development Research Centre, Ottawa.

Biljetina, R. (1987). Commercialization and economics. In *Anaerobic Digestion of Biomass* (eds. D. P. Chynoweth and R. Isaacson), pp. 231–55. Elsevier Applied Science, London.

Brown, N. L. and Tata, P. B. S. (1985). *Biomethanation*. ENSIC Review no. 17/18. Environmental Sanitation Information Center, Asian Institute of Technology, Bangkok.

ESCAP (1975). *Report of the Preparatory Mission on Bio-gas Technology and Utilization.* RAS/74/041/A/01/01, United Nations Economic and Social Commission for Asia and Pacific, Bangkok.

Ferguson, T. and Mah, R. (1987). Methanogenic bacteria. In *Anaerobic Digestion of Biomass* (eds. D. P. Chynoweth and R. Isaacson), pp. 49–63. Elsevier Applied Science, London.

Gnanadipathy, A. and Polprasert, C. (1992). Anaerobic treatment processes for high-strength organic wastewaters. *Proceedings of the National Seminar on Conventional and Advanced Treatment Techniques for Wastewater Treatment.* 7–8 February 1992, Bangkok, Thailand.

Gunnerson, C. G. and Stuckey, D. C. (1986). *Anaerobic Digestion: Principles and Practices for Biogas Systems.* UNDP project management report no. 5. World Bank. Washington, D.C.

Hao, P. L. C. (1981). The application of red mud plastic pig manure fermenters in Taiwan. In *Proceedings of the Regional Workshop on Rural Development Technology*, pp. 489–504. Seoul.

Henze, M. and Harremoes, P. (1983). Anaerobic treatment of wastewater in fixed film reactors—a literature review. *Water Sci. Technol.*, **15**, 1–101.

Hesse, P. R. (1982). *Storage and Transport of Biogas*. Project field document no. 23. Food and Agriculture Organization of the United Nations, Rome.

Lettinga, G., Hobma, S. W., Hulshoff Pol, L. W., de Zeeuw, W., de Jong, P., and Roersma, R. (1983). Design operation and economy of anaerobic treatment. *Water Sci. Technol.*, **15**, 177–95.

Lettinga, G. and Hulshoff Pol, L. W. (1992). UASB process design for various types of wastewaters. In *Design of Anaerobic Processes for the Treatment of Industrial and Municipal Wastes* (eds. J. F. Malina, Jr. and F. G. Pohland), pp. 119–45. Technomic Publishing Co., Inc., Lancaster, PA.

Lettinga, G., van Velsen, A. F. M., Homba, S. W., de Zeeuw, W., and Klapwijk, A. (1980). Use of the upflow sludge blanket (USB) reactor concept for biological wastewater treatment, especially for anaerobic treatment. *Biotech. Bioeng.*, **22**, 699–734.

McGarry, M. G. and Stainforth, J. (1978). *Compost, Fertilizer, and Biogas Production from Human and Farm Wastes in The People's Republic of China*. IDRC-TS8e, International Development Research Centre, Ottawa.

McInerney, M. J. and Bryant, M. P. (1981). Review of methane fermentation fundamentals. In *Fuel Gas Production from Biomass*, vol. 1 (ed. D. L. Wise), pp. 19–46. CRC Press, Boca Raton, FL.

Metcalf and Eddy, Inc. (1991). *Wastewater Engineering: Treatment, Disposal, Reuse*, 3rd edition. McGraw-Hill Book Co., New York.

Mosey, F. E. (1982). New developments in the anaerobic treatment of industrial wastes. *Water Pollution Control*, **81**, 540–52.

NAS (1977). *Methane Generation from Human, Animal, and Agricultural Wastes*. U.S. National Academy of Sciences, Washington, D.C.

Nonhebel, G. (1972). *Gas Purification Processes*. Butterworth, London.

Polprasert, C., Edwards, P., Pacharaprakiti, C., Rajput, V. S., and Suthirawut, S. (1982). *Recycling Rural and Urban Nightsoil in Thailand*. AIT Research Report no. 143. Asian Institute of Technology, Bangkok.

Price, E. C. and Cheremisinoff, P. N. (1981). *Biogas Production and Utilization*. Ann Arbor Science, Ann Arbor.

Smith, P. H., Bordeaux, F. M., Wilkie, A., Yang, J., Boone, D., Mah, R. A., Chynoweth, D., and Jerger, D. (1988). Microbial aspects of biogas production. In *Methane from Biomass: A Systems Approach* (eds. W. H. Smith and J. R. Frank). Elsevier Applied Science, London.

Tam, D. M. and Thanh, N. C. (1982). *Biogas Technology in Developing Countries: an Overview of Perspectives*. ENSIC Review no. 9. Environmental Sanitation Information Center, Asian Institute of Technology, Bangkok.

Tata Energy Research Institute (1982). *Biogas Handbook*, Pilot Edition. Documentation Centre, Bombay, India.

United Nations (1984). *Updated Guidebook on Biogas Development*. Energy Resources Development Series no. 27, United Nations, New York.

U.S. EPA (1979). *Design Manual for Sludge Treatment and Disposal*. EPA 625/1-79-011. U.S. Environmental Protection Agency, Washington, D.C.

Vigneswaran, S., Balasuriya, B. L. N., and Viraraghavan, T. (1986). *Anaerobic Wastewater Treatment—Attached Growth and Sludge Blanket Process*. Environmental Sanitation Review No. 19/20, Environmental Sanitation Information Center, Asian Institute of Technology, Bangkok.

Weiland, P. and Rozzi, A. (1991). The start-up, operation and monitoring of high-rate anaerobic treatment systems: Discusser's report. *Water Sci. Tech.*, **24**, 257–77.

Yang, P. Y. and Nagano, S. Y. (1985). Red mud plastic swine manure anaerobic reactor with sludge recycles in Hawaii. *Trans. Am. Soc. Agric. Eng.*, **28**, 1284–8.

Young, J. C. and Yang, B. S. (1991). Design considerations of full scale anaerobic filters. *J. Water Pollut. Control Fed.*, **61**, 1576–87.

Young, J. C. and McCarty, P. L. (1969). The anaerobic filter for waste treatment. *J. Water Pollut. Control Fed.*, **41**, R160–73.

EXERCISES

4.1 **a** You are to visit a biogas digester located at a farm or at a wastewater treatment facility, and find out the organic loading rate, HRT, and biogas (CH_4) production rate. Compare these data with those given in Table 4.11 or 4.14. Discuss the results.

b Also check with the biogas digester operator about the common problems of the digester and the remedial measures employed to solve these problems.

4.2 A farm has 100 cattle, each producing 40 L manure/day with 12.5 percent total solids (125 kg TS/m^3), 75 percent volatile solids (VS) and a C/N ratio of 15. The manure is to be used for biogas production together with rice straw, which has a C/N ratio of 80, moisture content of 50 percent, bulk density of 0.5 kg/L and N content of 0.9 percent dry weight. How much rice straw is needed to be mixed with the cattle manure to reach a C/N ratio of 25? At the VS loading of 2 kg/m^3-day and an HRT of 40 days, determine the volume of the biogas digester and the volume of dilution water required to mix with the manure.

4.3 A biogas digester produces 5 m^3 of biogas per day, which contains 35 percent of CO_2 by volume. The density of CO_2 is 1.80 kg/m^3 at 25 °C and 1 atmosphere pressure. Determine the daily requirement for quicklime (containing 95 percent CaO) to remove the CO_2.

4.4 A 1,000-pig farm plans to purchase 8-m^3 PVC bag digesters to produce biogas using pig wastes in its own stock. Each pig weighs 80 kg. The loading rate of the bag digester and HRT recommended by the producer are, respectively, 1.5 kg VS/m^3-day and 30 days for pig wastes diluted with water. Other data are available in Table 2.7. Determine the number of bag digesters that should be installed to treat the pig wastes and the amount of biogas to be produced if the biogas production rate is 0.3 m^3/kg VS added. If the density of the pig wastes is 1.1 kg/L, what is the volume of dilution water required?

4.5 Differentiate the hydraulic patterns and treatment mechanisms occurring in an anaerobic filter and an upflow anaerobic sludge blanket reactor.

5

Algae production

Algae are a diverse group of microorganisms that can perform photosynthesis. They range in size from microscopic unicellular forms, which are smaller than some bacteria, to multicellular forms such as seaweeds that may become many meters long. Unicellular algae are collectively called phytoplankton or single-cell protein (e.g. green algae, blue-green algae), and these are of interest in waste treatment and recycling processes because they are tolerant to changes in environmental conditions.

Botanists generally classify algae on the basis of: (i) their reproductive structures; (ii) the kind of products synthesized and stored in the cells; and (iii) the nature of the pigments in the chromatophores. There are seven phyla of algae as shown in Table 5.1, and examples of some planktonic algae are given in Figure 5.1.

In most algal species the cell wall is thin and rigid. Cell walls of diatoms are impregnated with silica, making them rather thick and very rigid. Walls of blue-green algae contain cellulose and are semi-rigid. The motile algae such as the euglena have flexible cell walls. The cell walls of most algae are surrounded by a flexible, gelatinous outer matrix secreted through the cell wall. As the cells age, the outer matrix often becomes pigmented and stratified, developing into a semi-rigid surface membrane.

Of the seven phyla shown in Table 5.1 the largest division of algae is the Chlorophyta or green algae. The photosynthetic ability of algae enables them to utilize sunlight energy to synthesize cellular (organic) material in the presence of appropriate nutrients. Thus the algae (and also the bacteria) are the major primary producers of organic matter in aquatic environments.

5.1 OBJECTIVES, BENEFITS, AND LIMITATIONS

With respect to algal mass culture in wastewater, efforts have been directed towards single-cell protein production for potential human and animal consumption. The desirable properties of algal single cell protein are:

1. high growth rate;
2. resistance to environmental fluctuations;

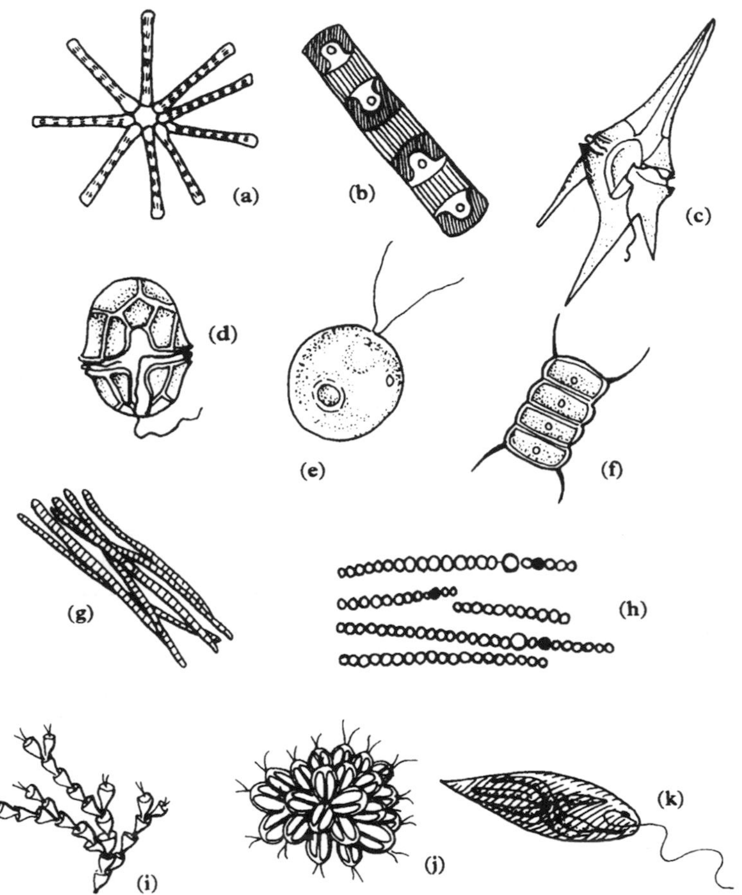

Figure 5.1 Examples of planktonic algae. Diatoms: (a) *Asterionella*; (b) *Skeletonema*. Dinoflagellates: (c) *Ceratium* and (d) *Peridinium*. Green algae: (e) *Chlamydomonas*; (f) *Scenedesmus*. Blue-green algae: (g) *Aphanizomenon* and (h) *Anabaena*. Chrysophytes: (i) *Dinobryon* and (j) *Synura*. Euglenoids: (k) *Euglena*

3. high nutritive value;
4. high protein content; and
5. ability to grow in wastewater.

The production of algal biomass from wastewater has the following objectives and benefits:

5.1.1 Wastewater treatment and nutrient recycling

The biological reactions occurring in algal ponds reduce the organic content and nutrients of the wastewater by bacterial decomposition and convert them into algal biomass by algal photosynthesis. Algal cells have high protein value, and subsequent harvesting of algae for human and animal consumption will be a financial incentive for wastewater treatment. The average production of algae is reported as 70 tons/(ha-year) or 35 tons/(ha-year) algal protein; as compared to

Table 5.1 Summary of the major phyla of algae[a]

Phylum	Common name and species	Pigmentation	Other characteristics
Cynophyta	Blue-green algae; *Spirulina*, *Oscillatoria*, *Anabaena*	Blue-green; phycocyanin, phycoerythrin chlorophyll *a* and *b*	Multicellular or unicellular but usually microscopic; some forms become unicellular in turbulent media; usually with gelatinous sheath
Euglenophyta	Euglenoids; *Euglena*	Grass green	Unicellular, motile; lacking cell wall
Chlorophyta	Green algae; *Chlorella*, *Oocystis*, *Scenedesmus*	Grass green Chlorophyll *a* and *b*	Unicellular, multicellular, a few macroscopic; cell wall of cellulose and pectins
Chrysophyta	Yellow-green or golden brown algae (diatoms); *Diatoma*, *Navicula*, *Asterionella*	Yellow-green to golden brown; xanthophylls and carotenes may mask the chlorophyll. Chlorophyll *a* and *c*	Microscopic, mostly unicellular; includes the large group of diatoms which have a cell wall containing silica
Pyrrophyta	Some are dinoflagellates; *Peridinium*, *Massartia*	Yellow-green to dark brown; xanthophylls predominant. Chlorophyll *a* and *c*	Unicellular, motile; cellulose cell wall
Phaeophyta	Brown algae; *Fucus*	Olive green to dark brown; fucoxanthin and other xanthophylls predominant. Chlorophyll *a* and *c*	Principally multicellular and marine (seaweeds); cellulose and pectin cell wall
Rhodophyta	Red algae; *Polysiphonia*	Red; phycocyanin, phycoerythrin. Chlorophyll *a*	Some unicellular, mostly multicellular; marine; cellulose and pectin cell wall

Adapted from Mitchell (1974), p. 75; reprinted by permission of Prentice-Hall Inc., Englewood Cliffs, New Jersey

[a] Chlorophyll, carotenes and xanthophylls are present in all phyla

the productivity of conventional crops, wheat 3.0 (360 kg protein), rice 5.0 (600 kg protein), and potato 40 (800 kg protein) tons/(ha-year) (Becker, 1981).

Almost all the organic wastes such as municipal wastewater, agricultural, and animal wastes can be treated by algal systems, resulting in considerable algal biomass yields.

5.1.2 Bioconversion of solar energy

Solar energy is the primary source of energy for all life, and is utilized by phytoplankton during photosynthesis and synthesis of new cells. Algae are phytoplankton, and they are the primary producers in the food chain with high productivity. Therefore algal production will be an efficient method for the conversion of solar energy.

5.1.3 Pathogen destruction

Wastewater usually contains pathogens which are harmful to humans, therefore pathogen destruction with waste stabilization is advantageous in waste recycling processes. A certain amount of pathogen destruction occurs in algal ponds due to the adverse environment prevailing in the pond. Adverse environmental conditions to pathogens are caused by:

1. diurnal variation of pH due to photosynthesis;
2. algal toxins excreted by algae cells; and
3. most importantly, solar radiation (UV light).

At present the main attractiveness of algal mass cultures is that they have great versatility for integration into multi-use systems for solving several environmental problems at the same time. Figure 5.2 shows possible applications for algal mass cultures, details of which are described in section 5.4.

As algal mass culture systems must be large and exposed to the outdoor environment, the bioengineering problems involved are greatly magnified. Specific problems relating to culture mixing, nutrient availability and addition, species control, algal separation, and harvesting have been of major concern. These problems are outlined below.

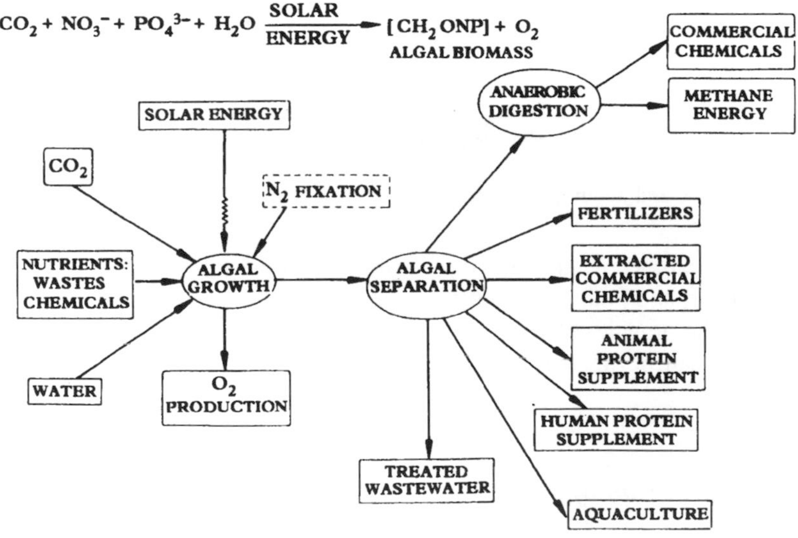

Figure 5.2 Possible applications of algal mass cultures (from Goldman, 1979a; reproduced by permission of Pergamon Books Ltd)

HARVESTING

Algae are very small (ranging from 1 to several μm), difficult to harvest, and require skillful operation. The existing technologies of algal harvesting are normally complex and expensive, and will be discussed in section 5.3.

An economical method for the harvesting of algal cells from an algal pond is still being studied.

ALGAE COMPOSITION

Except *Spirulina*, algae have thick cell walls which make the untreated algae indigestible to nonruminants. Therefore before feeding to animals, algae cells have to be ruptured, either by chemical alteration of the cell wall by acid treatment, or by mechanical or thermal treatment for cell wall destruction.

Another factor limiting the consumption of waste-grown algae is nucleic acid present in algae cells (4–6 percent), which may be harmful to humans (Becker, 1981).

To produce algal biomass for animal or human consumption, the blue-green algae, *Spirulina* (Figure 5.3), is a desirable species for culture because its cell wall is more digestible by nonruminants and its filamentous cells are easier to harvest. Mass culture of pure species of *Spirulina* requires certain environmental conditions (e.g. pH and nutrients) conducive to its growth without contamination from other microbes. Algal ponds or high-rate ponds treating wastewater will contain several algal and bacterial species essential for wastewater treatment. (The term 'high-rate' is commonly used for algal ponds because the algal growth rate in these ponds is several times greater than that occurring in conventional waste stabilization ponds whose objective is mainly waste treatment. Details of the high-rate ponds are given in section 5.2.) The waste-grown algae thus need to be processed further prior to being employed as human food or animal feed. These algae can be used for other purposes, as shown in Figure 5.2 and section 5.4.

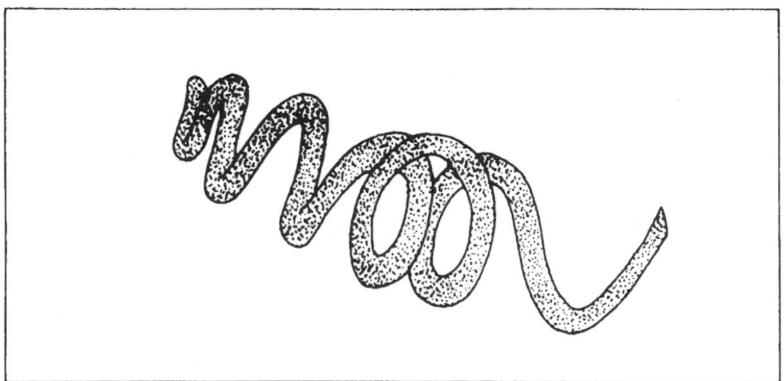

Figure 5.3 *Spirulina* 'jeejibai', showing tapering of end turns, (× 600)

CONTAMINATION OF TOXIC MATERIALS AND PATHOGENS

Possible contamination of algae with toxic substances (heavy metals, pesticides) and pathogens, which are common in wastewater, will reduce the potential value of the algal production.

5.2 ALGAL PRODUCTION AND HIGH-RATE ALGAL PONDS

There are three possible algal cultivation and processing systems, depending on the raw material used and the use for the obtained biomass (Becker, 1981):

1. A system in which a selected algal strain is grown using fresh water, mineral nutrients, and additional C sources. The algae produced in such systems are intended to be utilized mainly as food.
2. A system in which sewage or industrial wastewater is used as the culture medium without the addition of minerals and external C. In such a system the algal population consists of several species in the presence of large numbers of bacteria. The main purpose of this system is for wastewater treatment and the biomass produced is used as feed for animals, or as substrate for energy production.
3. A system where cultivation of algae is in an enclosed system (fermenter) under sunlight or artificial light, with cells being grown in a completely autotrophic medium.

System (2) is related to our objectives of waste treatment, recovery, and recycling and will be the main concern in this chapter.

The basic reactions occurring in an algal pond can be represented by eqs 2.1, 2.2 and 2.6, or the 'algal–bacterial symbiosis' previously cited in section 2.4. These reactions are schematically shown in Figure 5.4. Organic matter entering the system as wastewater or sludge is aerobically decomposed by bacteria, using oxygen photosynthetically produced by algae. The algae, utilizing solar energy and nutrients (or by-products) from bacterial oxidation, perform photosynthesis and synthesize new algal biomass. It is apparent from Figure 5.4 that the excess algae and bacteria biomass produced during the algal–bacterial symbiosis needs to be regularly removed from the system to maintain a constant biomass and efficient performance of the system.

5.2.1 High-rate algal pond (HRAP) systems and design criteria

The HRAP conventionally takes the form of a continuous channel equipped with an aerator–mixer to recirculate the contents of the pond. It is characterized by large area/volume ratios, and shallow depths in the range 0.2–0.6 m to allow sunlight to penetrate the whole pond depth. To minimize short-circuiting, baffles are normally installed in the pond to make the length/width ratio of the channel greater than

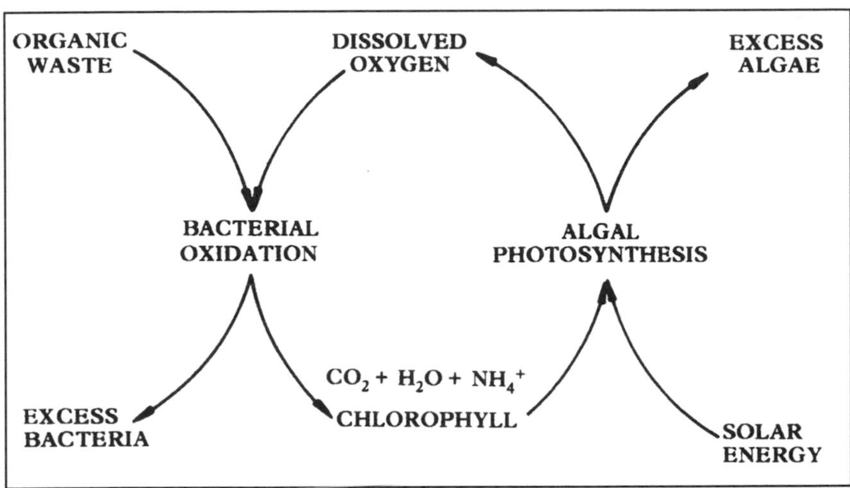

Figure 5.4 The cycle of oxygen and algal production in sewage treatment by photosynthesis (from Oswald and Gotaas, 1957; reproduced by permission of the American Society of Civil Engineers)

2 : 1. A diagram of an HRAP is shown in Figure 5.5. Depending on the mode of operation, sewage can be fed to the HRAP continuously or intermittently, i.e. 12 h/day during the sunlight period. HRAP is not sensitive to daily fluctuations in loading rate. Effluent overflow from HRAP, containing high algal suspension, normally goes into an algal separation unit. The effluent obtained, after the algae have been separated, is expected to have BOD_5 of 20 mg/L and DO of 0.5 mg/L. The effluent may be used for various purposes such as agricultural irrigation, industrial cooling, or recreation. Because of these advantages, HRAP has in recent years received increasing attention as a means both of treating wastewater and producing algal biomass.

Factors affecting the performance of HRAP and algal production include available C and nutrient sources, temperature, light intensity, mixing or agitation, pond depth, and hydraulic retention time (HRT). It is generally known that light intensity is important for photosynthesis and therefore for algal production. Temperature influences the biodegradation rate of the organic matter, and consequently the HRT to be designed for the HRAP.

In the context of photosynthesis, illuminance and irradiance are the two terms commonly used to express light intensity. Illuminance, or luminous intensity, is defined as luminous flux per unit area and bears a photometric unit of lux (lumen/m^2) or foot-candle (1 ft-candle = 10.764 lux), which can be measured by a lux meter or foot-candle meter. Irradiance, or radiant intensity, is defined as quantity of energy that is received on a unit area of surface over time and bears a radiometric energy unit of, for example, g-cal/cm^2-day (also called Langley/day), which can be measured by an actinometer or a pyranometer. Table 5.2 gives conversion factors for the most commonly used units of irradiance. It should be noted that illuminance and irradiance may not be directly correlated or convertible, depending on several factors such as location on the earth, latitude, season, and

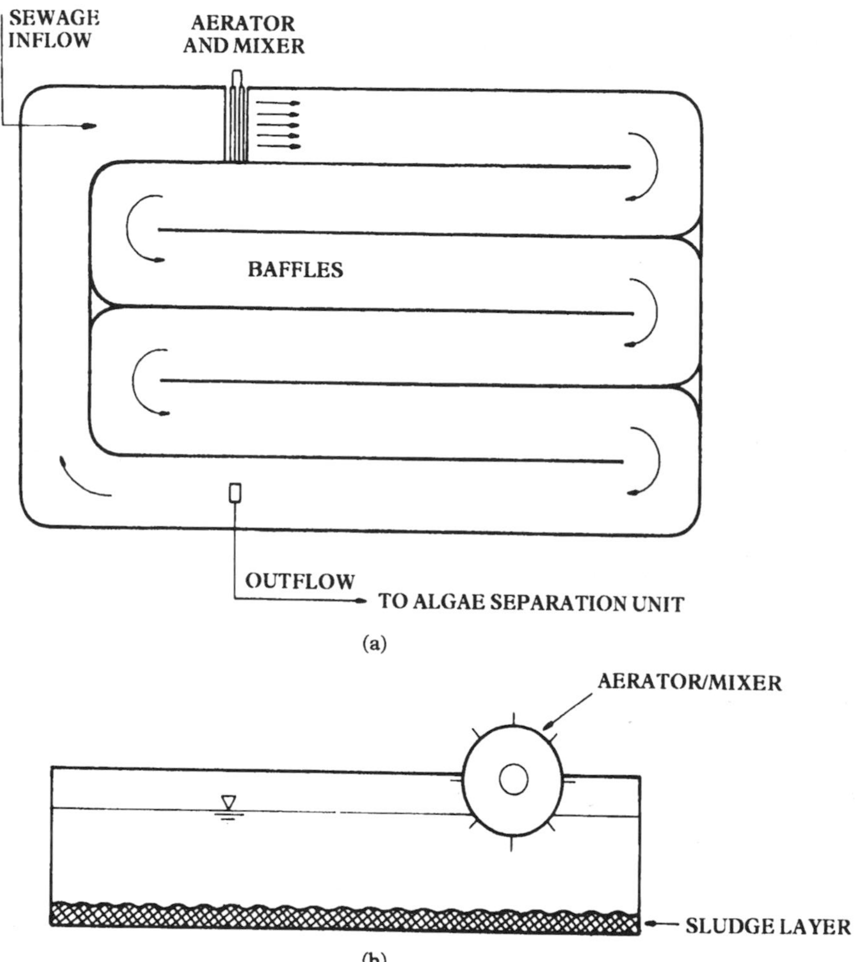

Figure 5.5 High-rate algal pond: (a) schematic plan, (b) schematic cross-section

other meteorological effects. Some reported energy equivalents of illuminance and irradiance in the visible range of daylight (400–700 nm wavelength) are shown in Table 5.3, where only ft-candle and g-cal/cm^2-day are used for the convenience of comparison.

Only solar radiation of wavelengths between 400 and 700 nm is available for photosynthesis by green plants and algae, coinciding with the range of wavelengths visible to the human eye. The daily amount of solar energy that reaches the

Table 5.2 Conversion factors for the most commonly used units of irradiance

Units	g-cal/cm^2-day	W/m^2 (or J/m^2-s)	Btu/ft^2-h
g-cal/cm^2-day	1	0.485	0.154
W/m^2 (or J/m^2-s)	2.06	1	0.317
Btu/ft^2-h	6.51	3.155	1

Table 5.3 Reported energy equivalents of illuminance and irradiance in the visible range of daylight (400–700 nm wavelength)

Energy equivalents	Reference
1 ft-candle = 0.0206 g-cal/cm^2-day	Kreith and Kreider (1978)
1 ft-candle = 0.0936 g-cal/cm^2-day	Wetzel and Likens (1979)
1 ft-candle = 0.0913 g-cal/cm^2-day	Golterman (1975)
1 ft-candle = 0.0933 g-cal/cm^2-day (daylight, full sun)	Coombs and Hall (1982)
1 ft-candle = 0.1119 g-cal/cm^2-day (blue sky light)	Coombs and Hall (1982)
1 ft-candle = 0.072 g-cal/cm^2-day	Bickford and Dunn (1972)

earth and water surface depends on astronomical, geographical, and meteorological factors. Nowadays, the data of illuminance and irradiance can be obtained from the meteorological data of the local weather bureau.

The engineering factors that can be controlled in the design and operation of HRAP are described below.

NUTRIENTS

Algae utilize ammonia as the principal source of N with which to build their proteinaceous cell material by photosynthetic reaction (eq. 2.6 or 2.7). At moderately long HRT of 3 or 4 days, when temperature and light are optimum, almost all the ammonia nitrogen appears in the form of algal cell material (Oswald and Gotaas, 1957). The N content of a waste material places an upper limit on the concentration of algal cells to be developed from it. As about 80 percent of N in the waste is recovered in the algal cells (C_a) and algal cells contain approximately 8 percent N, a relationship of $C_a = 10$ N has been proposed. This relationship suggests that wastewater containing 20 mg/L of N will yield an algal concentration of about 200 mg/L in HRAP, and that N in domestic sewage might become a limiting factor for algal growth if C_a exceeds 200 mg/L. To maintain efficient performance of an HRAP in waste stabilization and algal biomass production, the excess algal biomass has to be regularly removed or harvested, as previously mentioned.

Phosphorus (P), important to algal growth, is not expected to become a limiting factor because of the increased use of detergents in the home and in industry. Other nutrients, such as magnesium (Mg) and potassium (K), should be present in domestic sewage in sufficient quantities to support the growth of algae in an HRAP. The percentage contents of P, Mg, and K in algal cells are 1.5, 1, and 0.5, respectively (Oswald and Gotaas, 1957).

DEPTH

Pond depth should be selected on the basis of availability of light to algae. Oswald and Gotaas (1957) suggest that it may be approximated by Beer–Lambert law:

$$\frac{I_z}{I_i} = e^{-C_a \alpha z} \tag{5.1}$$

where I_i = measured light intensity at the pond surface, varying from 0 to 20,000 ft-candle

I_z = measured light intensity at depth z, ft-candle;

α = specific absorption coefficient, ranging from 1×10^{-3} to 2×10^{-3};

C_a = concentration of algae, mg/L;

z = pond depth, cm.

For practical design it should be assumed that all available light is absorbed; therefore, at the pond bottom the transmitted light, I_z, should be relatively small. If I_z is taken as equal to 1, eq. 5.1 can be written as:

$$z = \frac{\ln I_i}{C_a \alpha} \tag{5.2}$$

Equation 5.1 defines the effective depth for photosynthetic oxygen production in as much as there is no visible light, and hence no algal growth below the depth z.

To determine z for an HRAP, the values of I_i, C_a, and α have to be selected. I_i values can be obtained from meteorological data of the particular area which will vary with seasons, climate, and latitude, normally from a few hundred foot-candles to more than 10,000 foot-candles. C_a concentrations in HRAPs are between 200 and 300 mg/L. The value of α depends on the algal species and their pigmentation which in practice may be taken as 1.5×10^{-3}.

It appears from eq. 5.2 that C_a is the only controllable parameter which determines optimum pond depth for an HRAP. Theoretically, the depth for maximum algal growth should be in the range of from 4.5 to 5 inches (about 12.5 cm). Oswald (1963) carried out laboratory and pilot-scale experiments and found the optimum depth to range from 8 to 10 inches (20–25 cm). However, from a practical point of view, the depth should be greater than 20 cm (i.e. 40–50 cm) to allow for the sludge layer and for maintaining the needed HRT (Moraine *et al.*, 1979).

HYDRAULIC RETENTION TIME (HRT)

The optimum HRT for HRAP should be such that most nutrients are converted into algal cells. From theoretical analysis, Oswald and Gotaas (1957) derived the following relationship:

$$\text{HRT} = \frac{C_a \cdot h \cdot z}{1{,}000\, F \cdot I_0 \cdot T_c} \tag{5.3}$$

or

$$\text{HRT} = \frac{L_t \cdot h \cdot z}{1{,}000\, F \cdot I_0 \cdot T_c \cdot p} \tag{5.4}$$

where HRT = days

h = unit heat of algae, kg-cal/g of algae (for sewage-grown algae = 6 kg-cal/g);

F = efficiency of light energy conversion to chemical energy, usually = 0.1;
I_0 = the amount of visible solar energy penetrating a smooth water surface, varying from 0 to 800 g-cal/(cm^2-day);
T_c = temperature coefficient, values given in Table 5.4;
L_t = ultimate BOD or COD of influent wastewater, mg/L;
p = ratio of weight of O_2 produced and weight of algal cells synthesized, calculated from eq. 2.6, to be 1.58.

The terms C_a and z are as defined previously. Equation 5.4 indicates that HRT varies inversely with I_0. Therefore a strong I_0 may be associated with a reduced HRT. Oswald *et al.* (1953) indicated that, other factors being constant, the most favorable environment for algal growth in continuous cultures is obtained at relatively short HRT. Under these conditions algal growth is maintained in logarithmic phase, with cells being large, fat, rich in chlorophyll, low in carbohydrate, and rapidly producing protoplasm. For biological reasons the HRT should be larger than 1.8 days, which is a minimum time span for the generation of the algae in an HRAP (Oron and Shelef, 1982). Oswald *et al.* (1953) reported that the algal fraction of the suspended solids (SS) in effluent increased with increasing HRT, and at HRT greater than 4 days practically all of the volatile suspended solids (VSS) consisted of algal cells. Moraine *et al.* (1979) reported that algae constituted about 30–90 percent of the VSS in an HRAP in Israel, with 65 percent being typical. They also found longer HRT to favor a higher algal fraction in the HRAP water.

The maximum value for HRT of an HRAP should not exceed 8 days, since under these conditions the pond will be underloaded (lack of nutrients), causing a decrease in algae concentration. Therefore, for the maximization of the net production in an HRAP, the choice of the optimum value of HRT/z (subject to environmental and biological conditions) is of the utmost importance. From a practical point of view, HRT/z at a range of 6–12 day/m seems to be appropriate for all operational purposes (Oron and Shelef, 1982).

Table 5.4 Temperature coefficients for *Chlorella* cultures grown in pilot-scale plants

Mean temperature		Photosynthetic temperature coefficient, T_c
°C	°F	
0	32	—
5	41	0.26
10	50	0.49
15	59	0.87
20	68	1.00
25	77	0.91
30	86	0.82
35	95	0.69
40	104	—

BOD LOADING

BOD loading of HRAP should have an influence on algal yield because a too-high BOD loading can result in anaerobiosis and interference with the algal–bacterial symbiosis. Hsu (1970) did an experiment on an HRAP in Thailand using diluted, fresh Bangkok nightsoil as raw material to discover the optimum design parameters of the HRAP. He concluded that, for a tropical region such as Bangkok, the optimum pond depth, HRT, and BOD_5 loading should be 0.35 m, 1.5 days, and 300 lb/(acre-day) (or 336 kg/(ha-day)) respectively. Under these conditions BOD_5 removal efficiency was about 95 percent, the effluent BOD_5 was below 10 mg/L, and the algal yield was around 350 lb/(acre-day) (390 kg/(ha-day)). An experiment on a 200 m^2 HRAP by Edwards and Sinchumpasak (1981) using weak sewage (mean $BOD_5 = 45$ mg/L) resulted in the mean algal concentration of 94 mg/L or the mean algal yield of 157 kg/(ha-day); the organic load and HRT applied to this pond were 75 kg BOD_5/(ha-day) and 3 days, respectively.

MIXING AND RECIRCULATION

Mixing of the HRAP content is essential to prevent algal sedimentation and to provide interactions between the benthic deposits and free oxygen-containing supernatant. Mixing keeps the nutrients in active contact with the algal cell surface, leading to a stimulation of metabolic activity and a more effective utilization of incident light (Persoone *et al.*, 1980). In a large-scale HRAP, mixing can prevent thermal stratification and the development of anaerobic conditions at the pond bottom, and avoid photoinhibition due to the decrease of the duration of stay in the over-exposed layer, where irradiance might be too high for the algal cells (Soeder and Stengel, 1974).

On the other hand, mixing results in the suspension of sediments and reduces light penetration. Too much mixing is uneconomic for an HRAP operation. Moraine *et al.* (1979) found the intensified mixing to have an adverse effect on algal population stability, and they suggest a flow velocity of algal suspension in an HRAP to be 5 cm/s. Because the surface water in HRAPs is fairly homogeneous due to paddle wheel mixing, Green and Oswald (1995) suggested that the mixing linear velocity should be maintained near 15 cm/s due to the following reasons:

1. There are two distinct biological phases in a high-rate algal pond, an oxidative bacterial floc phase and the photosynthetic algal phase. At the flow velocity of 15 cm/s, the algal phase is suspended but the bacterial phase, being more sticky and hence heavier and more flocculant, stays near the bottom where its optimal pH is about 7.0 and it is protected from the higher pH surface water. Being near the bottom, the bacterial floc does not interfere with the penetration of light into the algal phase, permitting photosynthesis to proceed.
2. Maintaining a velocity of 15 cm/s requires only one-eighth as much energy as a velocity of 30 cm/s and only 1/64 the energy of a velocity of 60 cm/s.

Table 5.5 Design criteria for high-rate algal ponds

Pond depth (z), m	0.3–0.6
HRT, days	1.5–8
HRT/z, day/m	6–12
BOD loading, kg/ha-day	75–300
Mixing linear velocity, cm/s	5–15
Channel length/width ratio	>2

3. Delicate algal flocs that tend to form are not disrupted at 15 cm/s and hence are more settleable when transferred to the settling pond.

Based on the information given above, design criteria for HRAPs are given in Table 5.5. As the algal–bacterial symbiotic activities are greatly dependent on temperature, under tropical conditions, low HRTs and/or high organic loading rates can be employed for HRAP design and operation.

The principal advantages of pond recirculation are the maintenance of active algal and bacterial cells in the HRAP system and aeration of the influent wastewater. Most HRAPs have a configuration similar to that shown in Figure 5.5, in which recirculation is normally practiced in pond operation.

Although the major factors affecting HRAP performance are light intensity and temperature, the engineering parameters which can be manipulated to produce optimum HRT in year-round operation are pond dimensions, namely area and

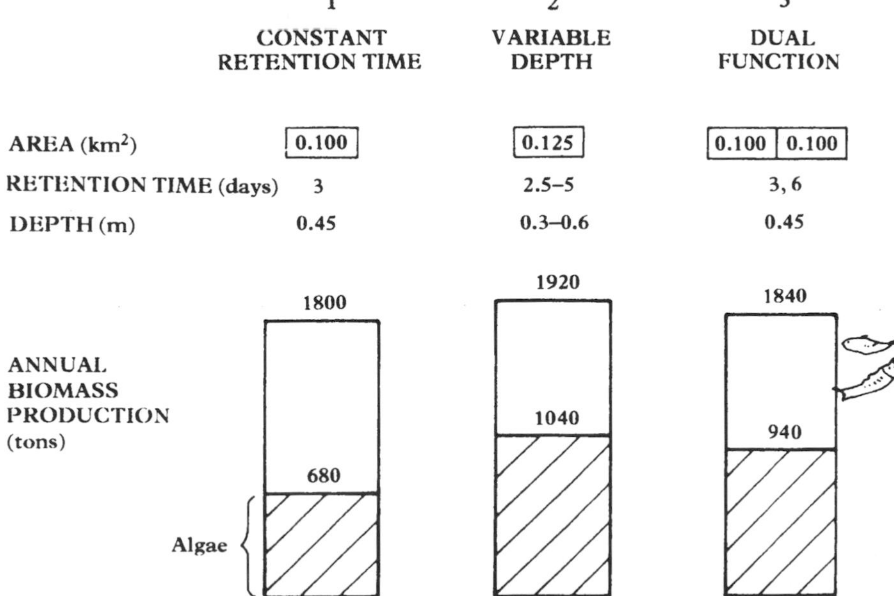

Figure 5.6 Estimated dimensions of HRAP and productivity for various operational strategies based on a community of 50,000 people (from Azov and Shelef, 1982; reproduced by permission of Pergamon Books Ltd)

depth (Azov and Shelef, 1982). Based on their HRAP research in Israel they have proposed three modes of pond operation in which a comparison of estimated pond dimensions and productivity based on a community of 50,000 people is given in Figure 5.6.

1. Constant-HRT operation, which is appropriate for tropical climates where seasonal variations in solar radiation and temperature are minimal. This method requires the least area, but also produces the least biomass.
2. Variable-HRT or variable-depth operation, which is recommended for moderate climates and can be economically achieved by varying pond depth at constant area using a variable-level overflow weir. During the summer period, when the temperature is high, the required HRT and consequently pond depth should be less; and vice-versa during the winter period. Azov and Shelef (1982) stated that determining the required changes in pond depth to produce optimal HRT is a matter of 'trial and error', depending on operational experiences and wastewater characteristics.

   This method of pond operation has a land area requirement 25 percent greater than method 1, but pond productivity is higher.
3. Variable-HRT operation using dual-function ponds, which might be of interest in agricultural locations but has doubled the land requirement of method 1. The ponds are operated solely for wastewater treatment during winter, while some can be converted into fish-rearing ponds during summer.

A recent development on HRAPs by Green and Oswald (1995), called Advanced Integrated Waterwater Pond System (AIWPS, Figure 5.7), was found very effective in wastewater treatment and algal production. The system consists of four ponds in series: a primary pond (advanced facultative pond, or AFP) with internal fermentation pits; a secondary shallow continuously mixed pond (high-

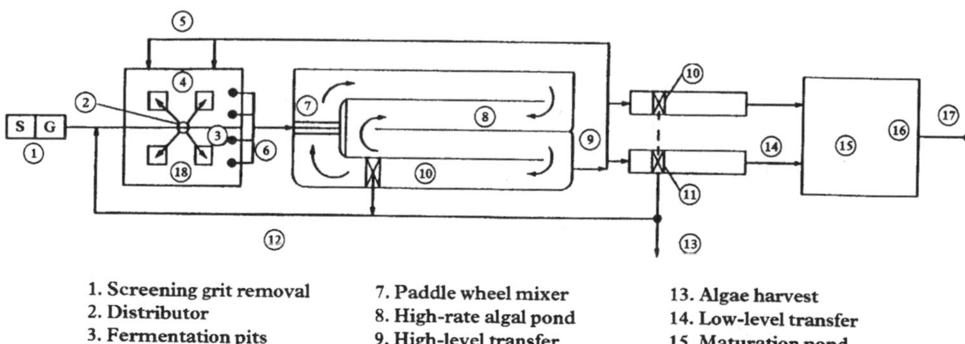

Figure 5.7 Advanced integrated algal–bacteria system for liquid waste treatment and oxygenated water, nutrient recovery and reuse (Oswald, 1991)

Table 5.6 Performance of AIWPS (Oswald, 1991)

(a) St Helena, California

Parameter	Influent sewage	AFP	HRAP	ASP	AMP	Percent removal
HRT (days)	0	20	10	5	N/A[a]	
BOD (mg/L)	223	17	9	6	7	97
COD (mg/L)	438	124	74	58	32	93
Total C (mg/L)	215	144	88	69	50	77
Total N(mg/L)	40	16	13	6	4	90
Total P (mg/L)	14	13	12	8	5	64

(b) Hollister, California

Parameter	Influent sewage	AFP	HRAP	ASP	AMP[b]	Percent removal
HRT (days)	0	32	10	7	—	
BOD (mg/L)	194	43	7	7	—	96
TVS (mg/L)	604	393	341	347	—	42
E. coli (MPN, 100 mL)	10^8	10^6	10^5	10^4	—	99,999

[a] N/A = not available

[b] Hollister's settling pond effluent is discharged to natural gravel percolation beds; there is no surface effluent

rate algal pond, or HRAP); a tertiary settling pond (advanced settling pond, or ASP); and a quaternary holding pond (advanced maturation pond, or AMP).

The AFPs are designed to retain the settleable solids in wastewater and to foster their long-term fermentation to gaseous products such as methane, CO_2, and N. The HRAPs are designed to produce algae and photosynthetic oxygen in sufficient, but not excessive, amounts to permit microbial oxidation of organic residuals. Algae in the HRAPs tend to raise the water pH and a pH of 9.2 for 24 h will provide an almost 100 percent kill of *Escherichia coli* and presumably most pathogenic bacteria (Oswald, 1991). It is not uncommon for HRAPs to reach pH levels of 9.5 or 10 during the day, so they have a high disinfection rate. The ASPs are designed to settle and remove algal–bacterial concentrates from the oxidized waste; and the AMPs are designed to hold the treated wastewater for seasonal applications of irrigation water for controlled discharges and further improve the disinfection. Table 5.6 shows the performance of AIWPS. The algal biomass production ranges from 80 to 220 kg (dry weight)/ha-day.

5.2.2 *Estimation of algal production*

A detailed literature review of photosynthetic algal yield was conducted by Goldman (1979b). He summarized that, at best, conversion efficiencies of less than 5 percent of total sunlight into algal biomass can be expected, leading to the upper limit in yields of 30–40 g (dry weight)/(m^2-day). Although a high temperature leads to high algal yields, respiratory and other decay processes are also influenced by this parameter, making temperature not nearly as important as sunlight in controlling algal productivity. To achieve maximum algal yield in an outdoor HRAP, consideration must be given also to the factors that affect algal growth, as described in section 5.2.1.

Some of the engineering models to predict algal productivity in outdoor mass culture or HRAP are given below.

GOLDMAN FORMULA

Based on the simple energy balance in photosynthesis and considering the characteristics of sunlight, Goldman (1979b) derived the following model in which the effect of algal decay was excluded:

$$P_a = 0.28 I_s \left[\ln \left(0.45 \frac{I_0}{I_s} + 1 \right) \right] \tag{5.5}$$

where P_a = algal productivity (dry weight), g/(cm^2-day);
I_s = saturation light intensity, g-cal/(cm^2-day);
I_0 is as defined in eq. 5.4.

The value I_s depends on temperature and is not the same for all cultures of algal species, but rather is a function of the physiological make-up of the particular cells in the culture. Goldman (1979b) stated that good values of I_s for different algal species from the literature are generally lacking. The data for I_s given in Table 5.7 only represent the general magnitude of I_s for different types of algae, and

Table 5.7 Summary of light saturation intensities (I_s) for different freshwater and marine microalgae

		I_s	
Species	Temperature (°C)	Illuminance (ft-candle)	Irradiance g-cal/(cm^2-day)
Freshwater			
Chlorella pyrenoidosa	25	—	51.8
	25	500	36.0
	26	—	82.1
Chlorella vulgaris	25	250	18.0
Scenedesmus obliquus	25	500	36.0
Chlamydomonas reinhardtii	25	500	36.0
Chlorella (7–11–05)	39	1,400	100.8
Marine			
Green algae	20	500	36.0
Diatoms	20	1,000	72.0
Dinoflagellates	20	2,500	180.0
Phaeodactylum tricornutum	18	—	82.1

Adapted from Goldman (1979b); reproduced by permission of Pergamon Books Ltd

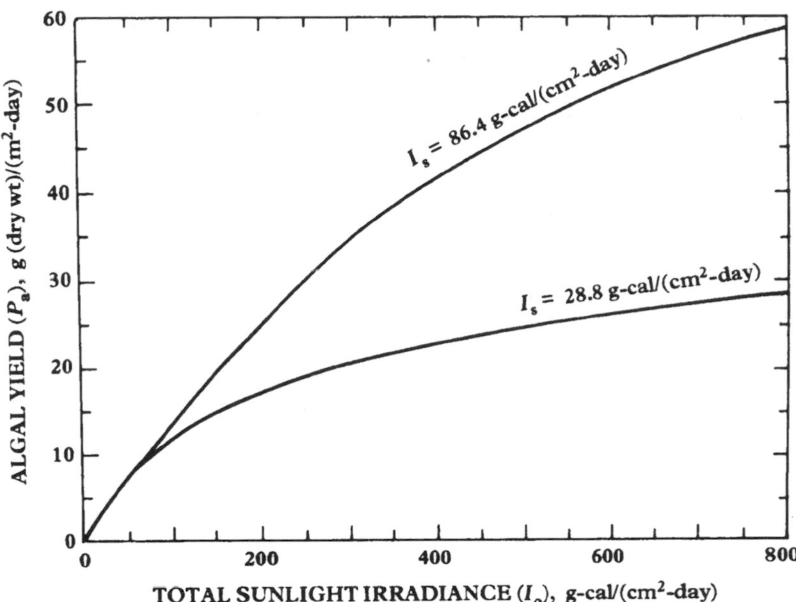

Figure 5.8 Algal yield (P_a) as a function of total solar irradiance (I_0) according to eq. (5.5) (adapted from Goldman, 1979b; reproduced by permission of Pergamon Books Ltd)

demonstrate the effect of temperature on the I_s values or the photosynthetic efficiency factor. The range of I_s values likely to be experienced with an outdoor HRAP is 29–86 g-cal/(cm^2-day) (Goldman, 1979b). The value of I_0 depends on the latitude and local weather conditions such as cloudy or clear sky, which can range from 0 to 800 g-cal/(cm^2-day).

The effects of I_s and I_0 on P_a values, based on eq. 5.5, are shown graphically in Figure 5.8. It is apparent from this figure that the algal productivity can be almost doubled by culturing the algal species with an I_s value of 86 g-cal/(cm^2-day) instead of an I_s value of 29 g-cal/(cm^2-day) at the higher light intensities (or higher I_0 values). However, P_a values do not linearly increase with I_0 at higher I_0 values because of the light saturation effect on algal growth. The large data compilation of maximum algal yields attained in outdoor HRAP systems (Goldman, 1979a) falls within the ranges shown in Figure 5.8.

ORON AND SHELEF FORMULA

To account for the effects of environmental and operational conditions of algal growth in HRAP, Oron and Shelef (1982) proposed the following empirical formula.

$$C_a = a \cdot (\mathrm{HRT}/z)^b \cdot I_0^\beta \cdot T^\gamma \tag{5.6}$$

where C_a = algal concentration, mg/L;
T = ambient temperature, °C;
a, b, β, γ = constants, dimensionless.

Units of HRT, z and I_0 are days, m and g-cal/(cm^2-day), respectively.

The numerical values of these constants were determined from the data collected from a pilot- and field-scale HRAP operating in Israel. By applying the nonlinear least-square method, the following equation was derived:

$$C_a = 0.001475(\text{HRT}/z)^{1.71} I_0^{0.70} T^{1.30} \tag{5.7}$$

The related algal productivity, P_a in g/(m^2-day) may be given by:

$$P_a = C_a \cdot (z/\text{HRT}) = 0.001475(\text{HRT}/z)^{0.71} I_0^{0.70} T^{1.30} \tag{5.8}$$

An experiment was carried out in Israel with domestic wastewater (average BOD_5 of 400 mg/L) and algae such as *Micractinium pusillum*, *M. quadriesetum*, and *Scenedesmus dimorphus* in an HRAP. Table 5.8 shows the predicted algal concentrations using eq. 5.7 and measured algal concentrations during six seasons. The results of this experiment show the applicability of eq. 5.7 for the assessment of algal concentration in HRAP.

Some drawbacks of eq. 5.7 are that it is applicable to influent BOD_5 ranging from 200 to 400 mg/L and to conditions in Israel where these experiments were carried out. However, the same approach can be applied to suit other local conditions, but with different values of a, b, β, γ.

THEORETICAL ESTIMATION

Equation 2.6 implies that 1 mole of algal cell (molecular weight 154 g) is synthesized with production of 7.62 mole O_2. Therefore, 1 g of algal cell is equivalent to

$$\frac{7.62 \times 32}{154} = 1.58 \text{ g } O_2, \text{ or the factor}$$

$$p = \frac{\text{weight of } O_2 \text{ produced}}{\text{weight of algal cell synthesized}} = 1.58,$$

as previously stated in eq. 5.4.

Table 5.8 Environmental and operational conditions and predicted algal concentration (dry weight) of the HRAP in Israel

	Solar radiation, I_0, (g-cal/(cm^2-day))	Ambient temperature (°C)	Effluent depth z, (m)	Retention time, HRT (days)	Measured algal concentration, C_a (mg/L)	Predicted algal concentration, C_a^{α} (mg/L)
Spring 1977	576	14.6	0.45	4.0	176	173
Summer 1977	679	24.2	0.45	2.9	229	215
Fall 1977	393	21.4	0.4	3.4	210	201
Winter 1977/8	318	13.8	0.45	3.9	95	101
Spring 1978	538	18.9	0.35	2.9	240	204
Summer 1979	649	25.4	0.25	2.0	285	321

[a] Algal concentration was predicted applying eq. 5.7

From Oron and Shelef (1982); reproduced by permission of the American Society of Civil Engineers

The oxygen demand of a waste is assumed to be met through photosynthetic oxygen production; the maximum biochemical oxygen demand, L_t in any time t, may be substituted for oxygen produced, and algal concentration C_a may be substituted for weight of algal cell synthesized. Therefore:

$$p = \frac{L_t}{C_a} \tag{5.9}$$

and algal production P_a, in g/(m^2-day)

$$= \frac{L_t}{p} \cdot \frac{z}{\text{HRT}} \tag{5.10}$$

where the units of z, HRT and L_t are m, days, and mg/L, respectively.

Oswald and Gotaas (1957) reported that, under environmental conditions suitable for photosynthetic oxygen production, the value of p is normally between 1.25 and 1.75. The deviation of value for p from theoretical value, 1.58, was due to various factors. In practice, all L_t cannot be oxidized even though enough O_2 is present; therefore a low value for p is observed. Or a strong wind can increase surface reaeration, so increasing the value of O_2 produced, therefore a high p value can be obtained.

Example 5.1 An agro-industry in Thailand is producing wastewater at a flow rate of 500 m^3/day and its COD or L_t is 400 mg/L. Determine the size of the HRAP to be used to produce algae from this wastewater. The following information is given:

h = unit heat of algae = 6 kg-cal/g;
p = 1.5;
F = efficiency of energy conversion = 0.1;
T_c = temperature coefficient = 0.9;
I_0 or I_i = visible solar energy irradiance = 400 g-cal/(cm^2-day) or 11,100 ft-candle, respectively;
α = specific light absorption coefficient = 1.5×10^{-3}.

From eq. 5.9 algal concentration in HRAP may be theoretically estimated.

$$C_a = L_t/p = 400/1.5 = 266.7 \text{ mg/L}$$

Pond depth can be estimated from eq. 5.2:

$$z = \frac{\ln I_i}{C_a \alpha} = \frac{\ln 11{,}100}{266.7(1.5 \times 10^{-3})} = 23.3 \text{ cm}$$

Select a z value of 25 cm.

HRT of the pond is estimated from eq. 5.3:

$$\text{HRT} = \frac{C_a \cdot h \cdot z}{1000\,F \cdot I_0 \cdot T_c}$$

$$\text{HRT} = \frac{266.7 \times 6 \times 25}{1000 \times 0.1 \times 400 \times 0.9} = 1.11 \text{ days}$$

To satisfy the minimum time span for algal generation in HRAP and to avoid cell wash-out from the system, select an HRT = 1.8 days.

$$\begin{aligned}\text{Volume of HRAP} &= \text{flow} \times \text{HRT}\\ &= 500 \times 1.8 = 900 \text{ m}^3\end{aligned}$$

$$\begin{aligned}\text{Surface area of HRAP} &= \text{pond volume/depth}\\ &= 900/0.25 = 3600 \text{ m}^2\end{aligned}$$

Choose two HRAP units, each with the following dimensions: length = 100 m, width = 18 m with six channels (or the width of each channel flow is 3 m).

From the above results the value of C_a could be recalculated using eq. 5.7 and taking $T = 30$ °C.

$$\begin{aligned}C_a &= 0.001475(1.8/0.25)^{1.71}(400)^{0.70}(30)^{1.30}\\ &= 238 \text{ mg/L}\end{aligned}$$

which is slightly lower than the theoretical value of 266.7 mg/L as determined from eq. 5.9.

The algal productivity is determined from eq. 5.10.

$$\begin{aligned}P_a &= C_a(z/\text{HRT}) = 266.7(0.25/1.8)\\ &= 37 \text{ g/(m}^2\text{-day)}\end{aligned}$$

The above P_a value is within the practical range predicted by eq. 5.5 as shown in Figure 5.8.

5.3 ALGAL HARVESTING TECHNOLOGIES

The algal biomass cultured in an outdoor HRAP is mainly in the microscopic unicellular form with concentrations in the pond water ranging from 200 to 400 mg/L. As shown in Figure 5.2, these algal cells need to be separated from the pond water prior to being used further. The HRAP effluent to be discharged to a water course may need to have algal cells removed so that the suspended solid content of the pond effluent is within the regulatory standard for discharge. Because of their microscopic size (generally less than 10 μm), the separation or harvesting of algal cells has been a major challenge and difficulty for environmental engineers and scientists developing an efficient harvesting technology which is economically viable. Some of the technologies that have been used to separate algal cells from effluents of conventional waste stabilization ponds and HRAP systems are presented below.

5.3.1 *Microstraining*

Microstrainers are low-speed (up to 4–7 rpm) rotating-drum filters operating under gravity conditions. The filtering fabrics are normally made of finely woven stainless steel and are fitted on the drum periphery. Mesh openings are normally in the range 23–60 μm, but microscreens with 1 μm mesh size made of polyester fabric have been developed (Harrelson and Cravens, 1982). Wastewater enters the open end of the drum and flows outward through the rotating fabric. The collected solids are continuously removed by high-pressure jets (located outside at the top of the drum) into a trough within the drum (Middlebrooks *et al.*, 1974; Metcalf and Eddy, 1991). Microstrainers are normally operated at hydraulic loading rates of 5–15 $m^3/(m^2\text{-day})$, drum submergence 70–75 percent of height or 60–70 percent of filter area, and drum diameters varying from 2.5 to 5 m depending on the design of the screen. The full-scale microstrainers with a 1 μm mesh size installed at the Camden waste stabilization ponds in South Carolina, U.S.A. (flow rate 7,200 m^3/day) were operated at hydraulic loadings between 60 and 120 $m^3/(m^2\text{-day})$.

Typical SS and algal removal achieved with microstrainers is about 50 percent, the range being 10–80 percent. Reed *et al.* (1988) suggested that microstrainers with 1 μm polyester screening media are capable of producing an effluent with BOD_5 and SS concentrations lower than 30 mg/L. However, the service life of the screen was found to be about $1\frac{1}{2}$ years, which is considerably less than the manufacturer's prediction of 5 years: this was probably due to operational and maintenance problems associated with this type of screen.

Problems encountered with microstrainers having large mesh openings include inadequate solid removal and inability to handle solids fluctuations. These problems may be partially overcome by varying the speed of drum rotation. In general, drum rotation should be at the slowest rate possible that is consistent with the throughput, and should provide an acceptable head differential across the fabric. The controlled variability of drum rotational speed is an important feature of the process, and the speed may be automatically increased or decreased according to the differential head best suited to the circumstances (Middlebrooks *et al.*, 1974).

5.3.2 *Paper precoated belt filtration*

Because conventional microstrainer fabric (mesh sizes greater than 23 μm) is too coarse for the unicellular algae grown in an HRAP, and extremely fine fabric media capable of effective separation of algal cells cannot be adequately cleaned, a new harvesting process developed in Australia, utilizing a paper precoated belt filter, seems to overcome these difficulties (Dodd and Anderson, 1977). A schematic diagram of the paper precoated belt filtration process is shown in Figure 5.9. The belt filter incorporates a coarse fabric belt on which a precoat of paper fibres is deposited, as in a paper machine, forming a continuously renewed filter medium. The paper precoat is continuously re-formed to provide a fresh filter medium having high trapping efficiency and good throughput characteristics.

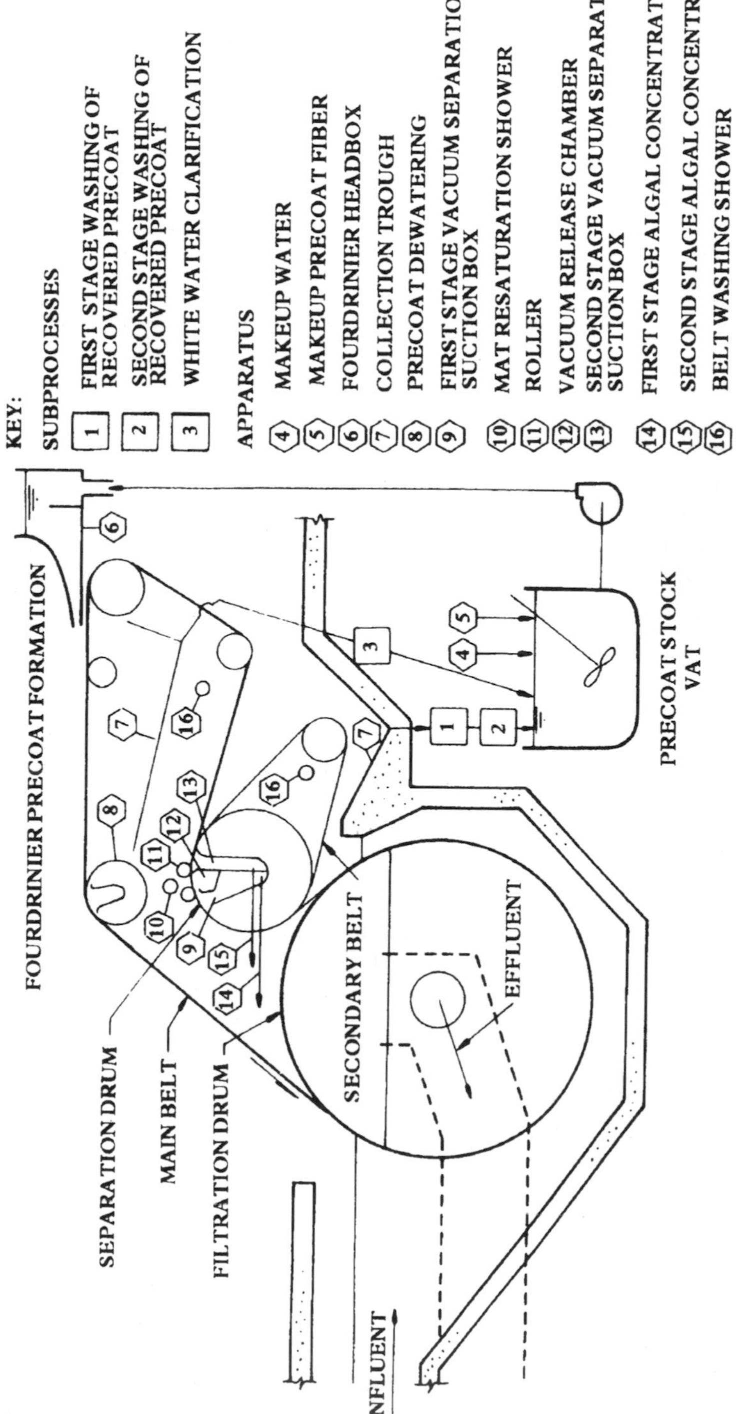

Figure 5.9 Schematic of a precoated belt filtration unit (after Dodd and Anderson, 1977; reproduced by permission of Pergamon Books Ltd)

According to Figure 5.9 the algal water or influent is filtered by the filtration drum. The filtered algal cells attached to the main belt are sandwiched between the main and secondary fabric belts and water, during application of vacuum to separate the algae from the precoat. The paper precoat and the algal cells are then removed from the belts by water showers (nos. 14 and 15 in Figure 5.9) which are called first-stage and second-stage algal concentrates. These algal concentrates are further washed to remove the algae, and the washed paper fibers are recycled to form new paper precoat.

From their experiments, Dodd and Anderson (1977) found the average SS of the first-stage algal concentrate (no. 14 in Figure 5.9) to be 1.18 percent, ranging from 0.81 to 1.49 percent. The average SS of the second stage algal concentrate (no. 15 in Figure 5.9) was 0.43 percent. The second-stage concentrate appeared to have flocculant characteristics which should be beneficial in the thickening of the algal concentrate to reduce the cost of further dewatering by centrifugation or filtration.

5.3.3 Flocculation and flotation

Flocculation is the process of floc formation through slow mixing so that the flocculent materials are large and heavy enough to settle in a sedimentation tank. In water and wastewater treatment flocculation is normally preceded by coagulation, in which coagulant materials such as alum, lime, ferric chloride, or polymers are added individually or in combination, and rapidly mixed to enhance floc formation. Flotation is a physical process in which solid particles float to the water surface through types of buoyant forces (such as dissolved-air flotation or foam flotation), and the floated particles can be skimmed off from the water surface, leaving the clear liquid in the lower portion of the flotation tank. The application of a coagulation process preceding flotation is expected to be beneficial to solids removal, as the floated particles will be larger, easier to entrap or absorb air bubbles, and will be buoyed up by dissolved air so that they can be effectively skimmed off from the water body.

COAGULATION–FLOCCULATION

Based on the above information it appears that higher efficiency for solid removal by flocculation or flotation can be achieved with the aid of coagulation. In the case of algal flocculation using alum as the coagulant, the pH range between 6.0 and 6.8 (6.5 optimum) gave good algal removal efficiency (Golueke and Oswald, 1965). The same result was also found by Batallones and McGarry (1970) when they studied the jar test using a fast mixing speed of 100 rpm for 60 s for coagulation and a final slow mixing speed of 80 rpm for 3 min for flocculation. They found the most efficient alum dose for algal flocculation to be between 75 and 100 mg/L, while Golueke and Oswald (1965) found the alum dose to be 70 mg/L.

Besides alum, other polyelectrolytes or polymers can also be used as coagulant aid materials. Only cationic polyelectrolytes should be used in algal flocculation because the algal cells act like a negative charge. Batallones and McGarry (1970)

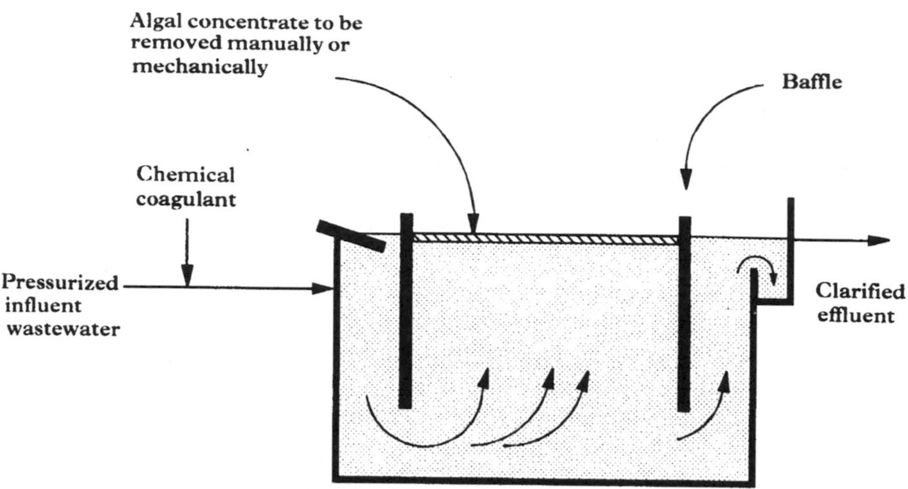

Figure 5.10 Schematic of a dissolved air flotation unit

found the cost of harvesting by alum alone at algal concentrations below 30 mg/L to be rather high and, to reduce the chemical cost, they suggested the use of alum in combination with some cationic polyelectrolyte, and reported that if polyelectrolyte (Purifloc-C31) was used to aid alum, the most economic doses were 40 mg/L of alum together with 2–4 mg/L of Purifloc-C31.

It is well known that the efficiency of coagulation–flocculation is dependent on several environmental parameters such as pH, alkalinity, temperature, turbidity, etc. Therefore laboratory or pilot-scale experiments on individual HRAP water should be conducted, wherever possible, to select the appropriate coagulant materials and other operating conditions for the coagulation–flocculation process.

DISSOLVED-AIR FLOTATION (DAF)

Theoretically there is a tendency for algal cells to float to the water surface due to the release of supersaturated oxygen generated during the photosynthetic period. To make the algal cells float more effectively, DAF is normally employed. In DAF systems, air is dissolved in the water under a pressure of several atmospheres, and the pressure is later released to atmospheric level in the flotation tank. For small-scale operation, part or all of the influent flow is pressurized, and is mixed with coagulant materials such as alum (Figure 5.10). For large-scale operation, and to improve the system performance, the clarified effluent is normally recycled to mix with the influent feed prior to being released into the flotation tank. In this case the recycled effluent is pressurized, semi-saturated with air, and mixed with the unpressurized influent water. DAF efficiency depends on various factors such as the air/solids ratio, dissolved air pressure, flotation time, types and dose of coagulant aids, and pH of the water. There are various types of DAF units produced by some manufacturers in Europe and the U.S.A. for full-scale application.

McGarry and Durrani (1970) conducted DAF experiments on HRAP water and found that at the alum dosages of 125–145 mg/L, pH 5–7, air pressure 35–50 psi gage (2.4–3.4 atmospheric pressures), and flotation time 6–10 min, an algal concentration of 8 percent was obtained in the overflow effluent. Bare *et al.* (1975) studied algal removal using a dissolved-air flotation process, with and without coagulant aids. They found that at 25 percent recycle and 3 atmospheric recycle pressure, the algal removal was 35 percent in the batch process without coagulants. The percentage removal increased to about 80 when ferric sulfate was used at 85 mg/L. The same result could be obtained when 75 mg/L of alum was used at the same operating conditions.

AUTOFLOCCULATION

Autoflocculation refers to the precipitation of algae and other particulate matter in a pond when the pH rises to a highly alkaline level. This phenomenon is related to the chemical make-up of the water, and, in particular, to the presence of calcium and magnesium carbonates (Middlebrooks *et al.*, 1974). As the algae remove CO_2 the pH rises to a point at which precipitation of magnesium hydroxides and calcium carbonate, along with algae, occurs, causing removal of the particulate matter.

A study by Sukenik and Shelef (1984) found the concentrations of calcium and orthophosphate ions in the medium to be important for autoflocculation. To attain autoflocculation within the pH range of 8.5–9.0 the culture medium (or HRAP water) should contain 0.1–0.2 millimol (mmol) orthophosphate and between 1.0 and 2.5 mmol calcium. Calcium phosphate precipitates are considered as the flocculating agent which reacts with the negatively charged surface of the algae and promotes aggregation and flocculation.

Removal of the precipitated algal cells requires another non-agitated basin for the cells to settle, or mechanical aeration has to cease for a few hours daily if cell precipitation is to occur in the HRAP. Removal of the precipitated cells may be conducted during night-time or early morning, when the photosynthetic activity is non-existent and the algal cells tend not to rise to the water surface.

Autoflocculation appears to be a simple method of algal harvesting, because it does not require sophisticated mechanical equipment or operations. However, its efficiency in algal removal is generally less than the aforementioned methods, and it requires a large area for construction of a settling basin. In addition a very warm and cloudless day is required to attain the high pH values (greater than 9) in HRAP water; in most locations this condition does not occur all year round.

5.3.4 Biological harvesting

The unicellular algae grown in HRAP are food for herbivorous fish (fish feeding on phytoplankton) and other macroinvertebrates. The culture of herbivorous fish in algal pond water to graze on the algae is an attractive means to produce protein biomass in the fish tissue. However, fish cannot effectively graze on all the algal cells, and will also discharge fish feces, causing the pond water to still contain high algal content and organic matter content. Details on fish production in wastewater can be found in Chapter 6.

5.3.5 Comparison of alternative algal harvesting methods

The literature on algal harvesting contains numerous studies and discussions of various harvesting methods (Golueke and Oswald, 1965; Middlebrooks *et al.*, 1974; Parker and Uhte, 1975; Benemann *et al.*, 1980). Besides the four harvesting methods listed above there are other methods that can be used to separate algal cells, such as centrifugation, fine-weave belt filtration, and sand filtration. However, there does not appear to be any method clearly superior for a 'typical' application. Owing to the very wide range of conditions and objectives encountered, the selection of an algal harvesting technology and its application should be approached as a distinct case, weighing the merits and drawbacks of each method for the particular circumstances.

Factors to be considered in the selection of a harvesting method include wastewater and HRAP effluent characteristics, treatment objectives in terms of harvested effluent quality for various water reuse applications or discharge requirements, algal product quality, capital and operating costs, energy requirements, level of operator skill required, and availability of equipment and chemicals. The specific physical properties of algae also directly affect the selection of harvesting method, e.g. the motile euglenoids may resist sedimentation or flotation because they can swim away from the process effluent, or the unicellular green algae will be too small for conventional microstrainers. It is apparent that laboratory scale or pilot-scale experiments with the particular HRAP water should be conducted, and the data thoroughly analyzed prior to the selection of an algal harvesting method.

5.4 UTILIZATION OF WASTE-GROWN ALGAE

Besides the benefits in wastewater treatment gained from waste-grown algae, these algae, depending on their characteristics, can be further used according to the applications shown in Figure 5.2.

5.4.1 Algae as food and feed

Tables 5.9 and 5.10 show the chemical compositions of various algal species and soya bean. The protein content of algae is about 50 percent, which is higher than soya bean, while other vitamins and minerals are present in desirable proportions. However, Waslien (1975) reported that there is an imbalance in the amino acid composition of the algal protein, notably in the absence of sulfur-containing amino acids, methionine and cystine, making the nutritional value of algae considerably less than conventional foods such as eggs and milk. Another problem with the

Table 5.9 Chemical composition of different algae compared with soya (percentage dry matter) (Becker, 1981)

Component	*Scenedesmus*	*Spirulina*	*Chlorella*	Soya
Crude protein	50–55	55–65	40–55	35–40
Lipids	8–12	2–6	10–15	15–20
Carbohydrates	10–15	10–15	10–15	20–35
Crude fiber	5–12	1–4	5–10	3–5
Ash	8–12	5–12	5–10	4–5
Moisture	5–10	5–10	5–10	7–10

Table 5.10 Chemical composition of algae (g/100 g dry matter)

Component	*Spirulina*	*Scenedesmus*
Total nitrogen	11.0	8.3
Non-protein nitrogen	1.5	1.05
Protein-nitrogen	9.5	7.25
Available lysine (g/100 g protein)	2.96	3.66
Ribonucleic acid (RNA)	2.90	4.4
Deoxyribonucleic acid (DNA)	1.00	1.6
Calcium	0.75	0.85
Phosphorus	1.42	1.9

From Becker, 1981; reproduced by permission of Turnep plc

algae is their high nucleic acid content, which is about 4 percent or more (see Table 5.10, the values of RNA + DNA). These nucleic acids, when being ingested in large quantities, can produce an unacceptably high level of uric acid (illness) in humans. In addition, except for *Spirulina* spp., most of the waste-grown algae have thick cell walls which are not digested well by humans and non-ruminant animals (such as poultry). Therefore these cell walls have to be ruptured, for example by heat or acid treatment or some mechanical means such as ball milling and crushing.

The thin-walled algae, such as *Spirulina*, thus appear to be a suitable species for culture as a human food supplement. However, as pointed out by Goldman (1979a), it is difficult to control algal speciation in outdoor cultures for sustained periods, except in unique chemical environments which rarely exist in nature. Owing to the rapid generation of algae, certain species tend to dominate through natural selection, regardless of the type of algae used as an inoculum. Invariably the thick-walled and multicellular freshwater species, such as *Chlorella*, *Scenedesmus*, and *Micractinium*, tend to become dominant over a period of time (Goldman, 1979b).

The direct use of algae for human food supplement may be unnecessary, if the algae can be fed to animals (pigs, poultry, and fish, etc.) and these animals are used as food for humans. Hintz *et al.* (1966) reported that the waste-grown algae (*Chlorella* and *Scenedesmus*) were 73 per cent digestible when fed to ruminant animals such as cattle and sheep, and were only 54 percent digestible when fed to pigs. The digestible energy content for the ruminants was 2.6 kcal/g algae. These algae were found to supply adequate protein to supplement barley for pigs. Alfalfa–algae pellets, when fed to lambs, resulted in higher weight gains than alfalfa pellets alone.

Algae are basically not palatable to most livestock, but this may be overcome by pelletizing the processed algae with the usual feed of the particular animal, such as steam barley in the case of cattle. In general the waste-grown algae appear to have potential as livestock feed because of the high content of protein and other valuable substances (Tables 5.9 and 5.10).

The use of waste-grown algae as feed for herbivorous fish (tilapia) was reported by Edwards *et al.* (1981b). Extrapolated fish yields approaching 20 tons/(ha-year) were obtained in the 4 m^3 concrete pond system based on 3-month growing periods under ambient, tropical conditions. A linear relationship was established between fish yields and mean algal concentration in the fish ponds, in which an algal concentration of 70 mg/L in the pond water was considered to be high enough to produce good fish growth. Higher algal concentrations were not recommended, as they might lead to zero dissolved oxygen concentrations in the early morning hours.

5.4.2 *Algae for fertilizer*

Algae may be used directly or indirectly as fertilizer in agriculture. In direct use, algae are cultured in HRAP and then irrigated to crops. This method is simpler, but requires more time because the algal cells need to be decomposed in the soil first. In terms of indirect use, algae are harvested and composted and then applied to soil as fertilizer. The application of algal-laden water to crops should be undertaken with due respect to public health considerations and guidelines, as proposed in Table 2.27.

As algae are capable of fixing N from atmosphere, they have an important role in agriculture. It was reported that a rice field inoculated with the N-fixing algae *Tolypothrix tenvis*, after 4 years, produced a 128 percent greater crop than uninoculated field. Furthermore, the plant from the inoculated field contained 7.5 lb N per acre (8.4 kg N/ha) more than the uninoculated field. Using algae as fertilizer will also improve the soil's water-holding capacity, which is an advantage to crop yield (Alexopulos and Bold, 1967).

5.4.3 *Algae for energy*

The fuel characteristics of dry algae (average heat content 6 kg-cal/g) are similar to those of medium-grade bituminous coal, and suitable for use as an energy source (Benemann *et al.*, 1980). Algae can be used together with other organic additives to produce biogas, from anaerobic digestion.

5.4.4 *Algae as a source of chemicals*

A significant amount of lipids is present in algal cells (Table 5.11) which can be used for many industrial purposes such as manufacturing surfactants, grease,

Table 5.11 Total lipids in algae harvested from high-rate algal ponds in Israel (Aaronson *et al.*, 1980)

Major species	Total lipids (percentage of dry weight)	Neutral lipids (percentage of dry weight)
Chlorella	22.6	10.9
Euglena	11.0	8.9
Micractinium	17.4	6.0
Oocystis	19.9	13.5
Scenedesmus	22.2	10.4

Reproduced by permission of Elsevier Science Publishers B.V.

textiles, food additives, cosmetics, and pharmaceuticals. The lipids of microalgae often contain large amounts of neutral lipids, mainly as glycerides, and this might serve as a source of glycerol (Aaronson *et al.*, 1980).

Microalgae may also serve as a source of steroids. The concentration of steroids in algae is variable, but significant amounts may be found in some algae. Algae may also contain up to 0.2 percent of dry weight as carotenoids (Paoletti *et al.*, 1976). Some medicinal products have been isolated from algae (Volesky *et al.*, 1970).

5.5 PUBLIC HEALTH ASPECTS AND PUBLIC ACCEPTANCE

5.5.1 Public health risks

Production of waste-grown algae involves some public health risks due to:

1. contamination of the pathogens normally present in raw waste, and
2. accumulation of heavy metals and other toxic compounds in algae cells.

As discussed in section 5.1, pathogen die-off in the algal ponds can take place due to ultraviolet light, algal toxins produced by algae cells, competition with other microorganisms, inhibitory environments such as high pH during photosynthesis, and sedimentation with sludge. However, a complete removal of pathogens never exists in practice, because they may be protected by self-shading of algae and clumping with sewage solids.

Edwards *et al.* (1981a) found the microorganism removal in their HRAP unit located in Thailand to be about 1 order of magnitude. The ranges of bacterial densities in the HRAP effluents were 1.0×10^5–1.3×10^{10}, 1.6×10^5–2.4×10^6, and 7.0×10^4–9.2×10^5 no./100 mL, for the standard plate count bacteria, total coliforms and fecal coliforms, respectively. According to Table 2.27 this HRAP effluent is not suitable for unrestricted irrigation (e.g. irrigation of edible crops), but may be used for restricted irrigation such as irrigation of trees and other industrial crops.

Care should be taken regarding occupational risks to people working with HRAP units. The possibility of contamination to these people is high, particularly during algal harvesting, which may lead to the spread of pathogens to other people by carrying them on the body or clothes. The possibility of swallowing pathogens during the working period cannot be ruled out. Therefore it is always advisable to use protective covers, such as masks, gloves, etc., when working in an algal pond system.

The possibility of algal contamination with toxic substances and heavy metals also has another adverse effect on public health. Heavy metals and pesticides may be concentrated in algae by the process of bioaccumulation, inducing impact on other consumers in the food chain through biomagnification (as discussed in section 2.5). The concentration of toxic substances in algae is expected to be higher than in wastewater discharged into the pond. Therefore, to avoid such risks, treatment of wastewater up to the allowable concentration of heavy metals and other toxic substances should be done prior to the feeding of this wastewater into an HRAP. The waste-grown algae to be used for human food or animal feed should be regularly monitored for the presence and concentrations of these substances.

5.5.2 Public acceptance

Unprocessed freshwater algae tend to have a strong smell and taste similar to those detected in natural waters which are undergoing eutrophication. Algal texture is also slimy and uninviting, making direct dietary use unlikely for humans and animals (except for herbivorous fish). Therefore the conversion of algal biomass into acceptable forms, or into palatable material (such as by pelletization) is important, and some encouraging results on the use of pelletized algae as animal feed have been reported (section 5.4).

Considering the results on acceptability and nutritive value of waste-grown algae, the prospect of their direct use as human protein supplement is relatively remote. Although the strain *Spirulina* seems to be potentially useful as human food or animal feed, the main problem is that, whereas species control is relatively simple with terrestrial crops so that the best-suited species are grown, it is almost impossible to contain algal speciation in outdoor HRAP cultures (Goldman, 1979a). Uses of HRAP and the waste-grown algae appear to be limited to solving specific environmental problems such as wastewater treatment and other applications shown in Figure 5.2. These limitations include the upper limit in algal yield of 30–40 g/(m^2-day), the economics of algal harvesting, and the algal cell characteristics, which are key factors in determining the application of HRAP to a particular situation.

REFERENCES

Aaronson, S., Berner, T., and Dubinsky, Z. (1980). Microalgae as a source of chemicals and natural products. In *Algae Biomass, Production and Use* (eds. G. Shelef and C. J. Soeder), pp. 575–601. Elsevier/North-Holland Biomedical Press, Amsterdam.

Alexopulos, C. J. and Bold, H. C. (1967). *Algae and Fungi*. Macmillan Company, New York.

Azov, Y. and Shelef, G. (1982). Operation of high-rate oxidation ponds: theory and experiments. *Water Res.*, **16**, 1153–60.

Bare, R. W. F., Jones, N. B., and Middlebrooks, E. J. (1975). Algae removal using dissolved air flotation. *J. Water Pollut. Control Fed.*, **47**, 153–69.

Batallones, E. D. and McGarry, M. G. (1970). Harvesting of algae through chemical flocculation and flotation. Research Program Report No. 4. Asian Institute of Technology, Bangkok.

Becker, E. W. (1981). Algae mass cultivation—production and utilization. *Proc. Biochem.*, **16**, 10–14.

Benemann, J., Koopman, B., Weissman, J., Eisenberg, D., and Goebel, R. (1980). Development of microalgae harvesting and highrate pond technologies in California. In *Algae Biomass, Production and Use* (eds. G. Shelef and C. J. Soeder), pp. 457–95. Elsevier/North-Holland Biomedical Press, Amsterdam.

Bickford, E. D. and Dunn, S. (1972). *Lighting for Plant Growth*. Kent State University Press, Ohio.

Coombs, J. and Hall, D. O. (eds.) (1982). *Techniques in Bioproductivity and Photosynthesis*. Pergamon Press, Oxford.

Dodd, J. C. and Anderson, J. L. (1977). An integrated high-rate pond algae harvesting system. *Prog. Water Technol.*, **9**, 713–26.

Edwards, P. and Sinchumpasak, O. (1981). The harvest of microalgae from the effluent of a sewage fed high rate stabilization pond by *Tilapia nilotica*. Part 1: Description of the system and the study of the high rate pond. *Aquaculture*, **23**, 83–105.

Edwards, P., Sinchumpasak, O., Labhsetwar, V. K., and Tabucanon, M. (1981a). The harvest of microalgae from the effluent of a sewage fed high rate stabilization pond by *Tilapia nilotica*. Part 3: Maize cultivation experiment, bacteriological studies, and economic assessment. *Aquaculture*, **23**, 149–70.

Edwards, P., Sinchumpasak, O., and Tabucanon, M. (1981b). The harvest of microalgae from the effluent of a sewage fed high rate stabilization pond by *Tilapia nilotica*. Part 2: Studies of the fish ponds. *Aquaculture*, **23**, 107–47.

Goldman, J. C. (1979a). Outdoor algal mass culture—I. Applications. *Water Res.*, **13**, 1–19.

Goldman, J. C. (1979b). Outdoor algal mass culture—II. Photosynthetic yield limitations. *Water Res.*, **13**, 119–36.

Golterman, H. L. (1975). *Physiological Limnology*. Elsevier Scientific Publishing Co., Amsterdam.

Golueke, C. G. and Oswald, W. J. (1965). Harvesting and processing of sewage-grown planktonic algae. *J. Water Pollut. Control Fed.*, **37**, 471–98.

Green, F. B. and Oswald, W. J. (1995). Engineering strategies to enhance microalgal use in wastewater treatment. *Water Sci. Technol.*, **31**, 9–18.

Harrelson, M. E. and Cravens, J. B. (1982). Use of microscreens to polish lagoon effluent. *J. Water Pollut. Control Fed.*, **54**, 36–42.

Hintz, H. F., Heitman, H., Jr, Weir, W. C., Torell, D. T., and Meyer, J. H. (1966). Nutritive value of algae grown on sewage. *J. Anim. Sci.*, **25**, 675–81.

Hsu, S. C. (1970). Factors affecting algal yields from high-rate oxidation ponds treating sewage. Thesis No. 325, Asian Institute of Technology, Bangkok.

Kreith, F. and Kreider, J. F. (1978). *Principles of Solar Engineering*. Hemisphere Publishing Corp., Washington.

McGarry, M. G. and Durrani (1970). Flotation as a method of harvesting algae from ponds. Research Program Report No. 5. Asian Institute of Technology, Bangkok.

Metcalf and Eddy, Inc. (1991). *Wastewater Engineering: Treatment, Disposal and Reuse*, 3rd edition. McGraw-Hill, New York.

Middlebrooks, E. J., Porcella, D. B., Gearheart, R. A., Marshall, G. R., Reynolds, J. H., and Grenney, W. J. (1974). Techniques for algae removal from waste stabilization ponds. *J. Water Pollut. Control Fed.*, **46**, 2676–92.

Mitchell, R. (1974). *Introduction to Environmental Microbiology.* Prentice-Hall, Englewood Cliffs, N.J.

Moraine, R., Shelef, G., Meydan, A., and Levi, A. (1979). Algal single cell protein from wastewater treatment and renovation process. *Biotechnol. Bioeng.*, **XXI**, 1191–207.

Oron, G. and Shelef, G. (1982). Maximizing algae yield in high-rate oxidation ponds. *J. Env. Eng. Div.—ASCE*, **108**, 730–8.

Oswald, W. J. (1963). High rate ponds in waste disposal. *Devel. Indust. Microbiol.*, **4**, 112–19.

Oswald, W. J. (1991). Introduction to advanced integrated wastewater ponding systems. *Water Sci. Technol.*, **24**, 1–7.

Oswald, W. J. and Gotaas, H. B. (1957). Photosynthesis in sewage treatment. *Trans. Am. Soc. Civ. Eng.*, **122**, 73–105.

Oswald, W. J., Gotaas, H. B., Ludwig, H. F., and Lynch, V. (1953). Algae symbiosis in oxidation ponds. III. Photosynthetic oxygenation. *Sewage Indust. Wastes*, **25**, 692–705.

Paoletti, C., Phushparaj, B., Florenzano, G., Capella, P., and Lercker, G. (1976). Unsaponifiable matter of green and blue-green algal lipids as a factor of biochemical differentiation of their biomass: I. Total unsaponifiable and hydrocarbon fraction. *Lipids*, **11**, 258–65.

Parker, D. S. and Uhte, W. R. (1975). Discussion: Technique for algal removal from oxidation ponds. *J. Water Pollut. Control Fed.*, **47**, 2330–2.

Persoone, G., Morales, J., Verlet, H., and De Pauw, N. (1980). Air lift pumps and the effect of mixing on algal growth. In *Algae Biomass, Production and Use* (eds. G. Shelef and C. J. Soeder), pp. 505–22. Elsevier/North-Holland Biomedical Press, Amsterdam.

Reed, S. C., Middlebrooks, E. J., and Crites, R. W. (1988). *Natural Systems for Waste Management and Treatment.* McGraw-Hill, New York.

Soeder, C. J. and Stengel, F. (1974). Physico-chemical factors affecting metabolism and growth rate. In *Algal Physiology and Biochemistry* (ed. W. D. P. Stewart), pp. 714–40. Blackwell Scientific Publications, Oxford.

Sukenik, A. and Shelef, G. (1984). Algal autoflocculation—verification and proposed mechanism. *Biotechnol. Bioeng.*, **XXVI**, 142–7.

Volesky, B., Zajic, J. E., and Knettig, E. (1970). Algal products. In *Properties and Products of Algae* (ed. J. E. Zajic), pp. 49–82. Plenum Press, New York.

Waslien, C. I. (1975). Unusual sources of proteins for man. *Crit. Rev. Food Sci. Nutr.*, **6**, 77–151.

Wetzel, R. G. and Likens, G. E. (1979). *Limnological Analysis.* W. B. Saunders, Co., Philadelphia.

EXERCISES

5.1 There are three formulas presented in this book for estimating algal productivity in high-rate algal ponds (HRAPs): the Goldman formula (eq. 5.5), the Oron and Shelef formula (eq. 5.8), and theoretical estimation (eq. 5.10). Discuss the merits and shortcomings of these formulas.

5.2 You are to obtain the meteorological data (for the past 12 months) which are important for algal production in high-rate algal ponds from the local weather bureau. Using these data and those reported in Chapter 2, determine the amount of algal cells that can be produced if all of your city's wastewater is to be used as the influent feed. Use the three formulas in section 5.2.2 for the calculation and discuss which of the three calculated results of algal production is most applicable to your city.

5.3 You are to survey the algal harvesting equipments currently available in the market in your country or elsewhere. Based on your inspection and information obtained from the equipment manufacturers, describe the equipment that appears most impressive with respect to: the harvesting technique employed, efficiency of algal harvesting, costs of investment, and operation and maintenance.

5.4 If human beings are to live in space or on the moon for an extended period of time, discuss the possibility of producing oxygen and algal single-cell protein from the wastes we produce. Estimate the amount of oxygen and algal protein that can be produced from a person's wastes using the data given in Chapter 2. Also sketch the bioreactor(s) to be constructed in a space shuttle and/or the moon to recycle the human wastes.

5.5 A food processing factory produces 400 m^3/day of wastewater with an average COD concentration of 400 mg/L. The factory wants to construct its own high-rate algal pond (HRAP) to produce algal single cell protein from this wastewater. As a consulting engineer, you arc asked to estimate the land area required and the annual algal production including the income that can be earned from the sale of the algal biomass. Submit the schematic drawings of the HRAP. The following information is given:

HRT for HRAP = 2–5 days
Depth of HRAP = 0.2–0.4 m
Average $I_0 = 400$ g-cal/(cm^2-day)
Mean temperature = 25 °C
Light saturation intensity (I_s) is to be estimated from Table 5.7.
Cost of algal harvesting = US\$0.2/m^3 of algal water
Algal density in HRAP = 300 mg/L
Price of algae = US\$30/kg (dry weight)

5.6 A small community has a 1-ha land available to treat its sewage with a COD of 300 mg/L. It plans to build high rate algal ponds (HRAPs) on this piece of land to treat the sewage. Determine the total treatment capacity of the HRAPs and the optimum pond depth. Also estimate the daily algal and protein production to be obtained from this piece of land. Additional information is given below:

I_0 or I_i = visible solar energy irradiance = 500 g-cal/(cm^2-day) or 13,900 ft-candle
α = specific light absorption coefficient = 1.5×10^{-3}
T_c = temperature coefficient = 0.8
F = efficiency of energy conversion = 0.1
p = 1.6
h = unit heat of algae = 6 kg-cal/g

6

Fish production

Among several waste recycling methods, the reuse of human and animal wastes for the production of algae and fish has been extensively investigated. Although the algal cells, photosynthetically produced during sewage treatment, contain about 50 percent protein, their small sizes, generally less than 10 μm, have caused some difficulties for the available harvesting techniques, which as yet are not economically viable (see Chapter 5). Apart from aesthetic reasons, one of the drawbacks concerning direct utilization of waste-grown algae as an animal feed, except *Spirulina*, has been the low digestibility of algal cell walls. Thus the culture of phytoplankton-feeding (herbivorous) fish in the same pond to graze on the algae or feeding the algal-laden water to herbivorous-fish ponds is attractive for the production of fish protein biomass, which is easily harvestable for animal feed or human food.

There are basically three techniques for reusing organic wastes in aquaculture: by fertilization of fish ponds with excreta, sludge, or manure; by rearing fish in effluent-fertilized fish ponds; and by rearing fish directly in waste stabilization ponds (such as in maturation ponds). A World Bank report (Edwards, 1985) cited several Asian countries, especially China, where excreta or animal manure is fed to fish ponds for fish production. Overhanging latrines are a common sight on fish ponds in Guangdong Province in southern China. In brackish pond waters in Taiwan, excreta is spread on the pond bottom when the ponds are empty during the winter, but in freshwater ponds, excreta is added at intervals, four to six times during the growing season. Most fish farmers also add supplementary feed such as agricultural by-products and grains to the fish ponds.

The second and third techniques of waste reuse in aquaculture have been practiced in developing and developed countries. There are about 2,500 ha of sewage-fertilized fish ponds in Calcutta, India; 270 ha in Hunan, China; 233 ha in Munich, Germany; 50–100 ha in Israel; and a smaller-scale operation in Hungary. When properly operated the productivity of fish ponds using wastewater has been found to be higher than that of inorganically fertilized ponds (Allen and Hepher, 1976).

Some reported yields of fish raised in waste-fed ponds in various countries are shown in Table 6.1.

Table 6.1 Fish yields in waste-fed ponds

Country	Type of waste	Yield	Remarks	Reference
Israel	Cattle manure	10,950 kg/(ha-year)	Extrapolated from experiment	Schroeder, 1977
	Duck manure and waste duck food	14,600 kg/(ha-year)	Extrapolation	Wohlfarth and Schroeder, 1979
	Wastewater, loadings = 25–45 kg BOD_5/(ha-day)	5,000 kg/(ha-year)	Extrapolation	Edwards, 1985
Philippines	Biogas slurry	8,000 kg/(ha-year)	—	Maramba, 1978
Indonesia	Livestock waste	7,500 kg/(ha-year)	—	Djajadiredja and Jangkaru, 1978
Poland	Sugar-beet wastes	400–500 kg/(ha-growing season)	—	Thorslund, 1971
Taiwan	Nightsoil	6,893–7,786 kg/(ha-year)	—	Tang, 1970
Calcutta, India	Wastewater	958–1,373 kg/(ha-year)	Indian carp	Edwards, 1985
China	Wastewater	6,000–10,000 kg/(ha-year)	—	Edwards, 1985
Munich, Germany	Wastewater, loadings = 30–77 kg BOD_5/(ha-day)	500 kg/(ha-growing season)	Common carp	Edwards, 1985
Hungary	Wastewater	1,700 kg/(ha-growing season)	Polyculture, Chinese carp and common carp	Edwards, 1985
U.S.A.	Wastewater effluent, 37%	126–218 kg/(ha-year)	Channel catfish	Colt *et al.*, 1975
	Wastewater	5,000 kg/(ha-year)	Silver carp, bighead carp	Henderson, 1983
Asian Institute of Technology, Thailand	Septage, loading = 150 kg COD/(ha-day)	5,000–6,000 kg/(ha-year)	Tilapia (extrapolation)	Edwards *et al.*, 1984
	Biogas slurry, loading = 100 kg COD/(ha-day)	3,700 kg/(ha-year)	Tilapia (extrapolation)	Edwards *et al.*, 1988
	Composted nightsoil, loading = 100 kg COD/(ha-day)	2,800–5,600 kg/(ha-year)	Tilapia (extrapolation)	Polprasert *et al.*, 1982

6.1 OBJECTIVES, BENEFITS, AND LIMITATIONS

The main objectives and benefits that can be gained from waste recycling through aquaculture are as follows.

6.1.1 Waste stabilization and nutrient recycling

Waste treatment is the primary objective of any waste recycling scheme and the inclusion of fish production recovers nutrients in the waste, such as N, P, and K. Addition of waste or waste by-products such as biogas slurry and compost to fish ponds resulted in increased fish yields (Polprasert *et al.*, 1982). Owing to the economic value of fish, part of the waste treatment costs can be recovered. This financial return will be an incentive for safe disposal of waste which will lead to better public health conditions, particularly among rural people where malnutrition and excreta-related diseases are common.

6.1.2 Upgrading effluent from waste stabilization ponds

Algae in suspension increase the suspended solids content of most waste stabilization pond effluents. (Waste stabilization ponds are a waste treatment process encompassing reactions of the algal–bacterial symbiosis as described in Chapters 2 and 5.) Effluent discharge without the removal of algae might cause algal blooms or increase organic and nutrient loads to the receiving waters. Introducing fish ponds in series with waste stabilization ponds or the culture of fish in the waste stabilization ponds was found to reduce algal and bacterial concentrations to a considerable extent when herbivorous fish species (tilapia and silver carp) were used (Schroeder, 1975). The layouts in Figure 6.1 could be used for the above purpose. In this way the death of fish in the fish ponds may also indicate the presence of some toxic materials and/or low dissolved oxygen (DO) levels in the effluent discharging into the receiving water.

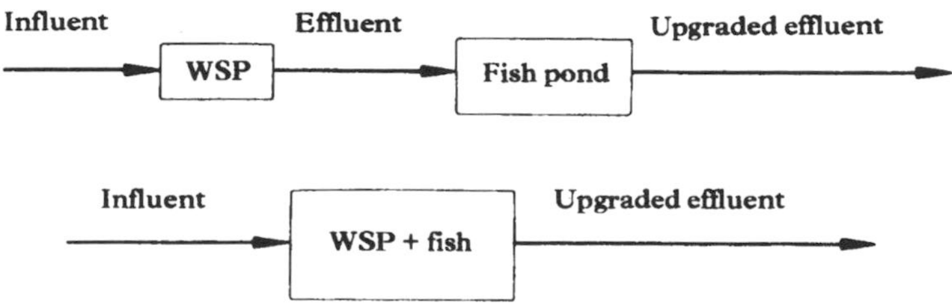

Figure 6.1 Layouts to upgrade the waste stabilization pond (WSP) effluents using fish

6.1.3 *Higher food conversion ratio*

Fish are cold-blooded animals and, unlike other farm animals such as cattle or poultry, do not have to spend a lot of energy for movement, because friction is less in water. Thus fish have a better food conversion ratio than any other farm animals. Table 6.2 compares the relative growth and feed utilization of farm animals and fish during a period of rapid growth. Food conversion ratio (FCR) is defined as:

$$\text{FCR} = \frac{\text{dry weight of feed given (kg)}}{\text{wet weight of fish gained (kg)}}$$

Therefore, the lower the FCR, the better the food conversion efficiency.

6.1.4 *Operational skill and maintenance*

Fish culture does not require highly skilled manpower for its operation. Fish harvesting is easier than algal harvesting. Fish processing is usually not required for human consumption. For other uses, such as animal feed, drying is generally sufficient.

However, fish production from wastes might have some or all of the following limitations.

LAND REQUIREMENT AND EXISTENCE OF A WASTE COLLECTION SYSTEM

Land requirement for fish production is quite large, as the organic loading to fish ponds has to be low. Land should be available at low cost and closer to the source of waste and natural watercourses (e.g. canals or rivers). The transportation of waste should be safe, easy, and cheap. If the waste is available in a concentrated form, such as septage, suitable sources of water to make up for evaporation and seepage losses should be available.

Table 6.2 Efficiency of feed utilization of various animal species per 1,000 g of feed intake (adapted from Hastings and Dickie, 1972)

Species	Live weight gain (g)	Food conversion ratio	Energy gain (kcal)	Protein gain (g)
Chicks	356	2.8	782	101
Pigs	292	3.4	1,492	30
Sheep	185	5.4	832	22
Steers	163	6.1	748	26
Channel catfish	715	1.4	935	118
Brown trout	576	1.7	608	75

AVAILABILITY OF SUITABLE FISH FRY

Fish species suitable for stocking in waste-fed ponds should be available locally at a low price. If the facilities for fish fry production are not available, or its production is not economically feasible locally, fish production is not possible.

PUBLIC HEALTH RISKS

Waste-grown fish have the potential of contamination with several kinds of pathogens that may be present in the waste itself (Table 2.22). Fish growing in a medium that has a high concentration of heavy metals or other toxic substances may eventually, through processes of bioaccumulation and biomagnification, contain high concentrations of these substances in their tissue.

MARKETING AND PUBLIC ACCEPTANCE

Unless the public is convinced that waste-grown fish are safe for consumption, fish production is bound to fail. In those areas where fish from the sea or rivers are available in sufficient quantity and cheap, marketing of waste-grown fish will be difficult for human consumption. However, other uses (i.e. as animal feed) or other recycling options have to be considered.

6.2 HERBIVORES, CARNIVORES, AND OMNIVORES

Feeding is one of the most essential factors that strongly influence the growth rate of fish. Moreover, similar to other animals, different fish species have their own special feeding habits. Fish can be classified into three groups on the basis of their feeding habits. Their morphologies are also different accordingly, mainly in respect of the specialized type of alimentary system. Thus the functional morphology of fish species can often indicate the type of food they eat, and those they are unable to eat. These three groups are characterized below.

6.2.1 Herbivorous fish

Herbivorous fish lack teeth, but possess fine gill rakers that can sieve microscopic plants from the water; they also lack a true stomach (i.e. a highly muscular, acid-secreting stomach) but possess a long, thin-walled intestine. They feed mainly on plants, including algae and higher plants. Typical examples are grass carp (*Ctenopharyngodon idellus*) (unlike other herbivores, its gut is very short), and silver carp (*Hypophthalmichthys molitrix*) (Figure 6.2a,b).

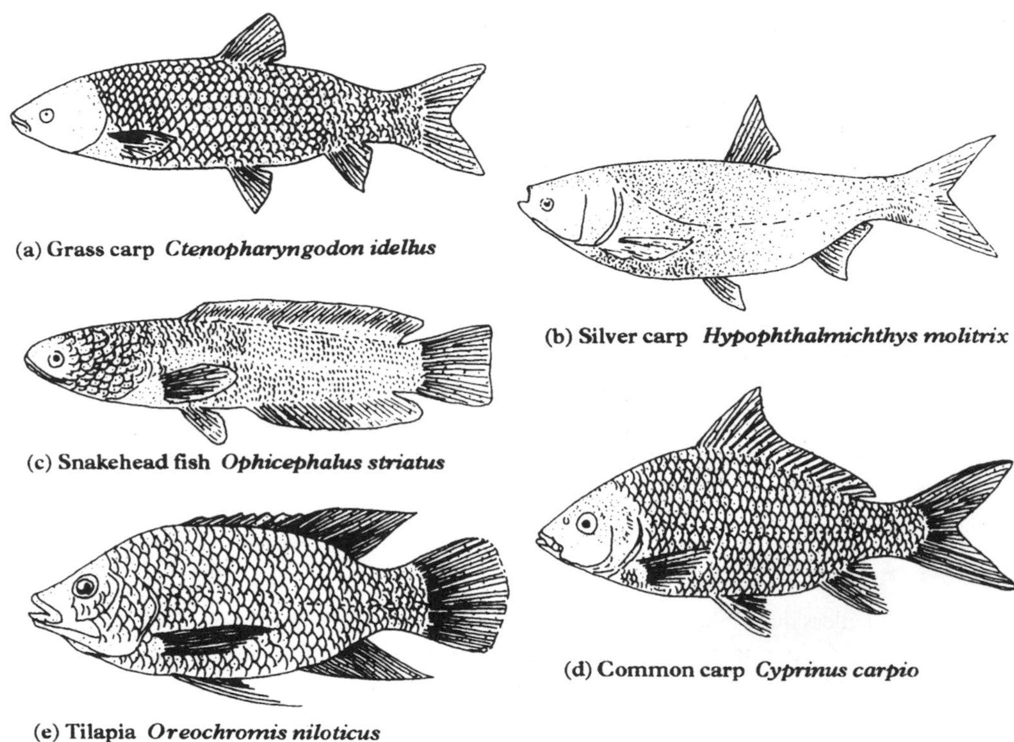

(a) Grass carp *Ctenopharyngodon idellus*

(b) Silver carp *Hypophthalmichthys molitrix*

(c) Snakehead fish *Ophicephalus striatus*

(d) Common carp *Cyprinus carpio*

(e) Tilapia *Oreochromis niloticus*

Figure 6.2 Examples of herbivores, carnivores, and omnivores

Under optimum conditions small grass carp, less than 1.2 kg, may eat several times their body weight of plant material daily (NAS, 1976). Thus, in addition to a source of protein, many herbivorous species such as grass carp and silver carp are often grown in weed-infested water for biological control of phytoplankton blooms and some aquatic weeds.

6.2.2 Carnivorous fish

These have teeth developed to seize, hold, and tear, and gill rakers modified to grasp, retain, rasp, and crush prey. There is a true flask-like stomach and a short intestine, elastic and thick-walled. These fish eat mainly zooplankton, insects, bacteria, trash fish, and other animals. Snakehead (*Ophicephalus striatus*) (Figure 6.2c) is mainly carnivorous, feeding on small fish, crustaceans, insects, and worms. Usually people prefer carnivorous fish to other fish for their high nutritive value and taste. But much higher investment is needed for carnivores feed and fertilizer than for that of other fish.

6.2.3 Omnivorous fish

These have alimentary systems more or less intermediate between those of the extreme herbivores and carnivores, but there is a range of types between the extremes. The alimentary canal is much longer, relative to body length. Most fish belong to this group, which consumes both animals and plants. For example, the Chinese common carp (*Cyprinus carpio*) and tilapia (*Oreochromis niloticus*) (Figure 6.2d,e) have a wide range of diet.

The terms herbivores and carnivores have relative meanings. Usually they indicate merely the type of predominant food consumed by the fish. Actually most fish are highly adaptable in their feeding habits and utilize the most readily available foods (Lagler *et al.*, 1962). Relatively few kinds are strict herbivores or carnivores, and perhaps none feed solely on one organism. Food habits may change as fish grow, accompanied by marked changes in the morphology of the alimentary system in early life. Fish such as the Bermuda angelfish (*Holacanthus bermudensis*) may even change their diet with season; it may be quite herbivorous in winter and spring, and become predominantly carnivorous in the summer and early fall.

Therefore, feeding materials will vary with the different fish species and the stages of development according to their feeding habits, so that fish can keep growing at higher rate.

Fish to be reared in waste-fed ponds should have the following characteristics:

1. tolerant to low DO level which can occur during the night or at dawn when photosynthetic oxygen production does not occur;
2. herbivorous or omnivorous in nature to feed on the waste-grown phytoplankton; and
3. tolerant to disease and other adverse environmental conditions.

Some species, such as tilapia (Figure 6.2e), Chinese carp, and Indian carp have been widely used in waste recycling practices. Organic wastes, including septage, could provide sufficient nutrients to promote the growth of phytoplankton, and consequently of zooplankton, which are the natural food of these fish (Polprasert *et al.*, 1984). Among these species, tilapia needs to be particularly mentioned, because it is widely used in waste recycling in tropical and subtropical areas. It feeds directly on algae and other primary aquatic vegetation (and on zooplankton as well). It grows rapidly and multiplies abundantly. Furthermore, it has better tolerance to low DO level and resistance to diseases (which often occur in fish ponds) than many other species such as carp.

The culture of carnivores is also of interest in waste recycling if fish for human consumption are desired. But carnivores have to be raised in separate ponds fed with pellets made from herbivores, omnivores, or trash fish (to allow the growth of herbivores or omnivores), and to safeguard public health because the carnivores do not have direct contact with organic wastes. Catfish, snakehead, and shrimp, which are of high market value, are the carnivores commonly reared.

6.3 BIOLOGICAL FOOD CHAINS IN WASTE-FED PONDS

A food chain comprises a series of organisms existing in any natural community through which energy is being transferred. Each link in such a chain feeds on, and obtains energy from, the ones preceding it. In a pond ecosystem there are generally three major groups of organisms present in the food chain, similar to the other marine and freshwater ecosystems. The three groups consist of: primary producers; primary, secondary, and tertiary consumers; and decomposer organisms (Figure 6.3).

At the beginning of the food chain the primary producers (algae and aquatic plants) represent the first trophic or energy level. They synthesize organic materials (or cell biomass) through photosynthesis (e.g. eq. 2.6), utilizing the nutrients present in the water and light energy. Next is the primary consumer group, mainly zooplanktons and herbivores, which consume the primary produ-

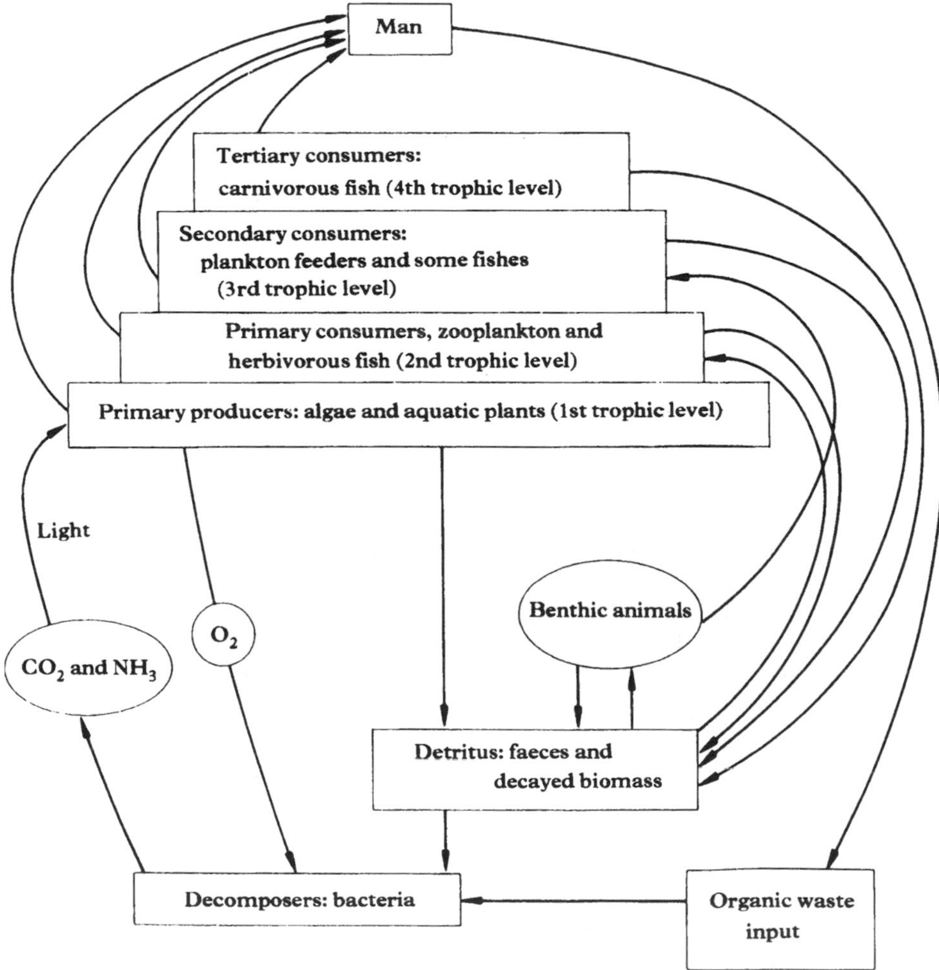

Figure 6.3 Food chain and other relationships in waste-fed ponds

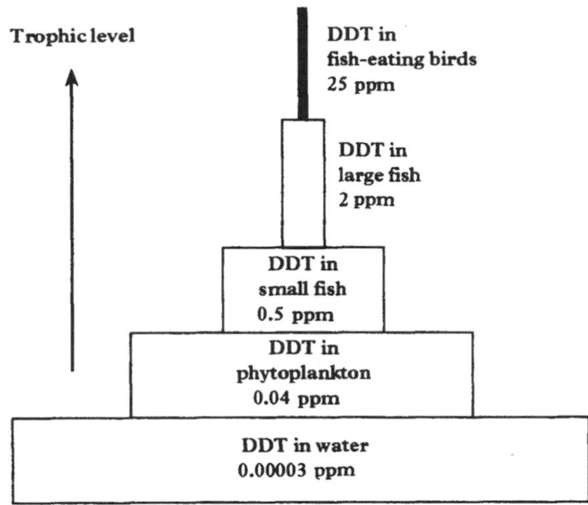

Figure 6.4 Bioaccumulation and biomagnification of DDT in a food chain (adapted from Campbell, 1990)

cers, then they are prey for secondary consumers such as small fish and other plankton-feeders. In this step some fish consume benthic animals which grow at the pond bottom. Some herbivorous fish, such as silver carp, consume phytoplankton directly, and can also take detritus. Tertiary consumers, such as snakehead, prey upon small fish. Depending on the type of fish stocked, they feed on phytoplankton and zooplankton and are primary consumers, secondary consumers, or tertiary consumers.

The waste material produced by fish and decaying biomass will settle to the pond bottom and be decomposed by bacteria (the decomposers), resulting in the release of nutrients such as CO_2 and NH_3 (eq. 2.1) required for the primary production.

When comparing food chains in a normal fish pond and a waste-fed fish pond, there are no wide differences between them. However, to allow the herbivorous fish to grow effectively and to maximize fish biomass production, carnivorous fish (tertiary consumers) are not normally stocked in waste-fed fish ponds where herbivores are being reared. Additionally, in waste-fed fish ponds there are more nutrients for primary producers due to the application of waste and its decomposition. The food chain depends on the primary productivity of the pond, which in turn depends on the nutrients and light. The subject of environmental requirements in waste-fed ponds will be discussed in section 6.5.

Another important consideration related to the food chain is problems caused by biomagnification and bioaccumulation. Biomagnification may be defined as the accumulation of toxic materials such as pesticides or heavy metals in an organism in any particular trophic level at a concentration greater than that in its food or the preceding trophic level, so that essentially animals at the top of the food chain accumulate the largest residues (see Figure 6.3). Bioaccumulation is the phenomenon in which toxic matter is in equilibrium at a higher concentration in tissues than in the surrounding aquatic environment. This depends on time of exposure,

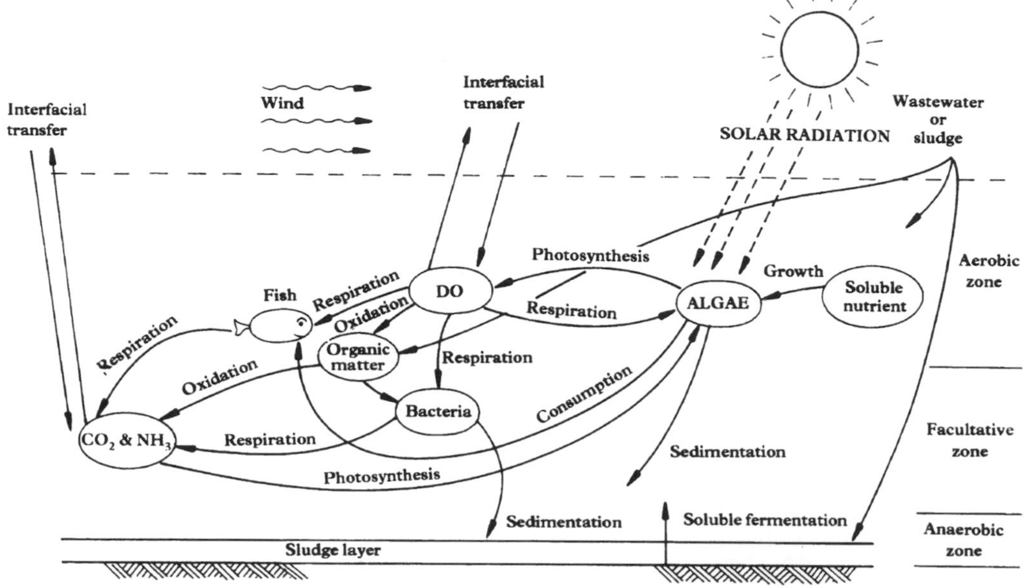

Figure 6.5 A simplified view of septage-fed fish pond dynamics (adapted from Bhattarai, 1985)

rate of uptake, metabolism within the organism, rate of excretion, potential for storage in tissues, and the physiological state of the organism. Therefore the effects of both factors should be investigated, especially when industrial or agricultural wastes containing high concentrations of heavy metals and/or toxic organic compounds are to be applied to fish ponds. A well-known example of bioaccumulation and biomagnification involves the pesticide DDT, which is now banned in the U.S. (Campbell, 1990). DDT concentration in a Long Island Sound food chain was magnified by a factor of about 10 million, from just 0.000003 ppm as a pollutant in the seawater to a concentration of 25 ppm in a fish-eating bird, the osprey (see Figure 6.4).

6.4 BIOCHEMICAL REACTIONS IN WASTE-FED PONDS

In a waste-fed fish pond that is functioning well, algae, bacteria, and fish form symbiotic relationships. The oxygen and food for fish are produced by the algae, and the waste is decomposed by the bacteria. Figure 6.5 depicts the interactions among these organisms in a pond fed with wastewater or sludge.

The biochemical reactions occurring in waste-fed fish ponds should be similar to those of the facultative waste stabilization ponds in which organic matter is decomposed by a combination of aerobic, facultative, and anaerobic bacteria. Three zones exist in the ponds: the first is an aerobic zone where aerobic bacteria and algae exist in a symbiotic reaction, i.e. the oxygen supplied partly by natural surface reaeration and from algal photosynthesis (eq. 2.6 or 2.7) is used by the bacteria in the aerobic decomposition of the organic matter (eqs 2.1 and 2.2); the

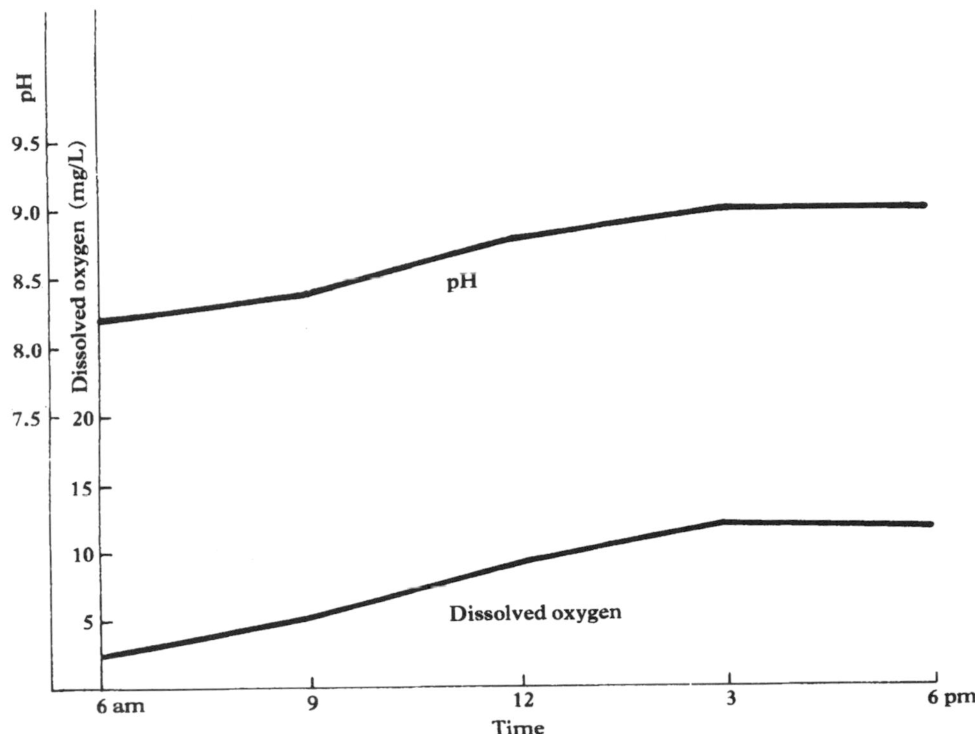

Figure 6.6 Semi-diurnal studies of the septage-fed fish pond system. Dissolved oxygen and pH were measured at 3-h intervals from 6 a.m. to 6 p.m. on four occasions. Mean of 15 septage fed ponds with fish stocking densities of 3, 6, 9, 12, and 15/m^2 in triplicate (Polprasert *et al.*, 1982)

nutrients and CO_2 released in this decomposition are, in turn, used by the algae (eq. 2.6 or 2.7). The second is an intermediate (facultative) zone which is partly aerobic and partly anaerobic, in which waste stabilization is carried out by facultative bacteria. The third is an anaerobic bottom zone in which the accumulated solids are decomposed by anaerobic bacteria. Fish normally lived in the aerobic and facultative zones where oxygen and food (algae) are present.

When organic waste is discharged into a pond the soluble and colloidal compounds that remain in suspension will be decomposed by the aerobic and facultative bacteria. The settleable solids will settle down to the pond bottom and, together with other decayed biomass that settles there, will form a sludge layer. Anaerobic reactions occurring in the sludge layer zone are similar to those described in Chapter 4, in which there will be a release of soluble organic by-products (such as amino acids and volatile acids), and gaseous by-products such as CH_4 and CO_2. As the pond depth is usually about 1 m these soluble by-products will dissolve in the water due to wind-induced mixing and fish movement, and will be further decomposed by the aerobic and facultative bacteria present in the pond layers above.

The above biochemical reactions normally result in diurnal changes in pH and DO in the pond water, as shown in Figure 6.6. The basic phenomena involved are that during the dark periods photosynthetic activity ceases to function, and the algal cells do not utilize the CO_2 released by bacterial decomposition of organic matter, resulting in the formation of carbonic acid and, consequently, a decrease in pH. Biomass respiration and the absence of photosynthesis during night-time contribute to a drop in DO. With the onset of the light period the algae consume CO_2 for photosynthesis, with the production of oxygen and more algal cells. This results in a gradual increase in pH and DO (up to supersaturation level) in the pond water. In Figure 6.6 the DO at dawn did not reach zero because the applied organic loading was not excessive. As fish are generally sensitive to low DO, organic loadings to be applied to fish ponds should be properly controlled so that the DO level at dawn, the critical period, does not drop to zero. Otherwise some kind of mechanical aeration needs to be provided to avoid fish suffocation.

6.5 ENVIRONMENTAL REQUIREMENTS AND DESIGN CRITERIA

6.5.1 Environmental requirements

To enhance fish growth in waste-fed ponds, various environmental parameters should be properly maintained or acquired, as follows:

LIGHT

Light should be of sufficient intensity and suitable duration during the day-time. It is the main factor in algal photosynthesis which results in the production of fish feed (algal cells) and oxygen for fish respiration. Generally, this requirement is met in tropical areas where depth of fish ponds is maintained at about 1 m to allow light to penetrate to the whole pond depth.

TEMPERATURE

Fish metabolic rate is directly correlated with water temperature. Huet and Timmermans (1971) reported that temperature has a considerable influence on the principal and vital activities of fish, notably their breathing, growth, and reproduction. Increased temperature will lower the DO in the water (according to Table 6.3) and also increase the metabolism of fish, which require more oxygen. The temperature tolerance limit within each species with individuals of different ages is the same, but the limit is different if they are acclimatized at different temperatures. Unfavorable temperatures at either end of the tolerance range produce stress, and prolonged exposure could result in lowered resistance and greater susceptibility to disease. Temperature does affect metabolism as well as food intake. Hickling (1971) reported that carp stop feeding at about 10 °C, and become torpid at about 5 °C; trout cease to feed at about 8 °C.

Table 6.3 Solubility of dissolved oxygen in water in equilibrium with dry air at 760 mmHg and containing 20.9 percent oxygen

Temperature (°C)	Chloride concentration (mg/L)				
	0	5,000	10,000	15,000	20,000
0	14.6	13.8	13.0	12.1	11.3
1	14.2	13.4	12.6	11.8	11.0
2	13.8	13.1	12.3	11.5	10.8
3	13.5	12.7	12.0	11.2	10.5
4	13.1	12.4	11.7	11.0	10.3
5	12.8	12.1	11.4	10.7	10.0
6	12.5	11.8	11.1	10.5	9.8
7	12.2	11.5	10.9	10.2	9.6
8	11.9	11.2	10.6	10.0	9.4
9	11.6	11.0	10.4	9.8	9.2
10	11.3	10.7	10.1	9.6	9.0
11	11.1	10.5	9.9	9.4	8.8
12	10.8	10.3	9.7	9.2	8.6
13	10.6	10.1	9.5	9.0	8.5
14	10.4	9.9	9.3	8.8	8.3
15	10.2	9.7	9.1	8.6	8.1
16	10.0	9.5	9.0	8.5	8.0
17	9.7	9.3	8.8	8.3	7.8
18	9.5	9.1	8.6	8.2	7.7
19	9.4	8.9	8.5	8.0	7.6
20	9.2	8.7	8.3	7.9	7.4
21	9.0	8.6	8.1	7.7	7.3
22	8.8	8.4	8.0	7.6	7.1
23	8.7	8.3	7.9	7.4	7.0
24	8.5	8.1	7.7	7.3	6.9
25	8.4	8.0	7.6	7.2	6.7
26	8.2	7.8	7.4	7.0	6.6
27	8.1	7.7	7.3	6.9	6.5
28	7.9	7.5	7.1	6.8	6.4
29	7.8	7.4	7.0	6.6	6.3
30	7.6	7.3	6.9	6.5	6.1

Reproduced from Whipple and Whipple, 1911; copyright 1911, American Chemical Society

DISSOLVED OXYGEN

DO is an important factor in the growth of fish because all metabolic activities of fish are dependent on the oxygen consumed during respiration. As shown in Figure 6.6, DO in a waste-fed pond has diurnal variation. Factors affecting variation of DO include organic loading, algal concentration, type of fish, fish stocking density, sludge accumulation, and temperature and chloride concentration (Table 6.3). Organic loading to a pond influences bacterial activity in waste stabilization and, consequently, oxygen utilization (eq. 2.1). Therefore, to design a waste-fed fish pond, organic loading and fish stocking density have to be properly selected so that the lowest DO to occur in the pond at dawn is within the tolerable range for fish.

Fish species have different rates of oxygen consumption, which vary with stage of life. Young fish need more DO than adult fish. Among the various species, tilapia are the most resistant to low DO. Among carp, common carp is more resistant than silver carp. Most of the data concerning the tolerance to low DO were derived from experiments conducted at constant DO; they thus have little relevance to waste-fed ponds where DO fluctuates widely. Fish may be much more resistant to low DO for a short period of time. Waste-fed fish ponds should be designed to have lowest DO levels of 1–2 mg/L at dawn, otherwise some kind of mechanical aeration (e.g. surface aerators) has to be provided during this critical period.

Experiments were conducted at the Asian Institute of Technology (AIT), Bangkok, to observe the effects of adding septage at loadings of 50, 150, 250, and 350 kg COD/(ha-day) to concrete ponds (the working dimensions of each pond were 2 × 2 × 0.9 m, length × width × depth). Septage loading to these ponds were conducted once daily, and the data reported in Figure 6.7 are mean values obtained during steady-state conditions (based on relatively constant COD concentrations in the pond water). The concentrations of algae and NH_3–N were found to increase with increasing septage loadings, while the levels of DO at dawn decreased at high septage loadings, according to the phenomena described in section 6.4. It appears from Figure 6.7 that under this ambient condition septage

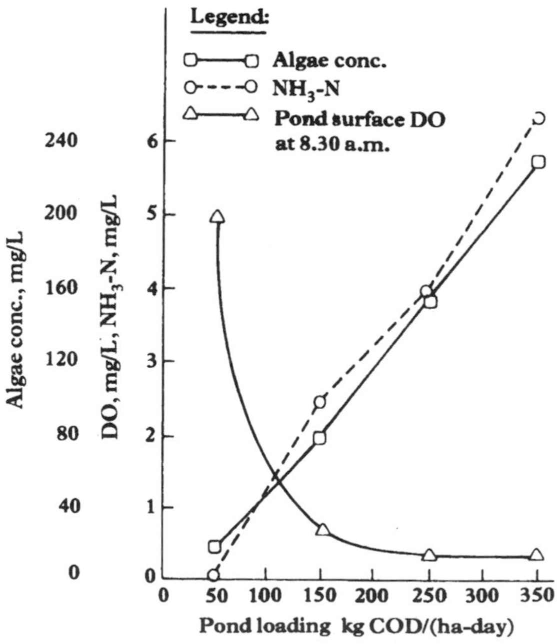

Figure 6.7 Effects of septage loadings on pond characteristics (no fish culture) (from Polprasert *et al.*, 1984; reproduced with permission of Pergamon Books Ltd)

loadings in the range 50–150 kg COD/(ha-day) should result in suitable DO levels at dawn for tilapia growth.

Sharma *et al.* (1987) investigated the variation with time of DO levels at dawn and COD concentrations in pond water fed with septage at various organic loadings. Their experiments were conducted at AIT campus, Bangkok, under ambient, tropical conditions. Because the pond system was static and non-flow-through, and septage loading was undertaken once a day, the total and filtered COD concentrations of the pond water increased gradually with respect to time (Figure 6.8a,b). The increase in COD consequently resulted in a decrease of DO at dawn, observed in Figure 6.8c.

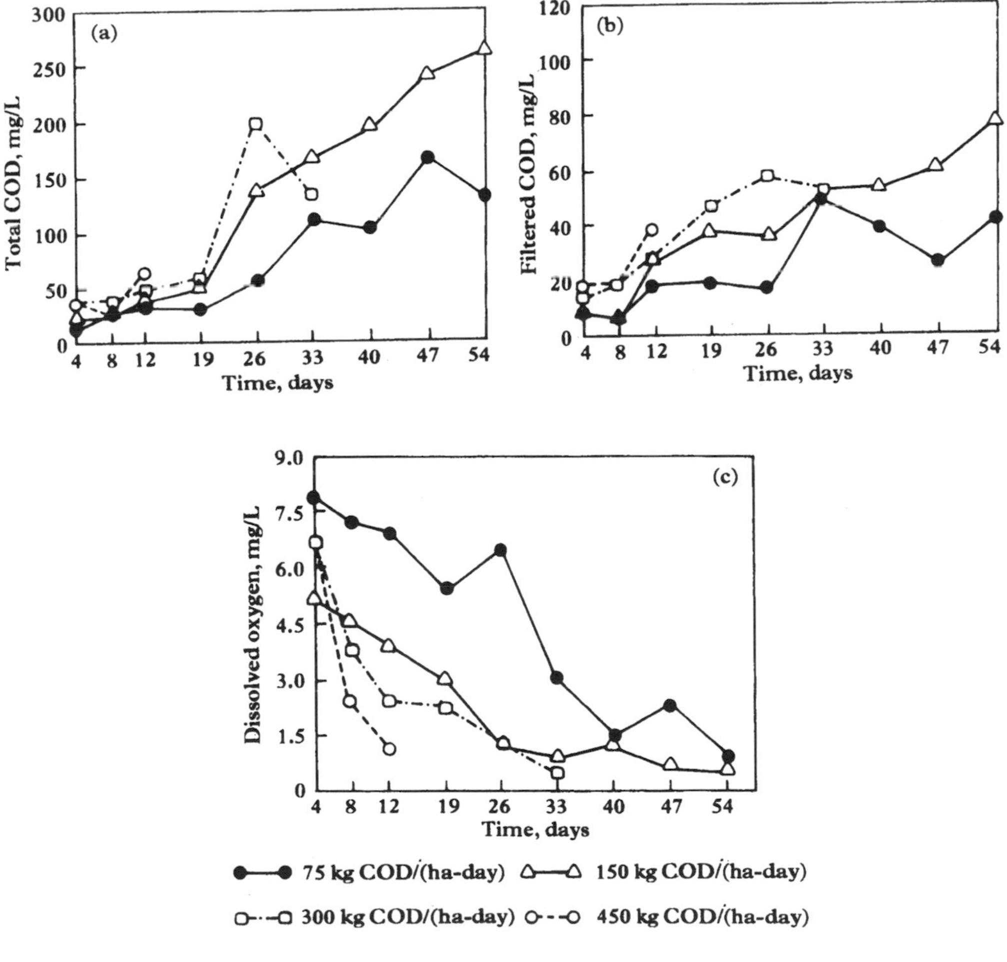

(a) Variation of total COD
(b) Variation of filtered COD
(c) Fluctuation of dissolved oxygen (DO) at dawn

Figure 6.8 Variations of COD and DO in septage-fed ponds at various COD loadings (Sharma *et al.*, 1987; reproduced by permission of Elsevier Science Publishers B.V.)

A portion of the COD increase could be due to the anaerobic decomposition of the settled septage (or sludge) occurring in the sludge layer which released, or resulted in the solubility of, some organic compounds to the pond water. These organic compounds were further decomposed by the facultative bacteria and the algal–bacterial symbiotic reactions taking place in the aerobic zone, hence a decrease in the DO level. The filtered COD concentrations of the pond loaded at 75 kg COD/(ha-day) became relatively stable (or attaining steady-state conditions) after about 40 days of operation (Figure 6.8b), similar to that of the DO level (Figure 6.8c).

Ponds fed with wastewater or algal-laden water will have a diurnal variation of DO similar to the septage-fed ponds. However, there will be less sludge accumulation at the bottom layer, and COD concentrations in the pond water should fluctuate less than those in the ponds fed with sludge.

AMMONIA CONCENTRATION

Un-ionized ammonia (NH_3) is toxic to fish, but the ammonium ions (NH_4^+) are not. Many laboratory experiments of relatively short duration have demonstrated that the acute lethal concentrations of NH_3 for a variety of fish species lie in the range of 0.2–2.0 mg/L (Alabaster and Lloyd, 1980). Un-ionized ammonia is more toxic when DO concentration is low. However, this effect is probably nullified in fish ponds as CO_2 concentrations are usually high when DO levels are low and the toxicity of NH_3 decreases with increasing CO_2 (Boyd, 1979).

The relationship between NH_3 and NH_4^+ is pH-dependent, as follows:

$$NH_3 + H^+ \rightleftarrows NH_4^+ \tag{6.1}$$

Equation 6.1 indicates that NH_3 formation is favored under high pH or alkaline conditions, as shown in Figure 6.9.

High concentrations of total ammonia ($NH_3 + NH_4^+$ or $NH_3 - N$) can occur following phytoplankton die-offs, but the abundant CO_2 production associated with such events depresses pH and the proportion of the total ammonia present as NH_3 (eq. 6.1). NH_3 also increases the incidence of blue-sac disease in the fry of freshwater fish when the eggs were cultured in water with high NH_3 content (Wolf, 1957). Considering safety factors the U.S. Committee on Water Quality Criteria (1972) has recommended that no more than 0.02 mg/L NH_3 be permitted in receiving waters. Sawyer *et al.* (1994) concluded that ammonia toxicity will not be a problem in receiving waters with a pH below 8 and NH_3–N concentrations less than 1 mg/L.

pH

Huet and Timmermans (1971) stated that the best water for fish cultivation is neutral or slightly alkaline, with a pH between 7.0 and 8.0. Reproduction

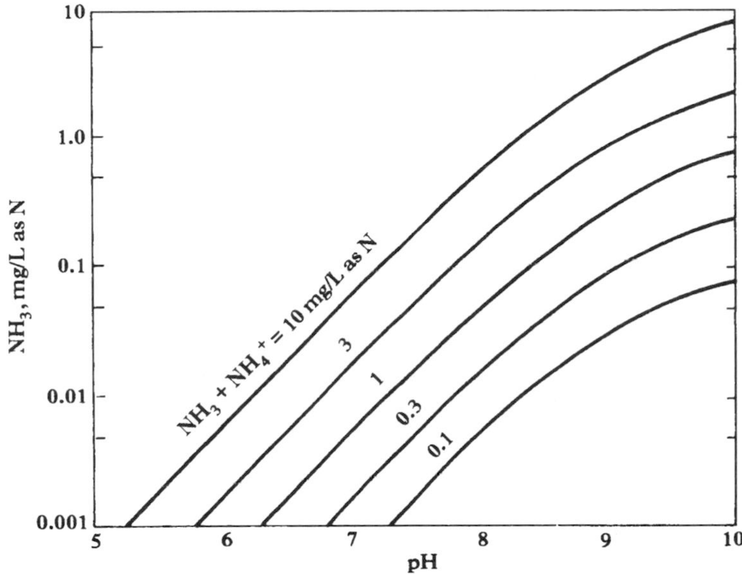

Figure 6.9 The effect of pH and ammonia nitrogen concentration ($NH_3 + NH_4^+$) on the concentration of free ammonia in water (from Sawyer *et al.*, 1994; reproduced by permission of McGraw-Hill Book Co.)

diminishes at pH values below 6.5, and growth rate becomes lower at pH range of 4–6.5.

CARBON DIOXIDE

The presence of free CO_2 may depress the affinity of fish blood for oxygen. Fish can sense small differences in free CO_2 concentration, and apparently attempt to avoid areas with high CO_2 levels. Nevertheless, 10 mg/L or more of CO_2 may be tolerable provided the DO concentration is high. Water with less than 5 mg/L of CO_2 is preferable. The most detrimental effect of free CO_2 results during the period of critically low DO (Boyd, 1979).

STOCKING

In addition to the limitations imposed by available oxygen, the density of fish population can increase to the point of growth inhibition. Overcrowding can result in a reduction in growth rate due to stress and competition for food, oxygen, and space, and in the poor food utilization caused by wastage. It is generally known that there is a limit to the density at which fish can be stocked in ponds; for beyond a certain point the advantage of growing a larger population of fish is cancelled by the slower growth of the fish, in spite of an excess of food.

Figure 6.10 shows some experimental results of mean weight of fish being cultured at different stocking densities (SD) in earthen ponds, but at a constant organic loading of 150 kg COD/(ha-day). Each pond had a dimension of $20 \times 10 \times 1$ m^3 (length $\times$ width $\times$ depth). These data clearly show the effects of SD on mean fish weight. After 6 months of operation the highest mean fish weight of 118 g was observed in the pond having a SD of 1 fish/m^2, while the lowest mean fish weight of 27 g occurred in the pond with a SD of 20 fish/m^2.

The data from Figure 6.10 were used in plotting Figure 6.11, to show the effects of SD on mean fish weight and total fish yield. It can be seen that although a lower SD could give a higher individual fish weight, it resulted in a lower total fish yield. The opposite took place in case of a high SD. This information implies that fish to be used as human food should be reared at a low SD to obtain table-size fish. Fish to be used as animal feed can be reared at a high SD to maximize total fish yield; these fish, small in size, may be used directly or processed further prior to being used as animal feed.

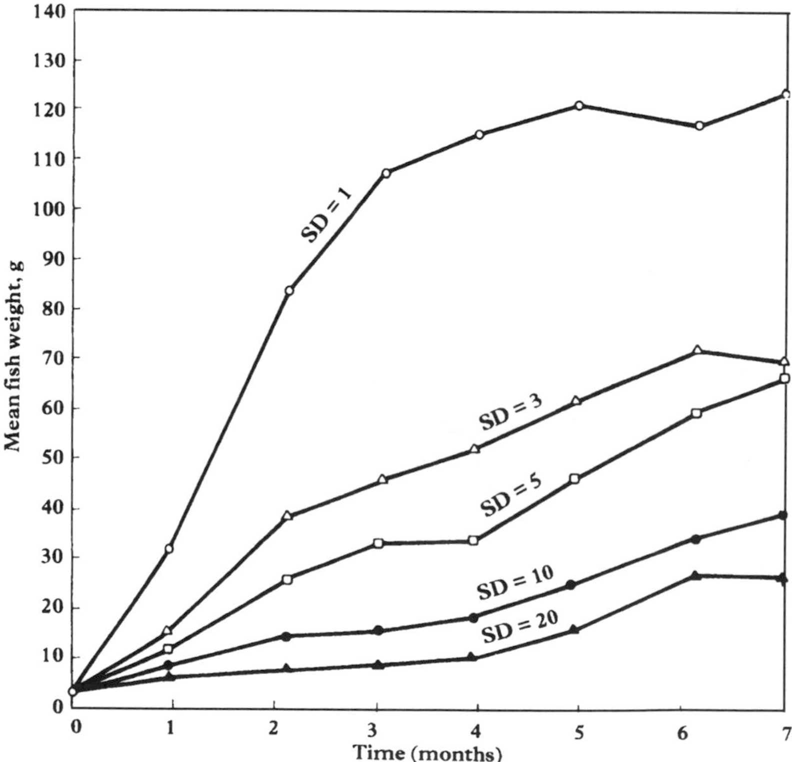

Figure 6.10 Mean individual fish weight (g) at monthly intervals in the septage-fed fish pond system. Each point is a mean of three ponds. Organic loading = 150 kg COD/(ha-day), stocking densities (SD) in number of fish/m^2 (Edwards *et al.*, 1984)

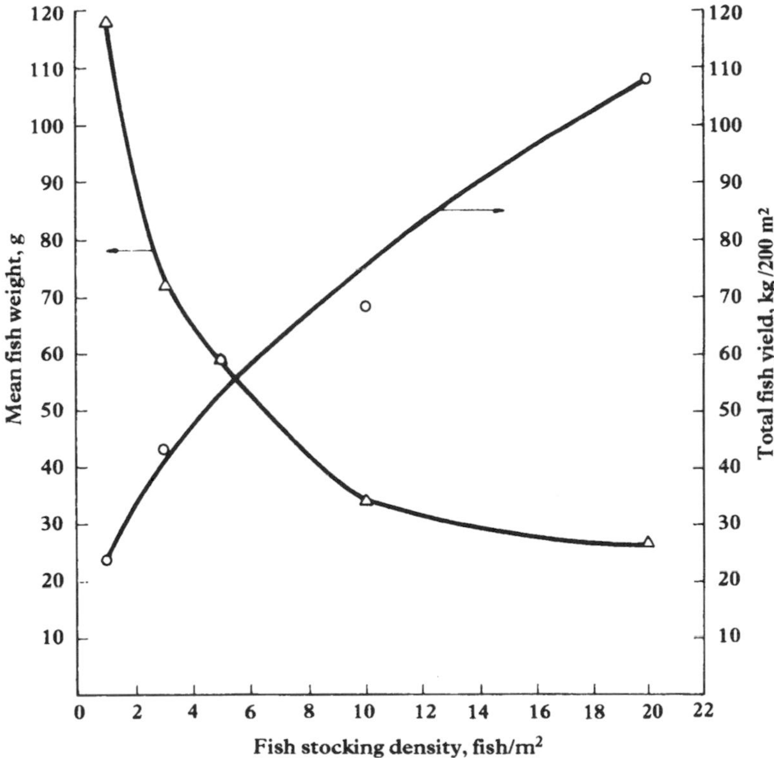

Figure 6.11 Effects of fish stocking density on mean fish weight and total fish yield after 6 months of pond operation. Data obtained from Figure 6.10 assuming no fish mortality and no spawning

HYDROGEN SULFIDE (H_2S)

Un-ionized H_2S, mainly from anaerobic decomposition of bottom sludge, is extremely toxic to fish at concentrations which may occur in natural waters. Results from various bioassay studies suggest that any detectable concentration of H_2S should be considered detrimental to fish production (Boyd, 1979).

Sulfide formation often occurs in anaerobic and facultative waste stabilization ponds due to the reduction of sulfate (SO_4^{2-}) under anaerobic conditions:

$$SO_4^{2-} + \text{organic matter} \xrightarrow[\text{bacteria}]{\text{anaerobic}} S^{2-} + H_2O + CO_2 \tag{6.2}$$

$$S^{2-} + 2H^+ \longrightarrow H_2S \tag{6.3}$$

The relationship between H_2S, HS^- and S^{2-} at various pH levels are shown in Figure 6.12. At pH values of 8 and above, most of the reduced S exists in solution

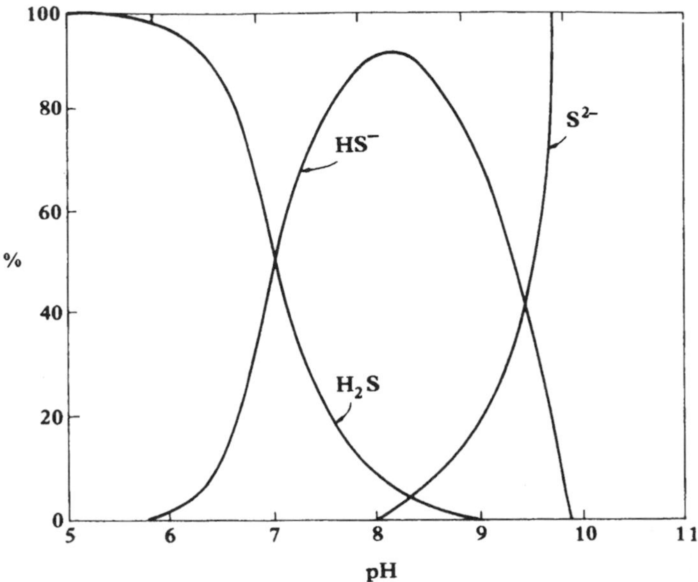

Figure 6.12 Effect of pH on hydrogen sulfide–sulfide equilibrium (after Sawyer *et al.*, 1994; reproduced by permission of McGraw-Hill Book Co.)

as HS^- and S^{2-} ions, and a small amount of un-ionized H_2S, the malodorous gas, is present. However, the formation of un-ionized H_2S is significant at pH levels below 8, which can cause detrimental effects to fish production and can produce problems of malodorous gas.

It is interesting to note that, although an increase in water pH to above 8 can avoid H_2S formation, this pH condition would enhance the formation of the un-ionized NH_3 compounds (see Figure 6.9), which are toxic to fish. Therefore organic loadings to fish ponds have to be properly controlled to avoid the occurrence of anaerobic conditions and DO depletion. Sludge deposits should be periodically removed from the ponds so that anaerobic reactions at the bottom layers are kept to a minimum.

HEAVY METALS AND PESTICIDES

These compounds, either individually or in combination, can produce acute toxicity to fish at concentrations as low as a few $\mu g/L$. The extent of toxic effects depends on several factors such as: water quality; species, age, and size of fishes; and antagonistic and synergistic reactions occurring in the pond water. Long-term effects of these compounds may include decline in growth rate and reproduction, and the enhancement of bioaccumulation and biomagnification in the food chains, as described in section 6.3.

6.5.2 *Design criteria*

ORGANIC LOADING, DO, AND FISH-YIELD MODELS

Almost all kinds of organic wastes can be fed to fish ponds. These wastes can be in any of the following forms: raw sewage; effluents from wastewater treatment plants or high-rate algal ponds; nightsoil and septage; biogas slurry; composted products; animal manures; and agro-industrial wastewaters. Fish may be reared directly in waste stabilization ponds. In any case, organic loadings to be applied to waste-fed fish ponds should be in the range where anaerobic conditions do not prevail in the ponds.

The data in Table 6.1 show that well-operated fish ponds fed with wastewater had organic loadings in the range of 25–75 kg BOD_5/(ha-day), while those fed with organic solids, e.g. septage or biogas slurry, had organic loadings from 50 to 150 kg COD/(ha-day). Data on fish yields vary widely depending on the mode of pond operation, climates, fish species, and stocking densities employed. However, these fish yield data were reported to be at least comparable with, or better than, in those ponds fed with conventional fish feed.

DO-at-dawn (DO_d) model

As the critical period of DO depletion in fish ponds occurs during night-time or in the early morning period, a model to predict this critical DO (or DO_d) would be useful for fish pond operators. Boyd (1979) proposed a mass-balance equation to estimate the amount of DO remaining at dawn.

$$DO_d = DO_{dusk} \pm DO_{df} - DO_f - DO_m - DO_p \tag{6.4}$$

where DO_d = DO concentration at dawn, mg/L;
DO_{dusk} = DO concentration at dusk, mg/L;
DO_{df} = gain or loss of oxygen due to diffusion, mg/L;
DO_f = DO used by fish, mg/L;
DO_m = DO consumed by sediment, mg/L;
DO_p = DO used by planktonic community, mg/L.

To determine DO_d from eq. 6.4, values of the parameters on the right-hand side have to be determined experimentally or from literature. Computer simulation programs have been developed by Boyd (1979) which could model the dynamics or fluctuation of DO_d in channel catfish ponds satisfactorily.

From a practical point of view, the DO_d model should be a simple one to enable fish pond operators to determine easily the occurrence of critical DO, so that appropriate measures (such as mechanical aeration or temporary discontinuation of organic waste feeding) can be undertaken. Bhattari (1985) proposed an empirical DO_d model for waste-fed ponds as follows:

$$\mathrm{DO_d} = 10.745\exp\{-(0.017t + 0.002L_c)\} \tag{6.5}$$

where exp = exponential;
t = time of septage loading, days;
L_c = cumulative organic loading to fish ponds up to time t, kg COD (per 200 m^3 pond volume).

Equation 6.5 was developed and validated with experimental data of Edwards *et al.* (1984), as shown in Figure 6.13 for the experiments with a septage loading of 100 kg COD/(ha-day). These experiments were conducted in Thailand with earthen ponds, each with working dimensions of 20 × 10 × 1 m, length × width × depth. These ponds were fed daily with septage at organic loadings of 50, 100, 150, 200, 250, and 300 kg COD/(ha-day), and tilapia stocking densities were varied at 1, 3, 5, 10, and 20 fish/m^2. The ponds were operated as non-flow-through, with no effluent overflow, but canal water had to be added periodically to make up for water losses due to evaporation and seepage. It is apparent that eq. 6.5 is applicable only to the above experimental conditions. Other types of fish pond conditions and operation will have different coefficient values for eq. 6.5, but the effects of t and L_c on $\mathrm{DO_d}$ might be similar.

The trend of $\mathrm{DO_d}$ decline with time as shown in Figures 6.8c and 6.13 is usually expected in fish ponds fed with organic wastes such as septage, sludge, or manure. This phenomenon is due to the pond hydraulics, which is static and without

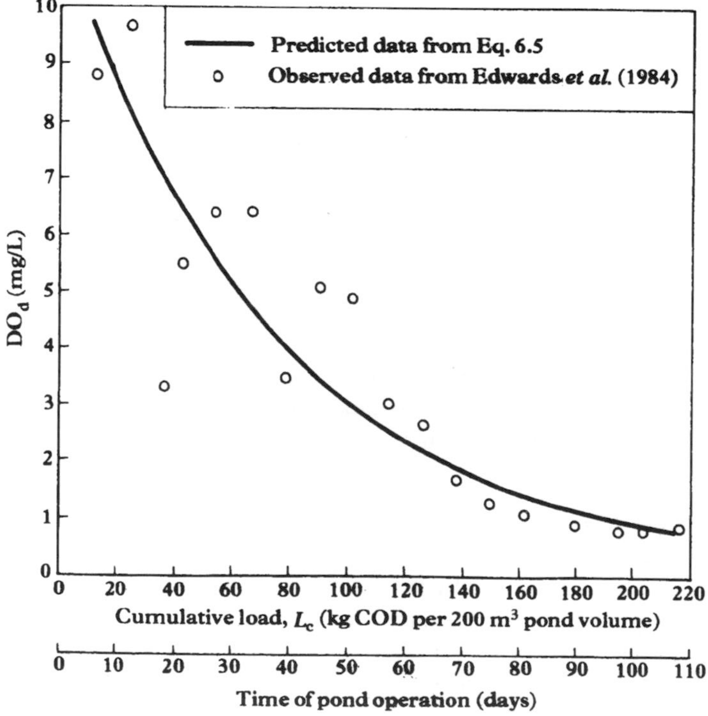

Figure 6.13 Cumulative organic load vs $\mathrm{DO_d}$ for 100 kg COD/(ha-day) loaded septage ponds

effluent overflow. All the loaded organic wastes stay in the ponds and exert an oxygen demand which keeps on increasing due to the accumulation of the residual organic compounds and solubilization of some organics from the sediment layer (see Figure 6.8a,b). After a period of pond operation the DO_d level seemed to reach a plateau in which only a small decline of DO_d was observed; an indication of a somewhat steady-state condition with respect to DO_d dynamics in the waste-fed ponds.

Tilapia growth model

Bhattarai (1985) attempted to develop an empirical model of tilapia growth in septage-fed fish ponds, using the data of Edwards *et al.* (1984). The model considered the effects of: fish SD, as evidenced from Figure 6.11; N loading, as a nutrient source for algal growth (or tilapia food); and time of fish culture. The effect of initial fish weight was not included because all the fish stocked had approximately the same initial weight of about 2.7 g. The developed model, assuming a constant temperature of 30 °C and neglecting spawning, is

$$W_t = 12.032(t \cdot N/SD)^{0.707} \tag{6.6}$$

where W_t = mean fish weight (g) at time t;
t = time of fish rearing, months;
N = total Kjeldahl N (organic N + NH_3 − N) loading, kg/(ha-day);
SD = fish stocking density, no./m^2.

Equation 6.6 is valid when N, t and SD are greater than zero, and the term $(t \cdot N/SD)$ can be interpreted as the applied weight of N per fish. Figure 6.14 shows

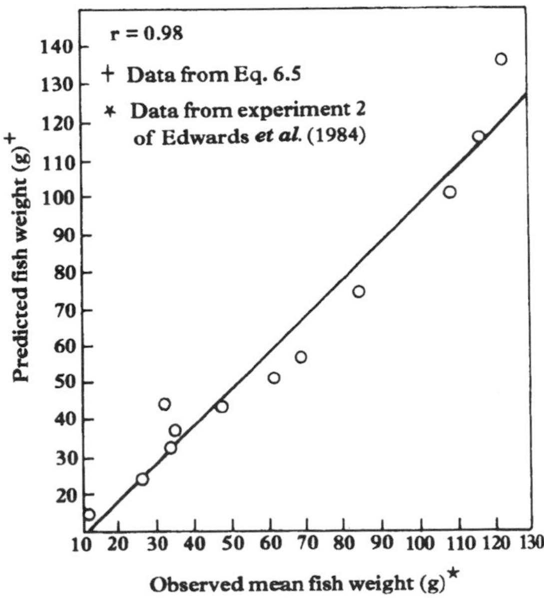

Figure 6.14 Observed vs predicted fish weight for septage-fed ponds; SD = 1, 5 fish/m^2; COD loading = 150 kg/(ha-day)

a typical relationship between the observed and predicted fish weight (using eq. 6.6) in which a good correlation is observed.

It should be pointed out that the development of eqs 6.5 and 6.6 and their validations were undertaken by using different sets of experimental data. However, it is apparent that, to ensure their applicability, these two equations should be validated further with other field-scale data, and with data from fish ponds having different modes of operation, fish species, and types of organic waste input.

FISH CULTURE AND STOCKING DENSITY

Polyculture of fish (such as herbivores and carnivores) is not advisable in waste-fed ponds because of public health considerations due to direct contact of all these fish with the wastewater, even though it utilizes all parts of the aquatic environment. It is not practical to design a waste-fed pond to rear different species of fish because their tolerance levels to low DO are different. Also the carnivores might prey upon the herbivores, thus hampering the potential growth of the herbivores.

Fish sex is an important factor when yield needs to be increased. Owing to genetic reasons, male tilapias generally grow faster than their female counterparts (Guerrero, 1982). Monosex male culture of tilapias gives faster growth and eliminates unnecessary reproduction which can take place during the culture period. Monosex tilapias can be obtained by: (i) manual separation of the sexes, which can be difficult when small fish fry is used; (ii) production of monosex broods through hybridization; or (iii) sex reversal with the use of sex steroids (such as androgen, methyltestosterone, and ethynyltestosterone) in the diet of fish fry during the period of gonadal differentiation (or the production of male or female organ). The effectiveness of these steroids was reported to be 93–100 percent (Nacario, 1987).

The tilapia species is most suitable for waste-fed ponds due to its hardiness. Other herbivorous fish, such as silver carp and bighead carp, are also promising species. Polyculture is recommended where benthic feeding fish such as mud carp are stocked together with omnivorous fish such as common carp.

Stocking density should be of primary concern in fish culture. As shown in Figure 6.11 and eq. 6.6, high stocking density gives high yield per pond volume, but with a smaller mean size of fish. To produce trash fish for animal feed, a stocking density of 10 or more fish/m^2 is recommended and the period of fish culture may be as short as 3 months. Ponds producing marketable-size fish, for possible use as human food, should be stocked at a lower density such as one or two fish/m^2 or even less, but the rearing period may be up to 1 year, or longer. However, several commercial fish farms are reportedly able to stock fish at densities greater than 50 fish/m^2 provided that the pond water quality (listed in section 6.5.1) and fish feed are maintained at the levels favorable for fish growth.

WATER SUPPLY

Water supply should be sufficient in quantity and of good quality, as ponds fed with solid or semi-solid organic wastes (e.g. animal manures or nightsoil) need a certain amount of water to compensate for losses through evaporation and seepage. Ponds fed with sewage or effluent of wastewater treatment plants sometimes need dilution water. The quality of water should be in the permissible range for fish growth.

POND SIZE

Generally a pond depth of 1–1.5 m is suitable for fish with the provision of free board of 0.3–0.6 m. In a rain-fed pond, if the organic input is in solid form, a deeper water level should be provided to compensate for the water loss during dry season. A polyculture fish pond is recommended to have 2–3 m depth with an area of manageable size; normally an area of 1,000–4,000 m^2 is suitable in the tropics. The width of ponds should not exceed 30 m to facilitate seining operations. Separate water supply and drainage systems should be provided in each pond.

POND ARRANGEMENT

Various pond arrangements can be made to suit the local conditions, as well as the type of organic wastes available. Figure 6.15 shows some possible arrangements of ponds with appropriate measures such as depuration to safeguard public health (see section 6.7). In some cases it might be necessary to drain the fish ponds during or after harvesting. Therefore additional pond area or water storage space should be provided for the drained water to prevent the fish pond water from contamin-

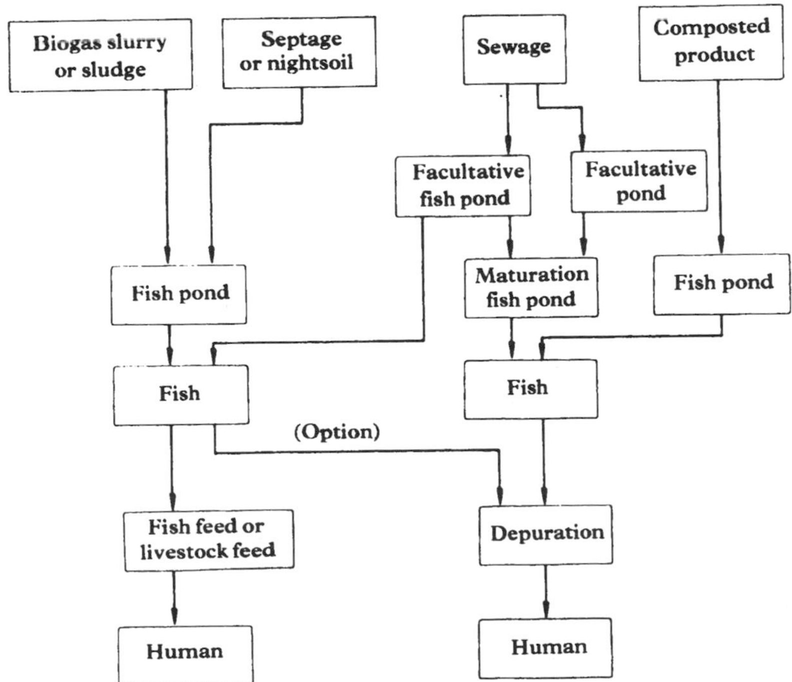

Figure 6.15 Some arrangements for waste-fed fish ponds

ating nearby water courses. Depending on water quality and local conditions the fish pond water can be used for irrigation. Fish ponds that have been drained should be sun-dried for 1–2 weeks to inactivate any pathogens inhabiting the pond bottom and side-walls. Bottom mud should be removed to maintain the working pond volume and to avoid excessive accumulation of sludge at the pond bottom.

Based on the above information and the data from Edwards (1992), design guidelines and performance data of waste-fed ponds are given in Table 6.4.

Example 6.1 Design a septage-fed pond from the given septage characteristics.

Quantity	10 m^3/day
BOD_5	5,000 mg/L
COD	25,000 mg/L
Org–N	1,500 mg/L
NH_3–N	500 mg/L
Total P	300 mg/L
Total solids	4.5 percent
Volatile solids	70 percent of total solids

Solution From data presented in Table 6.4 and Figure 6.7, use organic loading = 50 kg COD/(ha-day)

$$\text{Area requirement} = \frac{10 \times 25{,}000}{50 \times 1{,}000} = 50{,}000 \text{ m}^2 = 5 \text{ ha}$$

Table 6.4 Design guidelines and performance data of waste-fed fish ponds

Pond depth (m)	1–1.5
Pond width (m)	< 30
Pond area (m^2)	400–4,000
Minimum DO at dawn (mg/L)	1–2
NH_3 concentration of pond water (mg/L)	< 0.02
pH of pond water	6.5–9 (optimum 7.0–8.0)
Organic loading rate	
(kg BOD_5/ha-day)	25–75
(kg COD/ha-day)	50–150
Fish stocking density (fish/m^2)	0.5–50
Recommended fish species	Tilapia and carps
Culture period (months)	3–12
Fish yield (kg/ha-year)	1,000–10,000
Removal efficiencies	
Suspended solids (%)	70–85
BOD_5 (%)	75–90
Total nitrogen (%)	70–80
Phosphorus (%)	80–90

Select a fish pond size of 25.0 m (width) × 50.0 m (length) × 1.5 m (depth); water depth = 1.0 m

$$\text{Number of ponds} = \frac{50,000}{25 \times 50} = 40$$

Each pond is separated from the others to facilitate harvesting and drying operations.

Select tilapia (*Oreochromis niloticus*) with a stocking density of five fish/m^2. The products will be sold as trash fish to produce animal feed. This will reduce possible health hazard from direct consumption of these fish.

Number of fingerlings (about 0.5–5 g) used = 1,250 × 5 = 6,250 fish/pond.

Select a harvesting period of 3 months, and 10 days for pond drying (or 100 days in total) before subsequent operation.

Calculate fish yield from eq. 6.6:

$$W_t = 12.032(t \cdot \mathrm{N/SD})^{0.707} \tag{6.6}$$

$$\mathrm{N} = \frac{(500 + 1{,}500)}{1{,}000} \times \frac{10}{5} = 4 \text{ kg/(ha-day)}$$

$$t = 3 \text{ months}$$

$$\mathrm{SD} = 5 \text{ fish/m}^2$$

$$W_t = 12.032(3 \times 4/5)^{0.707} = 22.3 \text{ g/fish}$$

Assume mortality rate = 20 percent

$$\text{Estimated fish yield} = 6{,}250 \times 0.8 \times \frac{22.3}{1{,}000} = 112 \text{ kg/pond per 100 days}$$

$$\text{Extrapolated fish yield} = 112 \times \frac{365}{100} = 408.8 \text{ kg/(pond-year)} = \frac{408.8}{0.125} = 3{,}270 \text{ kg/(ha-year)}$$

$$\text{Estimated total fish yield} = 5 \text{ ha} \times 3{,}270 \text{ kg/(ha-year)} = 16{,}350 \text{ kg/year}$$

Calculate DO at dawn DO_d from eq. 6.5:

$$\mathrm{DO_d} = 10.745 \exp[-(0.017t + 0.002L_c)] \tag{6.5}$$

$$t = 90 \text{ days}$$

$$L_c = \text{cumulative COD loading in pond} = 50 \times 0.125 \times 90 = 562.5 \text{ kg/(1250 m}^3 \text{ pond volume)} = 90 \text{ kg/(200 m}^3 \text{ pond volume)}$$

$$\mathrm{DO_d} = 10.745 \exp\{-[(0.017 \times 90) + (0.002 \times 90)]\} = 1.94 \text{ mg/L.}$$

Tilapia should be able to survive at this DO_d level of about 2 mg/L; or organic loading to this fish pond can be a little more than 50 kg COD/(ha-day) because tilapia can tolerate DO level of about 1 mg/L.

6.6 UTILIZATION OF WASTE-GROWN FISH

Waste-grown fish may be used as a source of animal protein in human diet or as fish meal or feed for other animals.

6.6.1 Human consumption

The fish to be used for human consumption should be bigger in size, safe from pathogens and helminths, and have a nutritional value equal to or higher than that obtained from other sources of fish. Size can be increased by increasing the N loading, lengthening the rearing period, and using a lower stocking density. The following measures are believed to reduce the transfer of pathogens: raising carnivorous fish for human consumption using the herbivorous fish from waste-fed ponds as feed; pretreatment of waste input so that its microbiological quality meets the standards for aquacultural reuse stated in Table 2.28; depuration which is a natural process to remove some microorganisms from the fish body by putting contaminated fish in clean water for 1–2 weeks; and finally public education to wash, to remove intestines, and to cook the fish well before consumption. The nutritional value of waste-grown fish has shown to be better than that of those grown using other sources of feed. Most of the nutrition applied in the form of waste is converted to protein rather than fat, as reported by Moav *et al.* (1977) and by Wohlfarth and Schroeder (1979).

6.6.2 Fish meal or feed for other animals

This method of utilization is suitable for fish grown directly on waste, in order to lengthen the food chain and to reduce the possibility of transference of pathogens. The waste-grown fish can be used as feed for carnivorous fish or shrimps, which have a high market value when sold for human consumption. The biological value of protein in fish meal is 74–89 percent, which is quite high (Williamson and Payne, 1978) and suitable for feeding to pigs and chicken. For this purpose fish in waste-fed ponds can be reared at a higher stocking density and a shorter growing period to obtain a higher fish yield. Fish can be sun-dried, ground, and mixed with other foodstuffs to increase the value of fish meal.

Experiments were conducted at AIT to investigate the feasibility of using tilapia grown in septage-fed ponds as fish feed ingredients for carnivorous fish of high market value such as snakehead (*Channa striata*) and walking catfish (*Clarias batrachus* and *C. macrocephalus*) (Edwards *et al.*, 1987). Tilapia harvested from septage-fed ponds were either used fresh or processed as silage and mixed with other feed ingredients prior to feeding to the carnivores. A silage was produced by adding 20 percent carbohydrate (cassava) to minced tilapia (weight/weight) in the presence of *Lactobacillus casei*. It was found that growth of these carnivorous fish in ponds fed with tilapia fish meal, fresh tilapia or silaged tilapia was comparable with growth of those fed with marine trash fish (conventional fish feed), but poor

Table 6.5 Comparison of growth and food conversion ratios of carnivorous fish fed with different feeds (Edwards *et al.*, 1987)[a]

	Channa striata			*Clarias macrocephalus*		
	MWG (g)	DWG (g/day)	FCR	MWG (g)	DWG (g/day)	FCR
Marine trash fish feed	26.03	0.30	14.15	35.08	0.26	13.88
Tilapia meal feed	45.77	0.59	33.18	43.84	0.33	12.53
Tilapia silage feed	55.37	0.65	6.45	47.67	0.36	10.15
Fresh tilapia	52.81	0.62	10.99	51.43	0.39	12.87

[a] MWG = mean weight gain; DWG = daily weight gain; FCR = food conversion ratio

food conversion efficiencies were obtained because of excessive rate of feeding (Table 6.5).

Data in Table 6.5 were obtained by using a feeding rate of 20 percent body weight/day. In studies in which excellent FCRs of 2.5 or less were reported, a feeding rate of less than 10 percent body weight/day was used (Edwards *et al.*, 1987).

6.7 PUBLIC HEALTH ASPECTS AND PUBLIC ACCEPTANCE

As organic wastes such as sludge and animal manure can contain a large number of pathogens, fish-rearing in a waste-fed pond obviously carries a risk of contamination with these pathogens. In this regard fish are considered to be indicators of the sanitary condition of the fish pond water, in which the microbial flora present in the fish body directly reflects the microbiological condition of the water from which they are taken.

According to Feachem *et al.* (1983), three distinct health problems associated with fish culture in ponds enriched with human and/or animal waste are:

1. the passive transference of pathogens by fish which become contaminated in polluted water;
2. the transmission of certain helminths whose life cycles include fish as an intermediate host; and
3. the transmission of other helminths with a life cycle involving other pond fauna, such as snail hosts of schistosomes.

The first problem is a cause of concern worldwide, whereas the second and third apply only in areas where the helminths concerned are endemic.

The passive transference of pathogens occurs because fish can carry these pathogens, especially bacteria and viruses, on their body surfaces or in their intestines, which can later contaminate people who handle or eat these fish raw or partially cooked. Pilot-scale experiments on waste-fed ponds were conducted at the AIT using septage, composted nightsoil, and biogas slurry as organic wastes feeding into tilapia ponds (Polprasert *et al.*, 1982; Edwards *et al.*, 1984). Within the organic loadings up to 150 kg COD/(ha-day), the total coliform and fecal coliform bacteria and the *Escherichia coli* bacteriophages (a viral indicator) were found to be absent in fish organ samples such as blood, bile, and meat. These microorganisms were found at high densities in fish intestines (up to 10^9/mL), as normally expected. Levels of total coliforms and fecal coliforms in the fish pond water were 10^3–10^4/100 mL, comparable with those in the control ponds without septage feeding.

Fish have been shown not to be susceptible to the same enteric bacterial diseases as humans and animals (Janssen, 1970; Allen and Hepher, 1976). Therefore, several human bacterial pathogens which have been shown to be carried in and on fish (Janssen, 1970), may be carried passively, infecting only the gut surface which can be cleaned by depuration. Studies of human pathogens in salmonides reared in Arcata wastewater fish ponds, in California, showed bacteria causing human infections only in the gut (Allen *et al.*, 1976). Results from kidney, liver, and spleen samples indicated that none of the potential pathogens was present in these fish organs under the fish culture conditions in Arcata.

Cloete *et al.* (1984) reported that the skin, gills, and intestines of the waste-grown fish from a pond system treating cattle feedlot effluents contained large numbers of bacteria, including potential pathogens. However, similar bacterial numbers, including potential pathogens, were also associated with the skins, gills, and intestines of naturally grown fish. This suggests that the health risk involved in the consumption of waste-grown fish might not be substantially different from that of natural fish populations. In both cases the tissues and blood appeared to be sterile, which would contribute to a much-reduced health risk.

From the above information it appears that bacteria are normally present in fish intestines. However, if bacteria are present in very high concentrations in the fish pond water or in the fish body, the natural immunological barrier of fish could be overcome and the bacteria could invade the fish meat. Buras *et al.* (1985) defined a 'threshold concentration' as the minimum number of bacteria which, when inoculated into the fish, causes their appearance in the fish meat. From their experiments with tilapia and carp, a list of threshold concentrations is given in Table 6.6. Depuration experiments were found to be effective when the fish did not contain high concentrations of bacteria in their meat. With respect to the microbiological quality of fish pond water, Buras *et al.* (1987) experimentally found the 'critical concentration' of standard plate count bacteria to be 5×10^4/mL, in which bacterial concentrations higher than this critical concentration were found to cause their appearance in the meat of the fish reared in the ponds.

Hejkal *et al.* (1983) investigated the levels of bacteria and viruses in fish reared in experimental wastewater fish ponds in Arkansas, U.S.A. They found that even when levels of bacteria exceeded 10^5/100 g in the fish guts, very little penetrated into the fish muscle tissue; the maximum of 25 fecal streptococci/100 g was found in the fish meat. They finally suggested that while the fish do not accumulate bacteria in the muscle tissue, contamination of the muscle tissue during processing is difficult to avoid. Buras *et al.* (1985) proposed that, to prevent serious public health problems, the threshold values in Table 6.6 should be considered satisfac-

Table 6.6 Threshold concentrations of microorganisms inoculated to fish (adapted from Buras *et al.*, 1985)

Microorganisms	Threshold concentrations (no./fish)	
	Tilapia aurea	*Cyprinus carpio*
Bacteria		
E. coli	2.5×10^6	1.5×10^6
Clostridium freundii	9.3×10^3	—
Streptococcus faecalis	1.9×10^4	4.0×10^4
Streptococcus montevideo	1.8×10^4	3.7×10^4
Bacteriophages		
T2 virus	4.0×10^3	4.6×10^3
T4 virus	2.0×10^4	—

tory criteria for the design and management of fish ponds in which wastewater is used. In addition, as the experiments performed with Polio 1 LSc viruses suggested a very low threshold concentration, and as viruses have a low infective dose, their presence in the fish pond water or wastewater to be fed to the fish pond should be of a major concern.

The second public health problem is related to the helminths requiring fish as intermediate hosts. The major and common types of these helminths associated with waste-fed fish ponds include *Clonorchis* and *Opisthorchis* (Feachem *et al.*, 1983). The excreted eggs of these helminths, once discharged into a pond, will release miracidium (a larva) which, after being ingested by certain species of freshwater snails, will develop and multiply into free-swimming cercarial larvae. Within 1–2 days these cercaria have to find a fish as a second intermediate host. They penetrate under the fish scales and form cysts in the connective tissues. When the fish are eaten raw or partially cooked the cysts enter the human body and develop into worms in the digestive tract. In the human body female worms will lay eggs which will be released together with the feces into the environment. If the feces is discharged into a pond, this transmission cycle of the helminths could start again.

The third public health problem concerns schistosomiasis, in which certain species of freshwater snails serve as intermediate hosts for certain schistosome helminths. The schistosome eggs that are discharged into a pond will release miracidia. After ingestion by the snails these miracidia will develop and multiply into infective cercarial larvae that can penetrate directly into human skin. Once inside a human body, these cercaria will develop into mature worms and the female worms will lay eggs that will be excreted with the feces into the environment, and the transmission cycle can be repeated again.

It appears that cooking and eating habits will have great impact upon those people who handle and eat the waste-grown fish. Fish that are well cooked will have all the pathogens on or in the fish body killed. However, disease transmission during fish harvesting, cleansing, and preparation prior to cooking can occur, depending upon the types of the disease present. In areas where there are endemics or outbreaks of diseases related to fish ponds, organic wastes to be applied to fish ponds have to be properly treated to ensure destruction of these pathogens, or feeding of organic wastes to fish ponds should be discontinued temporarily. Fish to be used as animal feed will normally have to be processed further, e.g. through drying, silaging, and/or pelletization, which will result in die-off of most pathogens.

Another major problem relating to the consumption of fish raised on wastes is the public acceptability of the fish. As far as taste and texture are concerned, observations by various workers indicate that fish grown in well-treated domestic wastes are equal to, or even superior in taste or odor to, non-wastewater cultivated fish (Allen and Hepher, 1976). Fish grown in manure-fed ponds have a different taste and texture as they are much leaner, with only 6 percent fat, which is excellent compared with fish raised on high-protein feed pellets with 15 percent fat, and fish raised on grain with 20 percent (Wohlfarth and Schroeder, 1979).

Probably the most critical prerequisite before we can obtain public acceptance of waste-grown fish is public health concerns. All legitimate public health concerns must be adequately assessed and resolved, and adequate public health safeguards provided. Perhaps a way to overcome this public health problem is to treat the waste prior to use in fish culture to meet the WHO standards for aquacultural reuse (Table 2.28). Lime treatment of sludge can inactivate a large number of bacteria and viruses, but not helminthic ova. Waste stabilization ponds in series or sedimentation basins with a detention time of at least 10 days can settle out most of the helminthic ova, but care has to be taken with the second and third types of helminthic problems, described earlier. An increased destruction of pathogens would also take place if the waste was treated by a biogas or composting system before adding to the fish pond, as both processes, especially the latter, lead to an increased temperature which kills pathogens. An additional step could involve depuration, the maintenance of the fish in clean water for a week or two, which may eliminate any disease organisms surviving earlier treatments. The final step would be to have good hygiene at all stages of fish handling and processing, and to ensure that fish intestines are removed, and that the meat is thoroughly washed and cooked, before consumption. Another strategy could involve feeding the waste-grown fish to other animals which can be consumed by humans, so that the fish raised on waste are not consumed directly by humans. Both strategies would reduce the possibility of disease transfer.

There appears to be definite cultural differences concerning the consumption of fish reared on human wastes (Edwards, 1980). Chinese and Indians appear to have few objections to eating such fish, whereas Malays and Thais immediately reject such a suggestion. Muslims would also reject fish grown on swine waste, as it is religiously unclean to them. However, it should be pointed out that the fish grown in such ponds normally do not feed directly on the waste, but on the natural food, e.g. plankton, that develops as a result of the fertilizing effect of the waste, although there are some exceptions.

Public need will probably determine the degree of public acceptance of waste-grown fish in a particular area. Table 6.1 shows that the practice of fish production in waste-fed ponds, although at a low level, has been undertaken both in developed and developing countries. The social, economic, and political considerations that influence human behavior are very complicated. However, openness, candor, public education, and public involvement appear to be the keys to public acceptance. In spite of the many problems the future for waste-fish recycling systems seems promising. This is due to the ever-increasing demands both for waste disposal and food production in today's rapidly expanding populations.

REFERENCES

Alabaster, J. S. and Lloyd, R. (1980). *Water Quality Criteria for Freshwater Fish*. Butterworths, London.

Allen, G. H. and Hepher, B. (1976). Recycling of waste through aquaculture and constraints to wider application. *Proc. FAO Tech. Conf. Aquaculture*, Japan.

Allen, G. H., Bush, A., and Mortan, W. (1976). Preliminary bacteriological experiences with wastewater-fertilized marine fish ponds, Humboldt Bay, California. *Proc. FAO Tech. Conf. Aquaculture*, Japan.

Bhattarai, K. K. (1985). Septage recycling in waste stabilization ponds, Doctoral dissertation no. EV-85-1, Asian Institute of Technology, Bangkok.

Boyd, C. E. (1979). *Water Quality in Warmwaters Fish Ponds*. Craftmaster Printers, Alabama.

Buras, N., Duek, L., and Niv, S. (1985). Reactions of fish to microorganisms in wastewater. *Appl. Environ. Microbiol.*, **50**, 989–95.

Buras, N., Duek, L., Niv, S., Hepher, B., and Sandbank, E. (1987). Microbiological aspects of fish grown in treated wastewater. *Water Res.*, **21**, 1–10.

Campbell, N. A. (1990). *Biology*, 2nd edition. The Benjamin/Cummings Publishing Co., Inc., Redwood City, California.

Cloete, T. E., Toerien, D. F., and Pieterse, A. J. H. (1984). The bacteriological quality of water and fish of a pond system for the treatment of cattle feed effluent. *Agr. Wastes*, **9**, 1–15.

Colt, J., Tchobanoglous, G., and Wang, B. (1975). *The Requirement and Maintenance of Environmental Quality in the Intensive Culture of Channel Catfish*. Dept. of Civil Eng., University of California, Davis.

Djajadiredja, R. and Jangkaru, Z. (1978). *Small Scale Fish/Crop/Livestock/Home Industry Integration, a Preliminary Study in West Java, Indonesia*. Inland Fisheries Research Institute, Bogor, Indonesia.

Edwards, P. (1980). A review of recycling organic wastes into fish, with emphasis on the tropics. *Aquaculture*, **21**, 261–79.

Edwards, P. (1985). *Aquaculture: a component of low cost sanitation technology*. UNDP Project Management Report No. 3, World Bank, Washington, D.C.

Edwards, P. (1992). Reuse of human wastes in aquaculture—a technical review. *Water and Sanitation Report* No. 2, UNDP–World Bank Water and Sanitation Program, The World Bank, Washington, D.C.

Edwards, P., Pacharaprakiti, C., Kaewpaitoon, K., Rajput, V. S., Ruamthaveesub, P., Suthirawut, S., Yomjinda, M., and Chao, C. H. (1984). Reuse of cesspool slurry and cellulose agricultural residues for fish culture. AIT Research Report No. 166. Asian Institute of Technology, Bangkok.

Edwards, P., Polprasert, C., Rajput, V. S., and Pacharaprakiti, C. (1988). Integrated biogas technology in the tropics—2. Use of slurry for fish culture. *Waste Manage. Res.*, **6**, 51–61.

Edwards, P., Polprasert, C., and Wee, K. L. (1987). Resource recovery and health aspects of sanitation. AIT Research Report No. 205. Asian Institute of Technology, Bangkok.

Feachem, R. G., Bradley, D. J., Garelick, H., and Mara, D. D. (1983). *Sanitation and Disease—Health Aspects of Excreta and Wastewater Management*. Wiley, Chichester.

Guerrero, R. D. III. (1982). Control of tilapia reproduction. In *The Biology and Culture of Tilapias* (eds. R. S. V. Pullin and R. H. Lowe-McConnell), pp. 309–15, International Center for Living Aquatic Resources Management, Manila.

Hastings, W. H. and Dickie, L. M. (1972). Feed formulation and evaluation. In *Fish Nutrition* (ed. J. E. Halver). Academic Press, New York.

Hejkal, T., Gerba, C. P., Henderson, S., and Freeze, M. (1983). Bacteriological, virological and chemical evaluation of a wastewater-aquaculture system. *Water Res.*, **17**, 1749–55.

Henderson, S. (1983). An evaluation of filter feeding fishes for removing excessive nutrients and algae from wastewater. EPA-600/52-83-019. Ada, Oklahoma.

Hickling, C. F. (1971). *Fish Culture*, revised edition. Faber & Faber, London.

Huet, M. and Timmermans, J. A. (1971). *Breeding and Cultivation of Fish*. Fishing News (Book) Ltd., London.

Janssen, W. A. (1970). Fish as potential vectors of human bacterial diseases of fishes and shellfishes. Special Publication of American Fish Society, no. 5, 284–90.

Lagler, K. F., Bardach, J. E., and Miller, R. R. (1962). *Ichthyology: The Study of Fishes*. Wiley, New York.

Maramba, F. D. (1978). *Biogas and Waste Recycling. The Philippines Experience*. Maya Farms Division, Liberty Flour Mills Inc., Manila, Philippines.

Moav, R., Wohlfarth, G., and Schroeder, G. L. (1977). Intensive polyculture of fish in freshwater ponds. I—Substitution of expensive feeds by liquid cow manure. *Aquaculture*, **10**, 25–43.

Nacario, E. N. (1987). Sex reversal of Nile tilapia in cages in ponds. AIT master's thesis No. AE-87-35, Asian Institute of Technology, Bangkok.

NAS (1976). *Making Aquatic Weeds Useful: Some Perspectives for Developing Countries*. National Academy of Sciences, Washington, D.C.

Polprasert, C., Edwards, P., Pacharaprakiti, C., Rajput, V. S., and Suthirawat, S. (1982). Recycling rural and urban nightsoil in Thailand. AIT Research Report no. 143. Asian Institute of Technology, Bangkok.

Polprasert, C., Udom, S., and Choudry, K. H. (1984). Septage disposal in waste recycling ponds. *Water Res.*, **18**, 519–28.

Sawyer, C. N., McCarty, P. L., and Parkin, G. F. (1994). *Chemistry for Environmental Engineering*, 4th edition. McGraw-Hill, New York.

Schroeder, G. L. (1975). Some effects of stocking fish in waste treatment ponds. *Water Res.*, **9**, 591–3.

Schroeder, G. L. (1977). Agricultural wastes in fish farming—a commercial application of the culture of single-celled organisms for protein production. *Water Res.*, **11**, 419–20.

Sharma, H. P., Polprasert, C., and Bhattarai, K. K. (1987). Physico-chemical characteristics of fish ponds fed with septage. *Resources and Conservation*, **13**, 207–15.

Tang, Y. A. (1970). Evaluation of balance between fishes and available fish foods in multi-species fish culture ponds in Taiwan. *Trans. Am. Fish. Soc.*, **99**, 708–18.

Thorslund, A. E. (1971). *Potential Uses of Wastewaters and Heated Effluents*. European Inland Fisheries Advisory Commission Occasional Paper No. 5. FAO, Rome.

U.S. Committee on Water Quality Criteria (1972). *Water Quality Criteria*. Superintendent of Documents, Washington, D.C.

Whipple, G. C. and Whipple, M. C. (1911). Solubility of oxygen in sea water. *J. Am. Chem. Soc.*, **33**, 362.

Williamson, G. and Payne, W. J. A. (1978). *Animal Husbandry in the Tropics*, 3rd edition. Longman, London.

Wohlfarth, G. W. and Schroeder, G. L. (1979). Use of manure in fish farming—a review. *Agric. Waste*, **1**, 279–99.

Wolf, K. (1957). Blue-sac disease investigations, microbiology and laboratory induction. *Progressive Fish Culturist*, **19**, 14–18.

EXERCISES

6.1 What are the health problems associated with fish farming in wastewater-enriched ponds? What would you suggest to solve these problems?

6.2 Calculate the ammonia concentration in a water sample at pH = 8 and NH_3–N = 10 mg/L. The following equilibria exist:

$$NH_3 + H_2O \rightleftarrows NH_4^+ + OH^-$$

$$K_b = \frac{[NH_4][OH^-]}{[NH_3]} = 1.76 \times 10^{-5} \quad \text{(at 25° C)}$$

$$K_w = [H^+][OH^-] = 10^{-14} \quad \text{(at 25° C)}$$

The U.S. Committee on Water Quality Criteria (1972) has recommended that, to safeguard fish, no more than 0.02 mg/L NH_3 be permitted in receiving waters. Does the ammonia concentration in the above calculation meet the criteria? If not, suggest methods to make the water safe for the fish.

6.3 Compare the advantages and disadvantages of applying commercial fish meal or cattle manure into fish ponds for the purpose of raising tilapia for use as animal feed.

6.4 You are to visit a commercial fish farm and obtain the following information:

a fish species and stocking density
b type and amount of fish meal being fed to the fish ponds
c productivity of the fish ponds in unit of kg fish/(ha-year)

From the above data, estimate the organic loading rate (in kg BOD or COD/ha-day) and the percent fish mortality. Also suggest some improvements, if any, to be made to increase the fish productivity in this farm.

6.5 **a** A small community produces septic tank sludge (septage) with the following characteristics:

Quantity = 10 m^3/day
COD = 20,000 mg/L
Total Kjeldahl N = 1,800 mg/L

This septage is to be used to feed to fish ponds to produce tilapia with a mean weight of about 30 g for use as animal feed. Three fish harvests are planned annually, with 100 days for fish culture and 3 weeks for pond maintenance. Suppose tilapia can tolerate DO concentration at dawn of no lower than 1.5 mg/L. What should be the highest COD loading to be applied to the septage-fed fish pond(s)? Also determine the total land area required, stocking density, the number of fingerlings needed and the expected fish output for each cycle (120 days).

b Suppose that for efficient fish harvesting, the pond water is to be drained out, what should be done with this pond water and the pond sediment that needs to be removed?

6.6 What are the fish species that the people in your country like to eat? Would the people mind if the fish sold in the market are reared in waste-fed ponds? Suggest methods to make the waste-grown fish more acceptable to the people.

7

Aquatic weeds and their utilization

Aquatic weeds are prolific plants growing in water bodies, and can create a number of problems due to their extensive growth and high productivity. As they exhibit spontaneous growth they usually infest polluted waterways or water bodies, reducing the potential uses of these water courses. The problems of aquatic weeds have been magnified in recent decades due to intensive human use of the natural water bodies and pollution discharges into these water bodies. Eradication of the aquatic weeds has proven impossible, and even reasonable control is difficult. However, by turning these aquatic weeds to productive use—such as in wastewater treatment, in making compost fertilizer, and as human food or animal feed, etc.—some of the problems created by aquatic weeds may be minimized, and may provide incentives for people to harvest the aquatic weeds more regularly.

7.1 OBJECTIVES, BENEFITS, AND LIMITATIONS

7.1.1 Objectives

The main purposes of using aquatic weeds in waste recovery and recycling are waste stabilization and nutrient removal, and conversion of the harvested weeds into productive uses. Aquatic weeds provide a medium for bacteria to be attached and grow at their roots and stems, to stabilize the waste. The presence of weeds in the aquatic medium and subsequent harvesting enable nutrient removal from the wastewater. Even though stabilization of waste is a slow process in aquatic systems, removal efficiency is high, and can produce an effluent superior or comparable to that of any other treatment system. The productive uses of aquatic weeds are given below.

7.1.2 *Benefits*

Aquatic weeds can be used, directly or after processing, as soil additives, mulch, fertilizer, green manure, pulp and fiber for paper-making, animal feed, human food, organic malts for biogas production, and for composting. If properly designed the operation and maintenance of the aquatic weed system does not require highly skilled manpower.

7.1.3 *Limitations*

LAND REQUIREMENT

Waste treatment with aquatic weeds requires large areas of land at reasonably low cost. This will become one of the major limitations in urban areas, but in rural areas it will not be a problem.

PATHOGEN DESTRUCTION

The reliability of an aquatic system with regard to pathogen destruction is low because the inactivation mechanism is natural die-off during long detention time, and some of the aquatic plants provide suitable conditions for the survival of pathogenic microorganisms, e.g. preventing sunlight (or UV light) from inactivating the pathogens.

END-USES

Most of the end-uses of aquatic weeds are in agricultural and livestock rearing. So aquatic weed systems are suitable for rural areas.

7.2 MAJOR TYPES AND FUNCTIONS

Aquatic weeds may be divided into several life forms, a somewhat arbitrary separation as there are plants which may change their life form depending on stage of growth or depth of water. The major types are: submerged, floating and emergent weeds, as shown in Figures 7.1, 7.2, and 7.3, respectively. The most common plants and their scientific names are given in Table 7.1.

7.2.1 *Submerged type*

Weeds that grow below the water surface are called submerged. Often they form a dense wall of vegetation from the bottom of the water surface. Submerged species can only grow where there is sufficient light, and they may be adversely affected

Figure 7.1 Submerged aquatic weeds: (a) hydrilla (*Hydrilla verticillata*); (b) photograph of hydrilla; (c) coontail (*Ceratophyllum demursum*); (d) photograph of coontail

(d)

Figure 7.1 (*continued*)

by the onset of factors such as turbidity and excessive populations of planktonic algae, which decrease the penetration of light into the water. Thus these species are not effective for effluent polishing.

7.2.2 Floating types

According to Reimer (1984), there are two subtypes of floating weeds, i.e. floating unattached and floating attached. The roots of floating unattached plants hang in the water and are not attached to the soil. The leaves and stems are above the water, thus receiving sunlight directly. The submerged roots and stems are a good habitat for the bacteria responsible for waste stabilization. The aquatic weeds that cause some of the most widespread and serious problems, e.g. water hyacinth, duckweed, and water lettuce, fall into this category. As these plants are unattached, populations are seldom permanent and, if space permits, are liable to be moved by wind and water currents or, if confined in one place, initiate the accumulation of organic matter which decreases the depth of water until it is sufficiently shallow for the establishment of emergent swamp vegetation. Thus they are essentially primary colonizers in aquatic ecosystems.

Floating attached plants have their leaves floating on the water surface, but their roots are anchored in the sediment. The leaves are connected to the bottom of petioles (e.g. water lilies, Figure 7.2f and g), or by a combination of petioles and stems.

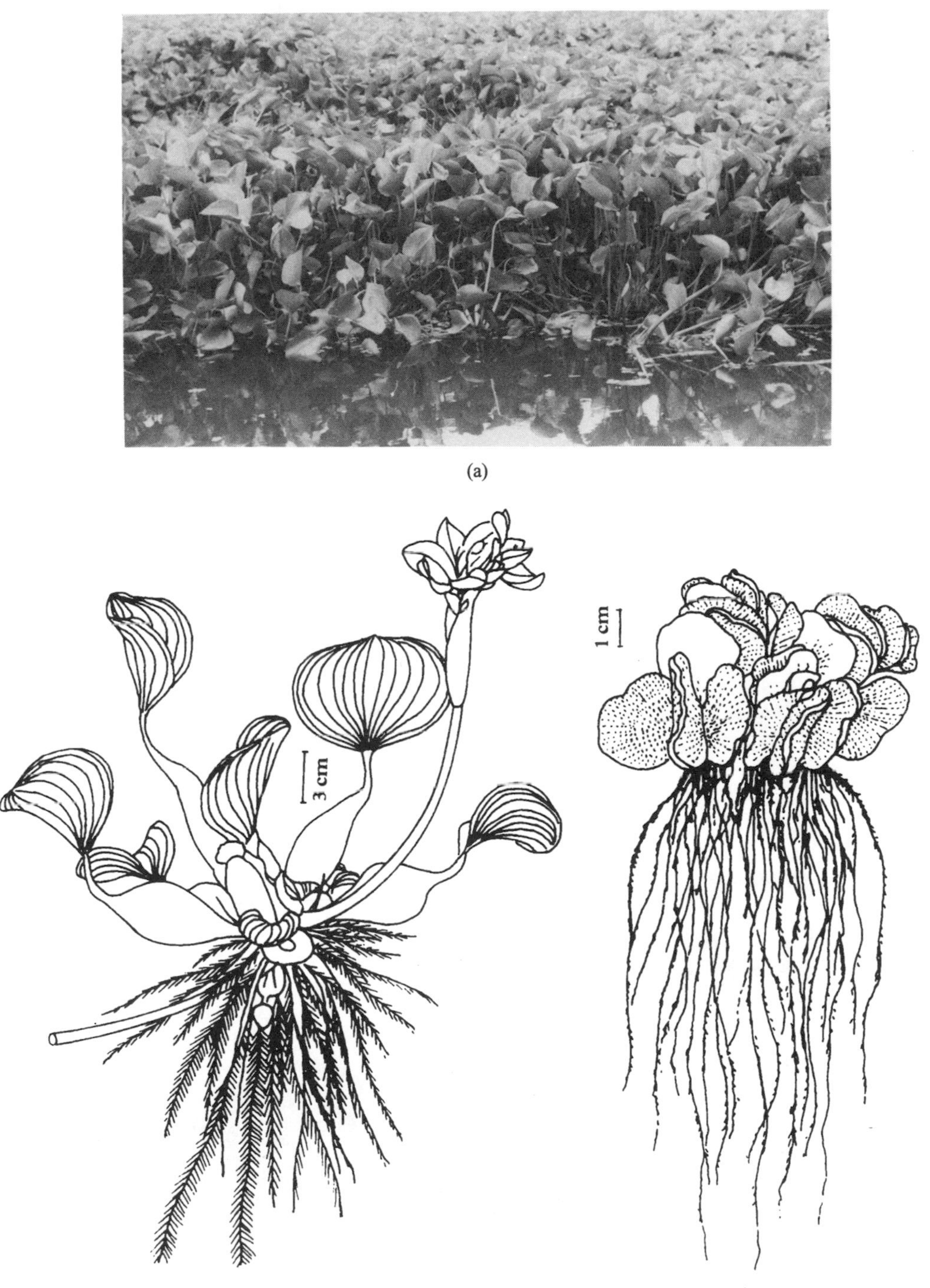

Figure 7.2 Floating aquatic weeds: (a) photograph of water hyacinth in a water body; (b) water hyacinth (*Eichhornia crassipes*); (c) water lettuce (*Pistia stratiotes*)

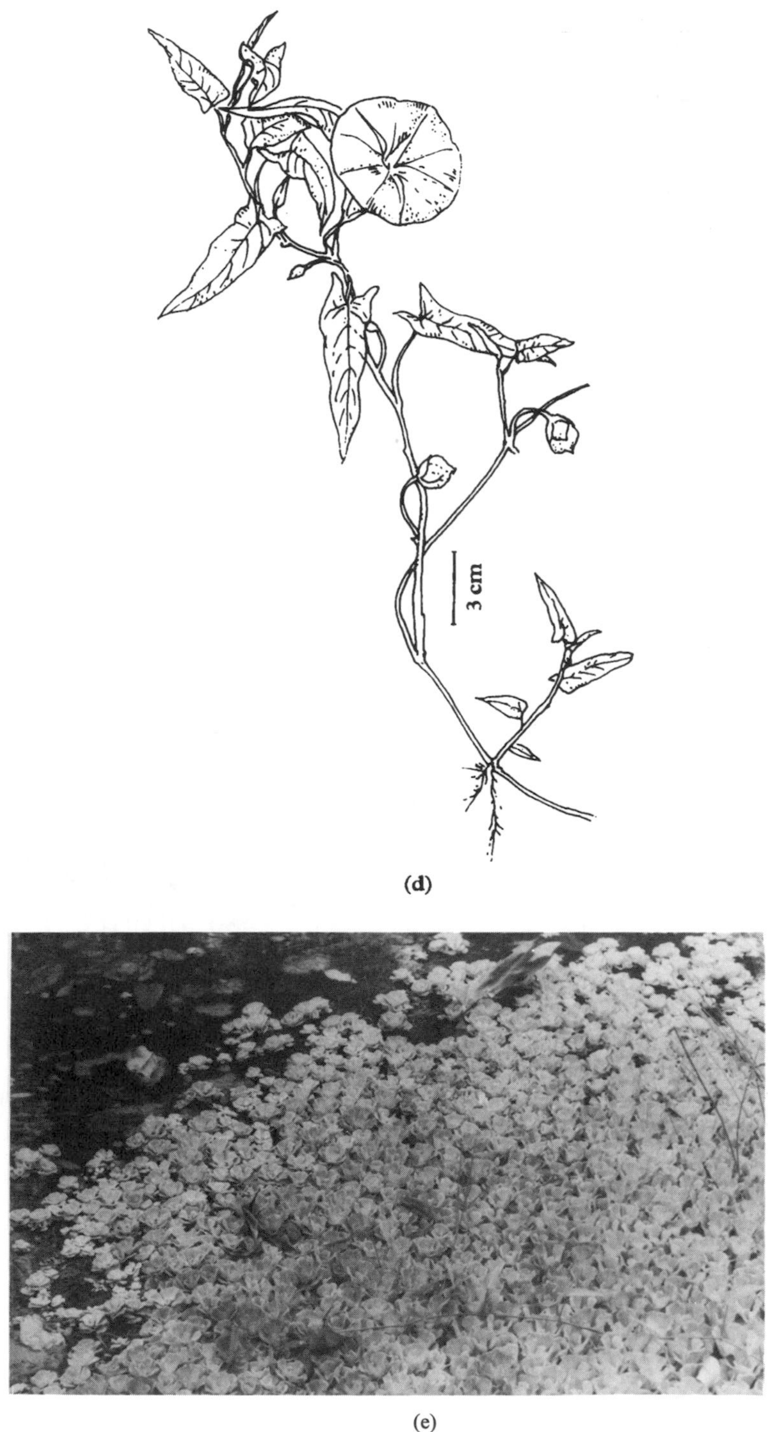

Figure 7.2 (*continued*) Floating aquatic weeds: (d) water spinach (*Ipomoea aquatica*); (e) photograph of water lettuce in a water body

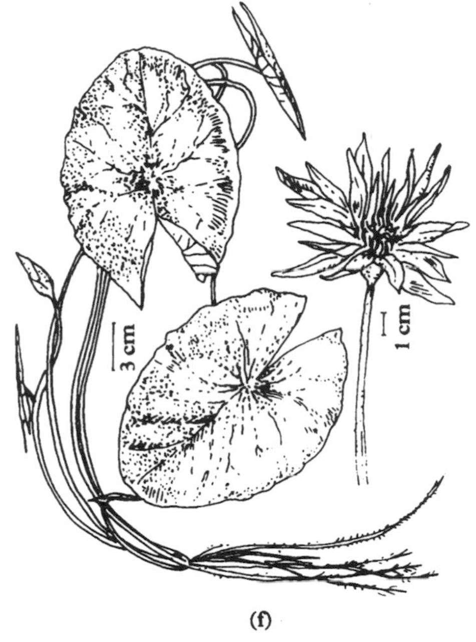

(f)

(g)

Figure 7.2 (*continued*) Floating aquatic weeds: (f) water lily (*Nymphaea nouchali*); (g) photograph of water lily

7.2.3 Emergent type

These are rooted weeds that extend above the water surface. Such attached aquatic plants usually grow well in a stable hydrological regime and are less likely to be a problem in situations where rapid or extensive fluctuations in water level occur. Specialized communities of emergent species may develop on a substratum of

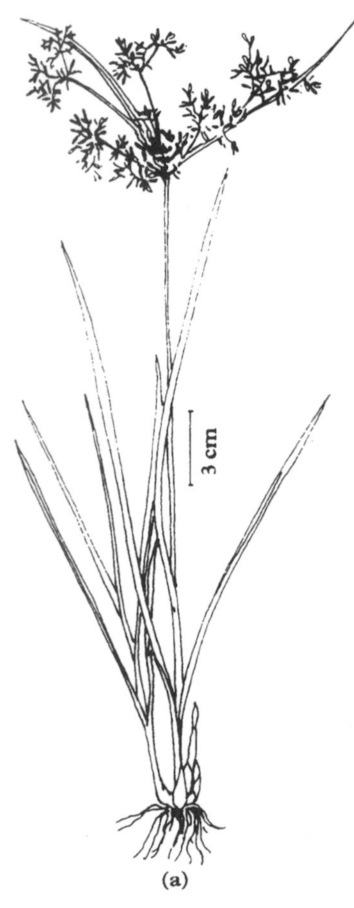

Figure 7.3 Emergent aquatic weeds: (a) bulrush (*Scirpus longii*); (b) photograph of bulrush; (c) cattail (*Typha latifolia*)

(d)

Figure 7.3 (*continued*) Emergent aquatic weeds: (d) cattail plants in a swamp

Table 7.1 Types and common names of aquatic weeds

Type	Common name	Botanical name
Submerged	Hydrilla	*Hydrilla verticillata*
	Water milfoil	*Myriophyllum spicatum*
	Blyxa	*Blyxa aubertii*
Floating	Water hyacinth	*Eichhornia crassipes*
	Duckweeds	*Wolfia arrhiga*
	Water lettuce	*Pistia stratiotes*
	Salvinia	*Salvinia* spp.
Emergent	Cattails	*Typha* spp.
	Bulrush	*Scirpus* spp.
	Reed	*Phragmites communis*

floating aquatic plants, especially where stands of the emergent species are particularly stable.

There is frequently a pronounced zonation of different life forms to different depths of water (Figure 7.4). The emergent type normally grows in shallow water and the submerged type in deeper water in which light can still penetrate to the bottom. The floating type is independent of soil and water depth.

Like any other plants, aquatic weeds require nutrients and light. The major factors governing their growth are:

1. ambient temperature
2. light
3. nutrients and substrate in the water
4. pH of water

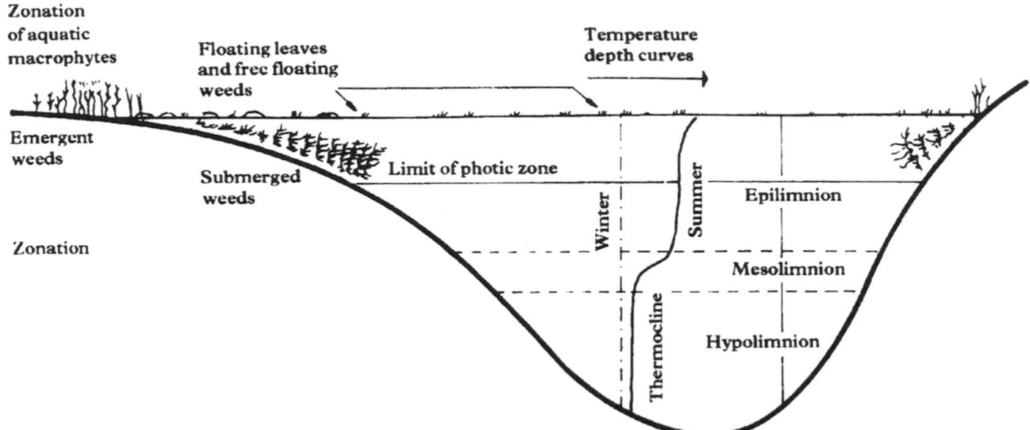

Figure 7.4 A diagrammatic representation of a lake showing zonation of aquatic weeds, the photic zone, and thermal stratification (adapted from Mitchell, 1974; © UNESCO 1974; reproduced by permission of UNESCO)

5. dissolved gases present in the water
6. salinity of the water
7. toxic chemicals present in the water
8. substrate and turbulence
9. water current in rivers and lakes
10. river floods
11. morphology of bodies of water.

While these factors modify the composition of the plant communities, they are in turn modified by the latter. This results in constant interaction between the plants and the environment. Furthermore, the environmental factors also interact with one another, resulting in a complex relationship between the environmental factors and the aquatic weeds.

7.3 WEED COMPOSITION

7.3.1 Water content

Aquatic weeds have a high water content, in general ranging from 85 to 95 percent. It varies from 90 to 95 percent for floating species, from 84 to 95 percent for submerged species, and from 76 to 90 percent for emergent species. The differences among the various life forms can be correlated to some extent with the amount of fiber present in the plant. Water supports the weight of submerged aquatic plants, so they do not develop fibrous stems. Floating plants and emergent plants require more skeletal strength in their aerial parts and so have more fiber than most of the submerged plants.

The low level of dry matter has been the major deterrent in the commercial use of harvested weeds. In order to obtain 1 ton of dry matter, 10 tons of aquatic weeds must be harvested and processed. For water hyacinth, which usually contains only 5 percent dry matter, 20 tons must be harvested and processed just to obtain 1 ton of dry matter. By comparison, terrestrial forages contain 10–30 percent solids, and are therefore, cheaper to harvest.

7.3.2 Protein content

For most species of aquatic weeds, about 80 percent of total N is in the form of protein. Aquatic weeds contain 8–30 percent of crude protein (on a dry-matter basis)—a range similar to that found in terrestrial crops. The considerable variations in crude protein content are due both to seasonality and environment. The crude protein of *Typha latifolia* decreased from 10.5 percent in April to 3.2 percent in July (Boyd, 1970) and that of *Justicia americana* from 22.8 percent in May to 12.5 percent in September (Boyd, 1974). In addition, the crude protein content of *Typha latifolia* from different sites varied from 4.0 to 11.9 percent (Boyd, 1970), and that of water hyacinth grown on a stabilization pond was 14.8 percent compared with 11.3 percent in samples from a lake (Bagnall *et al.*, 1974). This indicates that the crude protein content increases as the nutrient content of the water in which the plant is grown increases. According to Wolverton and McDonald (1976b), the crude protein of water hyacinth leaves grown in waste-water lagoons averaged 32.9 percent dry weight, which is comparable with the protein content of soya bean and cottonseed meal. This value is more than twice the maximum crude protein content of water hyacinth reported by Bagnall *et al.* (1974). Duckweeds have a crude protein content over 30 percent and have a better spectrum of amino acids, as regards to lysine and methionine, than do the other plants (Oron *et al.*, 1986).

Although the protein contents of various aquatic weeds differ greatly, the amino acid composition of the protein is relatively constant, nutritionally balanced, and similar to many forage crops. But the levels of methionine and lysine—generally considered as the limiting amino acids in plant protein—are lower than in terrestrial crops (Boyd, 1970).

7.3.3 Mineral content

The ash content of aquatic weeds varies with location and season. Sand, silt, and encrusted insoluble carbonates from the water account for much of the mineral content. Although silt can be washed off the plants, in practice it represents part of the chemical composition of the harvest. The amount of minerals varies from 8 to 60 percent of the harvest (dry weight), depending on the waterway's chemical content and turbidity (Table 7.2).

The amounts of P, magnesium, sodium, S, manganese, copper, and zinc in aquatic weeds growing in nature are similar to those in terrestrial plants. However, aquatic plants are often richer in iron, calcium, and K than land forages, and some

Table 7.2 Chemical composition of water hyacinth taken from several natural Florida, U.S.A., locations (composition reported as percentage dry weight)

Origin	Ash	C	N	C/N ratio	P	K	Ca	Mg	Na
Lake Istokpoga (Sebring)	24.4	18.0	1.08	16.7	0.14	1.00	0.73	0.38	0.15
Lake Eden Canal (SR 532)	19.4	28.8	0.86	33.5	0.09	1.95	0.46	0.31	0.23
Lake Thonotosassa	23.0	23.0	1.17	19.7	0.33	3.35	1.49	0.29	0.21
Waverly Creek (SR 60)	25.0	33.1	2.26	14.6	0.56	3.10	1.58	0.50	0.37
Arbuckle Creek	23.4	34.9	1.90	18.4	0.23	3.35	1.06	0.49	0.28
Lake Tohopekaliga (Kissimmee)	21.7	34.0	1.69	20.1	0.60	4.70	1.56	0.71	0.53
Lake Monroe (Sanford)	20.4	32.5	2.86	11.4	0.59	5.55	1.73	0.54	0.83
Duda Canal No. 1 (Belle Glade)	20.3	39.1	1.30	30.1	0.13	3.80	1.99	0.60	0.48
St Johns River (Astor)	20.1	36.4	2.33	15.6	0.51	6.50	1.43	0.51	0.63
W. R. Grace Landfill (Bartow)	19.0	36.4	1.86	19.6	0.59	2.72	1.99	0.56	1.54
Ponce de Leon Springs	18.5	37.5	1.74	21.5	0.33	5.40	2.34	0.50	0.47
Waverly Creek (SR 540)	18.5	38.1	1.76	21.6	0.32	4.85	1.45	0.55	0.67
Duda Canal No. 2 (Belle Glade)	17.5	37.8	1.66	22.8	0.15	4.70	2.28	0.69	0.57
Lake Alive (North of Florida)	17.3	38.6	1.17	33.0	0.40	3.66	2.41	0.69	0.40
Lake Apopka (Monteverde I)	15.8	38.8	1.22	31.8	0.14	4.26	2.07	0.54	0.41
St Johns River (Palatka)	15.8	38.0	1.82	20.9	0.16	3.44	1.83	0.73	0.86
Lake George	15.4	40.2	1.48	27.1	0.21	3.21	1.91	1.86	1.24
Lake Apopka (Monteverde II)	14.9	39.8	1.36	29.3	0.09	4.08	1.96	0.60	0.21
Lake East Tohopekaliga (St Cloud)	14.7	37.2	1.08	34.5	0.23	2.90	1.19	0.51	0.53
Mean	19.2	34.9	1.61	23.3	0.31	3.81	1.66	0.56	0.56
Standard deviation	3.2	5.9	0.50	7.0	0.18	1.30	0.53	0.14	0.36

From Taylor and Stewart (1978); reproduced by permission of Aquatic Plant Management Society, Inc.

species are known to concentrate such minerals. Element levels can be exceptionally high in aquatic plants grown in sewage or industrial and agricultural wastewaters. In fact, the low palatability of aquatic weeds to livestock has been attributed to a high mineral content (see subsection 7.6.2).

7.3.4 Miscellaneous

Some aquatic plants have carotene and xanthophyll pigment concentrations that equal or exceed those of terrestrial forages such as alfalfa. These pigments are important ingredients in poultry rations.

Pesticides have been found in aquatic plant samples collected when the waterways have been recently treated with herbicides or insecticides. Traces of cyanide, oxalate, and nitrate have also been found. However, no evidence of toxicity to mice, sheep, or cattle has yet been found in water hyacinth or hydrilla samples.

7.4 PRODUCTIVITY AND PROBLEMS CAUSED BY AQUATIC WEEDS

Aquatic weeds, especially rooted, emergent, and floating species, are some of the most productive freshwater ecosystems. Typical values for the net production of different types of aquatic vegetation from fertile sites (recorded in terms of unit weight per unit area of the earth's surface per unit time) are as follows (Westlake, 1963): lake phytoplankton 1–8, submerged weeds 3–18, and emergent weeds 27–77 tons of dry organic matter/(ha-year). At that time the highest net productivity recorded was for sugar cane, 85 tons dry matter/(ha-year).

The productivity of submerged weeds is usually low because the water reflects and absorbs some of the incident light, colored substances in the water absorb light, and the diffusion of carbon dioxide in solution is slow compared with its diffusion in air. The presence of phytoplankton in the water column also reduces the light available for submerged plants, and, in eutrophic waters, may be dense enough to cause the elimination of aquatic weeds. On the other hand, emergent weeds are particularly productive as they make the best use of all three possible states, with their roots in sediments beneath water and with the photosynthetic parts of the plant in the air. The mud surrounding the roots may be a good source of soluble nutrients which can diffuse to the roots via the pore water in the sediments. Light and carbon dioxide are more readily available in the air than in water. Thus they make the best of both aquatic and terrestrial environments.

Productivity of aquatic weeds has been reported in many ways, making comparison of results difficult. When evaluating aquatic weed for biomass production potential, growth rates measured as an increase in weight per unit area and time are the most useful. For water hyacinth, which is an exceptionally productive plant (it is a warm-water species with submerged roots and aerial leaves like emergent weeds), yields as high as 135 tons/(ha-year) have been predicted by some estimates (Westlake, 1963; Yount and Crossman, 1970; Wolverton and

McDonald, 1978). Growth rates of 600 kg dry matter/(ha-day) (which can be extrapolated to 199 tons dry matter/(ha-year) with a year-round growing season) have been recorded for water hyacinths grown on sewage effluent (Wolverton and McDonald, 1979). However, many of these optimistic projections are based on growth data from idealized conditions (high nutrient levels in the water and warm summer months) and, as such, are not realistic for annual yield projections in natural environments. More realistic projections place annual yields in a range of 35–90 tons/ha (DeBusk *et al.*, 1977; Reddy and Sutton, 1984). These yields can be obtained with minimum management inputs, and they far exceed the biomass yields of many subtropical terrestrial, saltwater, and freshwater plants (Wolverton and McDonald, 1978).

The growth rate for aquatic weeds can also be expressed as the specific growth rate independent of biomass or the units in which biomass is measured (dry weight). The values can be measured as fractional increase per day or percentage increase per day. In 1969, Bock calculated the daily incremental factor using the following formula (Mitchell, 1974):

$$N_t = N_0 \cdot x^t \tag{7.1}$$

where N_0 = initial number of plants,
N_t = final number of plants,
t = time interval in days,
x = daily incremental factor.

However, the specific growth rates are high at low plant density and decrease as plant density increases.

Growth of aquatic weeds can also be reported in terms of doubling time, the time taken for the material present to double itself. In 1974, Gaudet, while working with *Salvinia minima* and *S. molesta* under standardized culture conditions, calculated the doubling time using the following formula (Mitchell, 1974):

$$\text{Doubling time} = \frac{\ln 2}{(\ln N_t - \ln N_0)/t} \tag{7.2}$$

The notations are the same as used in eq. 7.1.

Under favorable conditions the area doubling time for water hyacinth ranges between 11 and 18 days, depending on the weather. Mitchell (1974) obtained doubling times for *S. molesta* of 4.6–8.9 days in culture solutions in the laboratory, compared with 8.6 days on Lake Kariba. Bagnall *et al.* (1974) and Cornwell *et al.* (1977) both reported doubling times of 6.2 days for water hyacinth grown on a stabilization pond receiving secondarily treated effluent. This value was about double the rate reported under natural conditions for the same species. The variations of doubling times reported in the literature are due to effects of weather, nutrients, growing season, and plant density.

Owing to their prolific growth, aquatic weeds can cause many problems, the major ones being listed below:

Figure 7.5 A drainage canal (klong) near Bangkok, Thailand, filled with water hyacinth, *Eichhornia crassipes*, and the emergent weed, bulrush, *Scirpus grossus*. The aquatic weeds block the canal and interfere with the use of the waterway for washing, drinking, and transport

1. water loss by evapotranspiration, ranging from 15 to 40 cm/month;
2. clogging of irrigation pumps and hydroelectric schemes;
3. obstruction of water flow (Figure 7.5);
4. reduction of fish yields and prevention of fishing activities;
5. interference with navigation;
6. public health problems, as aquatic weeds can become the habitat for several vectors;
7. retardation of growth of cultivated aquatic macrophyte crops, e.g. rice and water chestnuts;
8. conversion of shallow inland waters to swamps.

The problem of aquatic weed infestation is global, but is particularly severe in the tropics and subtropics where elevated temperatures favor year-round or long growing seasons, respectively. The annual world cost of attempts to control aquatic weeds is nearly US$2,000 million (Pirie, 1978).

Currently, the most serious problems associated with aquatic weeds are caused by water hyacinth, which is now more or less ubiquitous in warm waters. In the tropical and subtropical southeastern U.S.A. there is a serious water hyacinth problem; in Florida alone more than 40,000 ha are covered by the plant despite a continuous control program costing US$10–15 million annually. Subsistence-level farmers in the wet lowlands of Bangladesh annually face disaster when rafts of water hyacinth weighing up to 270 tons/ha are carried over their rice paddies by floodwaters. The plants remain on the germinating rice and kill it as the floods recede. In India, large irrigation projects have been rendered useless by plants that block canals, reducing water flow significantly.

Water hyacinth came originally from South America, where it causes few problems as it is kept in check by periodic flooding and changes in water level. The plants are flushed out as the water body enlarges due to seasonal flooding, and as the floods subside the aquatic plants are left stranded on dry land above the receding water level (Mitchell, 1976). The absence of natural enemies in their new environments has often been implicated as a causal factor in the rampant growth of water hyacinth and other aquatic weeds. Therefore, the absence of periodic flooding in artificial lakes and irrigation schemes may be the major contributing factor to the development of the aquatic weed problem. This problem is further exacerbated by eutrophication from human, animal, and agro-industrial wastes, and agricultural runoff (Figure 7.5). As new lakes and irrigation schemes are developed, the newly submerged soil and vegetation provide a rich source of nutrients favoring aquatic plant growth (Little, 1968).

Another problematical aquatic weed is the fern *Salvinia molesta*. On Lake Kariba, Africa, the largest man-made lake in the world, there was a steady increase in the area of the lake colonized by the fern following the closure of the dam in 1959, when 1,000 km^2 or 2.5 percent of the lake surface was covered. Since 1964 (2 years after the dam was opened in 1962) the area has fluctuated between 600 and 850 km^2 and is limited mainly by wave action, which has increased as the lake has reached full size (Mitchell, 1974). The same species is a serious threat to rice cultivation throughout western Sri Lanka, and covers about 12,000 ha of swamp and paddy fields (Dassanayake, 1976).

7.5 HARVESTING, PROCESSING AND USES

Harvesting aquatic weeds from the waterways and utilizing them to defray the cost of removal is one of the most successful approaches to aquatic weed management. It results in weed-free waterways while providing an extensive vegetation resource. This is especially advantageous in developing countries where forage and fertilizer are in short supply.

The removal and recycling of nutrients and other components are done by harvesting, processing, and utilizing the aquatic plants in which the nutrients are collected. The whole process is shown by a simple diagram given below:

Aquatic weeds → Harvesting → Processing → Product → Uses

A complete network of the possible processes used to convert aquatic weeds to a variety of products is shown in Figure 7.6. Harvesting and chopping are precursors to all of the other processes and all of the products. The chopped plants can be applied directly to land, composted to stabilize and reduce mass, digested to produce methane, pulped to produce paper, or pressed to reduce moisture content and produce a highly reactive protein–mineral–sugar juice. The juice can be separated to recover a high-quality food–feed protein–mineral concentrate, or can be digested to produce methane. The pressed fibers may be ensiled with appropriate additives or dried to produce a granular feed component.

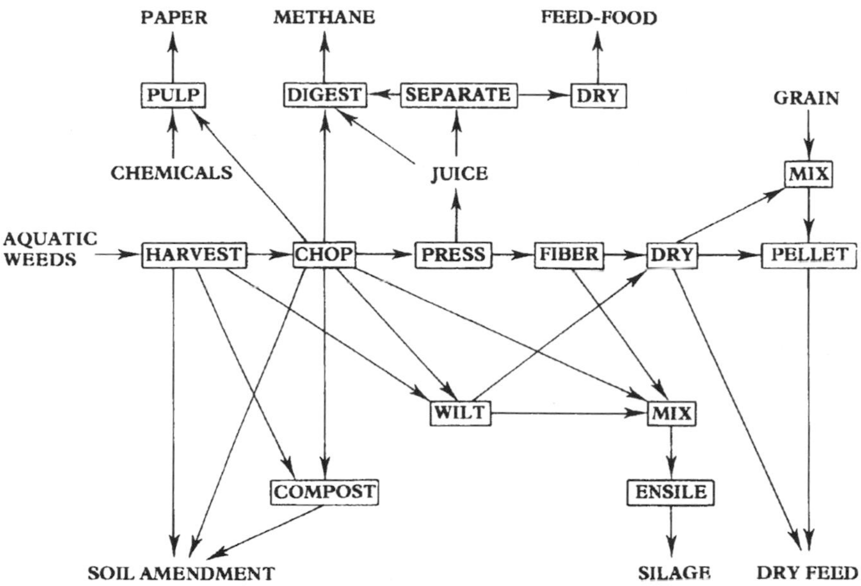

Figure 7.6 Different end-products of harvested aquatic weeds (from Bagnall, 1979; reproduced by permission of the American Society of Agricultural Engineers)

This section deals with all uses of harvested weeds except the use of weeds as food and feed, which is dealt with separately in section 7.6 (Food potential of aquatic weeds).

7.5.1 Harvesting

Harvesting can be accomplished manually or mechanically, depending on the amount of weeds to be harvested and the level of available local technology.

Large floating plants such as *Eichhornia crassipes* and *Pistia stratiotes* can be lifted from the water by hand or with a hay-fork. Smaller floating plants such as *Lemna* or *Azolla* can best be removed with small-mesh sieves or dip-nets. Submerged plants can be harvested by pulling rakes through the underwater plants. Emergent and floating-leaved plants can be cut at the desired height with knives, or in areas with loose bottom soil, pulled by hand. One person can harvest 1,500 kg or more fresh weight of plants per day from moderately dense stands of most species.

Large-scale harvesting is done with the help of various mechanical devices mounted on boats or barges. Harvesting can be carried out from a site at the water's edge or with a self-propelled floating harvester. Shoreside harvesting requires that the plants float to the harvester. Rooted species must be uprooted or mowed and then moved to the harvester by boat or by wind and current. Plants can then be lifted from the water by hand, crane, mechanical conveyor, or pump. On the other hand, mobile harvester sever, lift, and carry plants to the shore. Consisting basically of a large barge, on which a belt-like conveyor and a suitable cutting mechanism is mounted, these harvesters are mostly intended for harvesting submerged plants. Nowadays, some harvesters have also been designed to harvest floating plants or the mowed tops of submerged plants.

Harvesting frequency has to be chosen depending on the initial area of coverage of the plants (or stocking density) and the area doubling time. For example, if a plant is stocked to cover half of the pond area harvesting may be done to remove half of the plants at every interval of doubling time. Plants such as duckweeds, which are easily piled up due to wind or could not compete with algae for space and food, have to be stocked at a higher initial stocking density to prevent algal blooms.

After the weeds have been harvested their handling is another problem. Because of their enormous water content (which results in very low bulk density), the weeds are exceedingly difficult to handle and their transportation is very expensive. Thus, choppers are often incorporated into harvesting machinery designed for aquatic weeds. Chopping makes the plants much easier to handle and reduces their bulk to less than a fourth of the original volume, greatly simplifying transportation, processing, and storage.

The energy cost of combined harvesting–chopping amounts to about 0.34 kW-h/ton and the economic cost to about US$0.14/ton of fresh plants (Bagnall, 1979).

7.5.2 Dewatering

Because harvested or chopped aquatic weeds contain 80–95 percent water they should be dewatered prior to being used as animal feed or for other purposes. Although sun-drying can be employed, this must occur so rapidly that mold and decay do not ruin it, and this practice is not possible all year round.

About half the moisture is on the surface and some is loosely contained in the vascular system. This water can be removed relatively easily by lightly pressing the plant. The squeezed-out water contains only about 2 percent of the plant solids and often can be returned directly to the waterway without causing pollution. Even with half the water removed, aquatic vegetation is still much wetter than terrestrial grass. In order to reduce the moisture further, heavier pressing is required. This process can remove about 70 percent of the water content, yielding a product that is comparable with terrestrial forage grasses in moisture content.

Depending on plant species, process design and operating conditions, the water removed can carry with it 10–30 percent of the plant's solid matter, 15–35 percent of the protein, and up to 50 percent of the ash (for the most part, silt and mineral encrustations caught on the plants). Roller, belt, cone, and screw presses are the different types commercially available to dewater weeds, but they are usually in heavy, durable designs that are unnecessarily complex and cumbersome for pressing aquatic vegetation. Lightweight experimental screw presses (suitable for developing countries) have been designed. For example, a small screw press of a simple design has been constructed at the University of Florida (NAS, 1976). With its 23 cm bore this press weighs 200–250 kg. Complete with a power plant it can be carried by a truck, trailer or barge to remote locations, and can press 4 tons of chopped water hyacinth per hour. Such presses are compatible with a program of manual harvest and with the small-scale needs for animal feed in rural areas in developing countries. In addition, these presses can be manufactured locally.

Estimates of aquatic weed pressing costs range from US$5 to 10/ton of dry matter, depending on machine cost, machine use, production rate, amortization time, and labor costs (Bagnall *et al.*, 1974; NAS, 1976; Bagnall, 1979).

In some arid climates solar drying may be feasible, either by exposing a thin layer of the plants to the sun or by more sophisticated solar collector–heat transfer systems. In the U.S. a solar drying system for aquatic weeds is reportedly built at the National Space Technology Laboratories. Depending on the weather, it will be capable of drying 15–17 tons of newly harvested water hyacinth in 36–120 h (Figure 7.7). Drying of the aquatic weeds with conventional fuel-heated air is economically impractical because of the high moisture content.

7.5.3 Uses of aquatic weeds

Mineral fertilizers are expensive for many farmers in developing countries, yet there is a greater need now than ever before to increase food production. As an alternative, the 1974 World Food Conference, and more recently the Food and Agriculture Organization (FAO) of the United Nations, have stressed the urgency of reassessing organic fertilization. This includes the possible ways in which aquatic weeds can be used as organic fertilizers, namely as mulch and organic fertilizer, ash, green manure, compost, or biogas slurry.

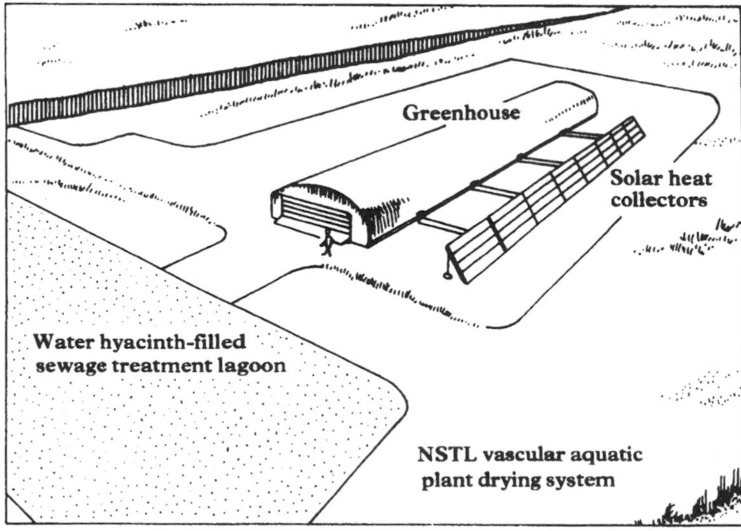

Figure 7.7 Solar energy being harnessed to dry aquatic weed at the National Space Technology Laboratories (NSTL) near Bay St Louis, Mississippi (NAS, 1976). *Note*: This system is connected to a lagoon full of water hyacinth growing in sewage effluent

MULCH AND ORGANIC FERTILIZER

Mulching involves the laying of plant material on the surface of the soil to reduce evaporation and erosion, to smother weeds and for temperature control. Both sand and clay soils need conditioning to make them productive. Working plant material into the soil improves its texture, and also, by acting as manure, improves the nutrient content. Several species of aquatic weeds that can be used as manure are *Pistia stratiotes*, *Hydrilla verticillata*, *Salvinia* spp., *Eichhornia crassipes* and *Azolla* (Frank, 1976; Gupta and Lamba, 1976; Little, 1968; Gopal and Sharma, 1981).

The high moisture content of aquatic weeds makes them suitable as a mulch. But the time and labor involved in harvesting and distributing even sun-dried material would preclude such a use, except on a small scale adjacent to the watercourse in which the aquatic weeds occur.

ASH

Although it has been suggested that the ash of water hyacinth can be used as a plant fertilizer, some reasons showing that this is not feasible are:

1. Burning of the plant to ash results in the loss of N and organic matter, which reduces the fertilization potential of the plant.
2. Burning of the plant requires previous drying, which involves additional labor and expenditure and is restricted to dry days of the year, while growth and distribution of the pest is most profuse in wet months.
3. Ash needs immediate bagging and storing in order to prevent it from being washed away by rain or carried away by wind.

Thus the cost of labor and energy required to obtain ash from aquatic weeds seems to exceed by far the value of the ash obtained as a fertilizer (Little, 1968).

GREEN MANURE

Green manure, in a strict sense, is plant matter cultivated for its fertilizer value to other crops. However, certain species of aquatic weeds which grow wild in rice fields and are ploughed into paddy, e.g. *Limnocharis flava* and *Sesbania bispinnosa*, are sometimes referred to as green manure. Thus the distinction between aquatic weeds which grow wild and are used as manure and fertilizer, and green manure which is cultivated, is not always maintained. Certain types of aquatic weeds are cultivated as green manure or biofertilizers to add N to the soil, and this practice is useful since it lessens dependence on commercial, inorganic fertilizer.

The cultivation of the fern *Azolla pinnata*, with its symbiotic N-fixing blue-green alga, was developed in Vietnam, where it is treated not as a weed but as a valuable crop (Ngo, 1984). Edwards (1980) cited literature which reported that the cultivation of the fern has spread to southern China. In both countries *Azolla* is produced to fertilize areas of rice and other crops. In northern Vietnam, just before or after rice transplanting, *Azolla* is scattered in the fields at the rate of 4–5 tons/ha, and in January and February it grows along with the rice. During this time, when the mean daily air temperature is 16–17 °C, the paddy is young, and available solar radiation and nutrients allow *Azolla* to grow normally, with the result that it completely covers the surface of the water. Towards the end of March, when the temperature rises to 22–24 °C, the paddy has grown up and shades most of the surface, resulting in the death of *Azolla* and release of P and K and newly acquired N to the paddy. The *Azolla* produces about 18 tons of fresh material/ha, and assimilates more than 100 kg N/ha in the 3–4 month growing period. A negligible portion of the fixed N is released when *Azolla* is growing, and it becomes available only on the death of the plant.

Before being applied to agricultural lands and fields, *Azolla* is produced in small, special fields where it receives appropriate fertilization and care. In these fields the *Azolla* biomass can double every 3–5 days in good weather. A 1 ha field fully covered with *Azolla* can produce 1–4 tons of fresh biomass per day, which contains 2.6–2.8 kg of pure N (Ngo, 1984). Within 3 months enough green manure to fertilize 4,000 ha of rice fields can be produced from 1 ha of *Azolla* fields. Thus, if year-round production of *Azolla* is maintained, 1 ha of *Azolla* can save 5–7 tons of costly imported chemical fertilizer (ammonium sulfate) per year. Ngo (1984) reported that increases in yield of rice by 30–50 percent, of sweet potato by 50–100 percent, and of corn by 30 percent, have been achieved in fields green-manured with *Azolla*.

Because *Azolla* is cultivated as green manure only in limited areas of Asia, there can be management problems in other areas. It grows in abundance in Japan, India, China, the Philippines, and Thailand, and tropical strains that grow at 30–35 °C are also found. Because these tropical, local varieties are more heat-resistant than the one found in northern Vietnam, they could be developed to become a high-yield and high N-content crop. However, experimentation is needed to determine the full potential of *Azolla* in tropical areas.

Attempts have also been made to use free-living, filamentous, N-fixing blue-green algae to improve fertility of rice fields in Japan and India, but they are still in an experimental stage.

COMPOSTING

One of the most promising methods to utilize aquatic weeds is to use them to make compost. Composting requires about 60–70 percent moisture content (see Chapter 3) and this can be achieved by a few days of wilting in the sun, a great saving over the other drying methods in terms of cost and equipment. The weeds also generally contain adequate nutrients (C/N ratio between 20 and 30, see Table 7.2), which favors the growth of microbes that produce compost. As previously stated in Chapter 3, the nutrient contents of compost products are usually several times less than inorganic fertilizers. However, when applied to land, about 25–30 percent of inorganic fertilizer nutrients can leach to groundwater or are not available to the crops. The compost N is largely in organic form (or cell biomass) and is released into the soil gradually. It is thus available throughout the growing season.

Compost can be used as a soil conditioner, and as an organic fertilizer to raise phytoplankton in fish ponds. The phytoplankton production was found to be directly related to nutrient content of the composts, and field trials using composted water hyacinth in fish ponds resulted in a considerable yield of tilapia (a herbivorous fish) as shown in Table 3.10 (Polprasert, 1984).

PULP, FIBER, AND PAPER

Aquatic weeds have found use in producing fibers and pulp for making paper. In Romania *Phragmites communis* (the common reed) has been used to make printing paper, cellophane, cardboard, and various synthetic fibers. Wood pulp is mixed with the reed pulp to increase the tear strength and the density of paper. In addition, the raw weed and pulp mill wastes yield a variety of other products, notably cemented reed blocks, compressed fiber board, furfural, alcohol, fuel, insulation material, and fertilizer (NAS, 1976). For centuries common reed has been used in peasant crafts, thatching, fences, and windbreaks. Its stems are used in basketwork, as firewood, fishing rods, weavers' spool, and as mouthpieces for musical instruments.

Cattail is another weed that can be used as a source of pulp, paper, and fiber. Its leaves and stems are suitable for paper-making, and the paper obtained is fairly strong but difficult to bleach. Bleaching is, however, not essential for the production of wrapping paper, which is in great demand nowadays. Cattail leaves yield a soft fiber that can be used in mats, baskets, chair seats, and other woven articles. Because the leaves swell when wet, they are good for caulking cracks in houses, barrels, and boats. Studies in Mexico have shown that woven cattail leaves coated with plastic resins have potential as place mats, building walls, and roof tiles. The resulting product is as strong as fiberglass (NAS, 1976).

At the University of Florida, attempts were carried out to make paper from water hyacinth (Bagnall *et al.*, 1974). Different pulping materials and conditions were used, but it failed to produce suitable pulp. Reasons for the failure were:

1. very low fiber yield;
2. excessive shrinkage of the paper on drying, and subsequent wrinkling;
3. brittleness of the paper;
4. paper turning very dark on drying;
5. poor tear properties.

Because of low fiber yield it was felt that the use of water hyacinth for pulp would not be profitable even if it were delivered to the pulp plant at no cost.

However, dried water hyacinth petioles (leaf stalks) are woven into baskets and purses in the Philippines (NAS, 1976). In Thailand, water hyacinth leaves are used as cigar wrappers and for preparing plastic moulded materials such as furniture, electrical insulation board, radio cabinets, etc. (Gopal and Sharma, 1981).

BIOGAS AND POWER ALCOHOL

Chopped water hyacinth alone, without dewatering or mixed with animal manure or human waste, can be anaerobically digested to produce methane gas (see Tables 4.7 and 7.3). Nutrients such as P and K are provided in adequate proportions by water hyacinth. Digestion takes about 10–60 days, and requires skilled supervision. Table 7.3 shows the production of several volatile fatty acids during anaerobic digestion of a mixture of cattle dung and water hyacinth. The highest biogas production and methane content were achieved during the digestion period of 30–60 days. It is estimated that water hyacinth harvested for 1 ha will produce approximately 70,000 m^3 of biogas.

Each kg of water hyacinth (dry weight) yields about 370 L of biogas with an average methane content of 60 percent, and the fuel value of about 22,000 kJ/m^3. Temperature has a marked effect on biogas production. The biogas was produced quickly and had higher methane content (69.2 percent) at 36 °C than 25 °C (59.9 percent). In the same study, Wolverton *et al.* (1975) noted that the rate of methane production was higher in plants contaminated with nickel and Cd than in uncontaminated plants.

The use of dried water hyacinth as fuel in Indian villages is common. Gopal and Sharma (1981) reported that, in 1931, Sen and Chatterjee were the first to demonstrate the possibility of using water hyacinth for generation of power alcohol and fuel gas. Sen and Chatterjee described three methods of utilizing the plant. In one of the methods the plant is saccharified by acid digestion and subsequent fermentation, yielding 100 kg potassium chloride, 50 L ethanol, and

Table 7.3 Production of volatile fatty acids and methane from laboratory fermentation of cattle dung + water hyacinth (1 : 1 mixing ratio) (Gopal and Sharma, 1981)

Duration (days)	pH	Temperature (°C)	Volatile fatty acids	Average gas produced (L/day)	Methane (%)
1–4	7.0–7.5	30	Acetic, propionic, and butyric acids	0.95	0
5–13	6.5	27–30	Acetic, propionic, butyric acids and ethanol on 7th day	1.22	3–8
14–28	6.5–7.0	26–29	Acetic, propionic, butyric, isobutyric, valeric, and isovaleric acids	0.81	10–60
29–49	7.0	27–28	Propionic acid isovaleric acids	6.01	57–62
50–60	8.0	26–28	—	4.31	60–64
1–60	8.0–8.5	26–28	From cattle dung alone, acetic, propionic, and butyric acids	3.81	50–60

200 kg residual fiber of 8,100 kJ calorific value. In another method the plant is gasified by air and steam to produce 1,150 m^3 of gas (equivalent to 150 kJ/m^3), 40–52 kg ammonium sulfate, and 100 kg potassium chloride. The gas obtained at 800 °C comprises 16.6 percent hydrogen, 4.8 percent methane, 4.1 percent carbon dioxide, 21.7 percent carbon monoxide, and 52.8 percent N. The third method employs bacterial fermentation which produces 750 m^3 gas with a calorific value of 22,750 kJ/m^3. It comprises 22 percent carbon dioxide, 52 percent methane, and 25 percent hydrogen.

The slurry which is the by-product of biogas digestion can be used as organic fertilizer for fish ponds and crops. Field experiments were conducted at the Asian Institute of Technology, Bangkok, in which biogas slurry from four 3.5 m^3 biogas digester units fed with water hyacinth and nightsoil, was added to fish (tilapia) ponds. Extrapolated yields of tilapia were found to be 2,800–3,700 kg/(ha-year) and this scheme was considered to have potential for rural waste recycling in developing countries (Polprasert *et al.*, 1982; Edwards *et al.*, 1988).

These initial studies have generated great interest in biogas and the recovery of fuel from aquatic weeds, especially for rural areas in developing countries. As many developing countries have an inexhaustible supply of aquatic weeds, this potential energy source deserves further investigation before being commercially exploited.

7.6 FEED AND FOOD POTENTIAL

As plants are a starting point of the food chains they are naturally the source of all food for animals and humans. The major pathways involving the use of aquatic weeds in food production are shown in Figure 7.8. The pathways involving composting, mulching, green manuring, ash, and biogas digestion were previously discussed in section 7.5. The remaining pathways are discussed in this section.

7.6.1 Feed for herbivorous fish

There are many species of herbivorous fish that feed on aquatic weeds (Mehta *et al.*, 1976; NAS, 1976; Edwards, 1980), thus converting them to valuable food. They basically fall into three categories:

Grazers—if they eat stems and foliage.

Mowers—if they devour the lower portions of aquatic plants and thus cut them down.

Algal feeders—if they consume algae.

Algal feeders are not considered, as they consume single-cell-protein algae, which is outside the scope of this chapter.

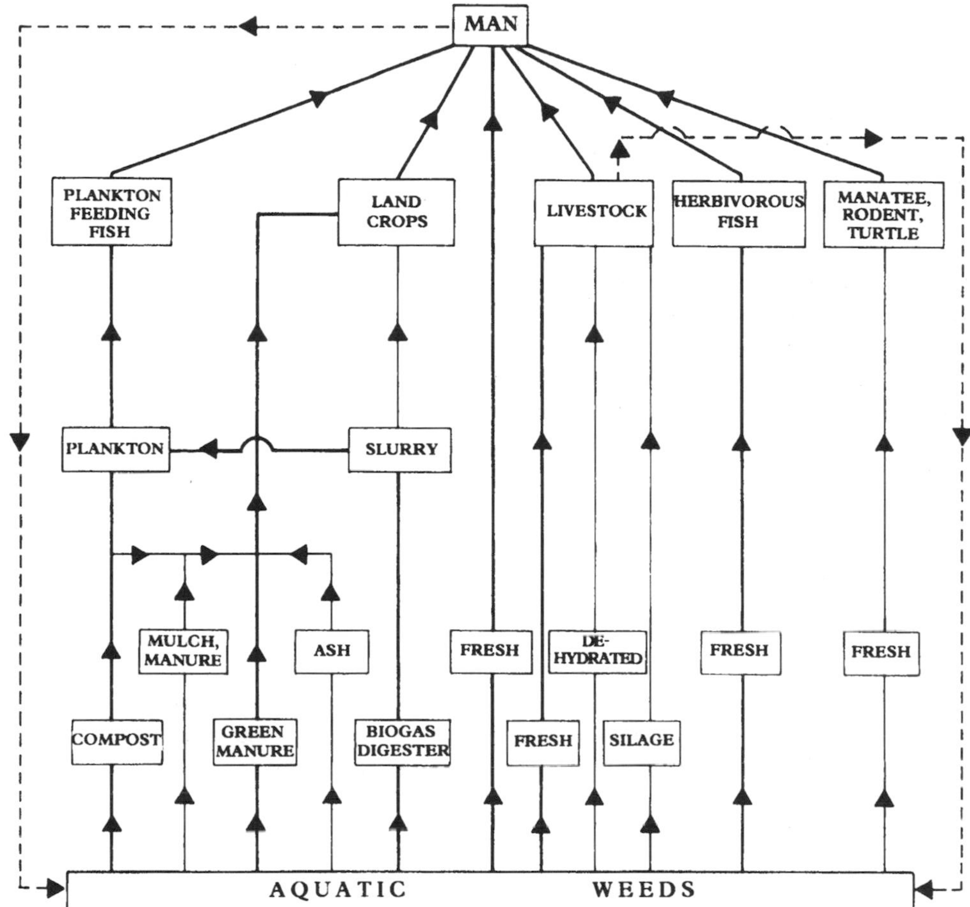

Figure 7.8 A schematic diagram of the major pathways involving aquatic weeds in food production. *Note*: (1) Pathways which may have the greatest potential at present are in a heavier solid line. (2) The broken line indicates that the recycling of livestock and human wastes could play an important role in food production (from Edwards, 1980; reproduced by permission of the International Center for Living Aquatic Resources Management)

CHINESE GRASS CARP

The most promising species for the consumption of aquatic weeds is the Chinese grass carp or white amur, *Ctenopharyngodon idellus* (see Figure 6.2a). It is a fast-growing fish that feeds voraciously on many aquatic weeds and grows to weigh as much as 34 kg. It exhibits tolerance to a wide range of temperatures, to low dissolved oxygen level, as well as to brackish water (Mitchell, 1974).

Being a grazer, the grass carp feeds primarily upon submerged plants, but also eats small floating plants such as duckweed. Submerged weeds consumed by the fish include hydrilla, water milfoil, chara, *Elodea canadensis*, *Potamogeton* spp., *Ceratophyllum demursum*, and most other submerged plants (Mehta *et al.*, 1976; NAS, 1976). Overhanging terrestrial plants and bank grasses are consumed when preferred plants are not available. Small fish, less than 1.2 kg, may eat several times their body weight of plant daily, while large fish consume their body weight daily (NAS, 1976). Thus rapid weed control is achieved if more than 75 fish/ha are present. Stocking rates lower than 75 fish/ha reduce the rate of weed control, but result in larger, more valuable food fish. Grass carp cultured for food yields 164 kg/ha in temperate zones and 1,500 kg/ha in tropical weed-infested water (NAS, 1976).

The main difficulty with grass carp is that it is restricted to southern China and does not reproduce in captivity. Recent studies, however, have shown that spawning can be induced by the injection of fish pituitary hormone, and in Arkansas, U.S.A., artificial spawning is routine. Another problem is that water hyacinth, *Salvinia*, and other floating weeds prevalent in the tropics are not preferred by the grass carp, and it is therefore not a satisfactory agent for controlling them. Neither is it good for controlling tall, hard, emergent aquatic weeds like cattails.

OTHER HERBIVOROUS FISH

Besides the grass carp there are several other species of fish used in the dual role of a source of food and as weed-control agents (Mitchell, 1974; NAS, 1976; Edwards, 1980). Of the different species, *Tilapia* spp. is the most commonly used. This tropical lowland fish is common in Africa and Asia, cheap and easy to breed, grows rapidly, and is a voracious feeder of aquatic weeds. As its flesh is much valued as food, peasant farmers in Africa and Asia culture it in small-scale subsistence and commercial enterprises. Several species of tilapia fish have been grown in different parts of the world, especially Africa and Southeast Asia.

Two South American species, *Metynnis roosevelti* and *Mylossoma argenteum*, both known as 'silver dollar', attack a variety of submerged weeds, especially pond weeds. Dense growths of weed are rapidly removed at stocking densities of 1,200–2,500 fish/ha. Little is known of their potential yield or value as food, although they occur in large numbers and are sought and relished by people along the Amazon River (NAS, 1976).

In addition, a number of other species, namely silver carp, tawes, common carp, tambaqui, pirapitinga, American flagfish, and goldfish are being tested for use as weed control agents.

Considerable research is required to identify techniques for their spawning, culturing, and management; their value as food; and their sensitivity to adverse water quality.

CRAYFISH

Crayfish, or freshwater lobsters, are an under-exploited food source. They are produced commercially in some European countries and a few areas of the U.S.A. A few tribes in New Guinea use them extensively as their major protein source. Boiled in salt water they are a delicacy and a gourmet's delight.

There are over 300 known species and a few are exclusively herbivorous. Such varieties appear promising for aquatic weed control and utilization. *Orconectes causeyi*, a species native to the western U.S.A., has been used experimentally for weed control and was effective against pond weeds. *Orconectes nais* has been shown to control aquatic weeds in Kansas, U.S.A. Beyond that, little is known about weed control by the herbivorous species of crayfish (NAS, 1976). In general, crayfish are thought of more in terms of an available crop associated with aquatic weeds, not as a weed-control agent. Red crayfish (*Procambarus clarkii*) are widely farmed in California and Louisiana in flooded rice fields, and live mainly on aquatic weeds that grow among the rice. The crayfish is too small to eat the rice seedlings at planting, and by the time the crayfish mature the rice plants are too tall and fibrous to be eaten.

Before crayfish are introduced into new areas their effect on rice production should be studied carefully. Imported crayfish have become a problem in Japan and in Hawaii, where rice paddy dikes have been weakened by their burrowing. Some species of crayfish eat tender shoots of newly germinated rice and should be avoided.

7.6.2 Livestock fodder

Aquatic weeds can be used as feed for livestock after suitable processing. Excessive moisture content in fresh plants restricts the ability of animals to obtain adequate nourishment. The palatability of feed processed from aquatic weeds compares poorly with that of most other conventional feeds. A good feed must contain adequate levels of protein, fat, carbohydrate, vitamins, and mineral nutrients for satisfactory growth. The feed should have a fairly low fiber content so that most of the organic matter is highly digestible even to non-ruminant animals.

The approximate compositions of some aquatic weeds which appear suitable as feed are compared with that of alfalfa hay, as given in Table 7.4. As can be seen, the composition of dried samples of many species showed that they were inferior to alfalfa hay for use as livestock feed, but some species were as suitable as, or better than, dried alfalfa. On average, aquatic weeds contained less crude protein, somewhat more ash and fat, and slightly less cellulosic fiber than alfalfa. Submerged and floating plants usually have higher values for crude protein and ash than emergent plants. However, the amino acid composition of protein in aquatic weeds is similar to many forage crops (section 7.3.2).

Figure 7.9 compares the crude protein content of fresh and dried aquatic weeds with that of fresh and dried alfalfa hay. Aquatic weeds compare favorably on a dry mass basis, but not on a fresh mass basis.

Even though nutritional values of aquatic weeds compare favorably (when dried) with alfalfa, the cost of artificial drying, grinding, formulating with other feed to improve palatability, and pelleting makes the cost of feed from aquatic weeds expensive (Frank, 1976; NAS, 1976). Nevertheless, water hyacinth can support the growth of livestock if it is partially dried and properly supplemented, and if the animals are accustomed to it.

Table 7.4 Approximate composition (percentage dry weight) of some aquatic plants and alfalfa hay

Species	Ash	Crude protein[a]	Fat[b]	Cellulose[c]
Eichhornia crassipes[d]	18.0	17.1	3.6	28.2
Pistia stratiotes[d]	21.1	13.1	3.7	26.1
Nelumbo lutea[e]	10.4	13.7	5.2	23.6
Nuphar advena[e]	6.5	20.6	6.2	23.9
Nymphoides aquatica[f]	7.6	9.3	3.3	37.4
Potamogeton diversifolius[f]	22.7	17.3	2.8	30.9
Najas guadalupensis[f]	18.7	22.8	3.8	35.6
Ceratophyllum demersum[f]	20.6	21.7	6.0	27.9
Hydrilla verticillata[f]	27.1	18.0	3.5	32.1
Egeria densa[f]	22.1	20.5	3.3	29.2
Typha latifolia[g]	6.9	10.3	3.9	33.2
Justicia americana[g]	17.4	22.9	3.4	25.9
Sagittaria latifolia[g]	10.3	17.1	6.7	27.6
Alternanthera philoxeroides[g]	13.9	15.6	2.7	21.3
Orontium aquaticum[g]	14.1	19.8	7.8	23.9
Alfalfa hay	8.6	18.6	2.6	23.7

[a] Nitrogen × 6.25
[b] Ether-extractable material
[c] Cellulose values are slightly lower than values for crude fiber
[d] Floating species
[e] Floating-leaved species
[f] Submerged species
[g] Emergent species
Notes:
1. Each value is the average of 3–15 samples
2. All samples represent plants which were in a lush, green stage of growth
From Boyd (1974). *Aquatic Vegetation and its Use and Control*; © UNESCO 1974; reproduced by permission of UNESCO

None of the feeding tests reported in the literature produced evidence of toxins in aquatic weeds. Potentially toxic substances such as nitrates, cyanides, oxidates, tanning, and discoumarins are all present in aquatic weeds, but they also occur in many terrestrial forages, so that in general aquatic plants are no more hazardous to livestock than conventional forages (NAS, 1976). Boyd (1974), however, reported a concentration of tannins of 10 percent or more of the dry weight in some species of aquatic weeds, which would greatly impair the digestibility of their protein.

SILAGE

A promising technique to eliminate the expense of artificially drying aquatic weeds is to convert them to silage (Frank, 1976; NAS, 1976). Ensiling aquatic weeds could become very important in the humid tropics where it is difficult to sun-dry plants to make hay. According to NAS (1976), water hyacinth silage can be made with 85–90 percent moisture content as the fiber retains water, and thus

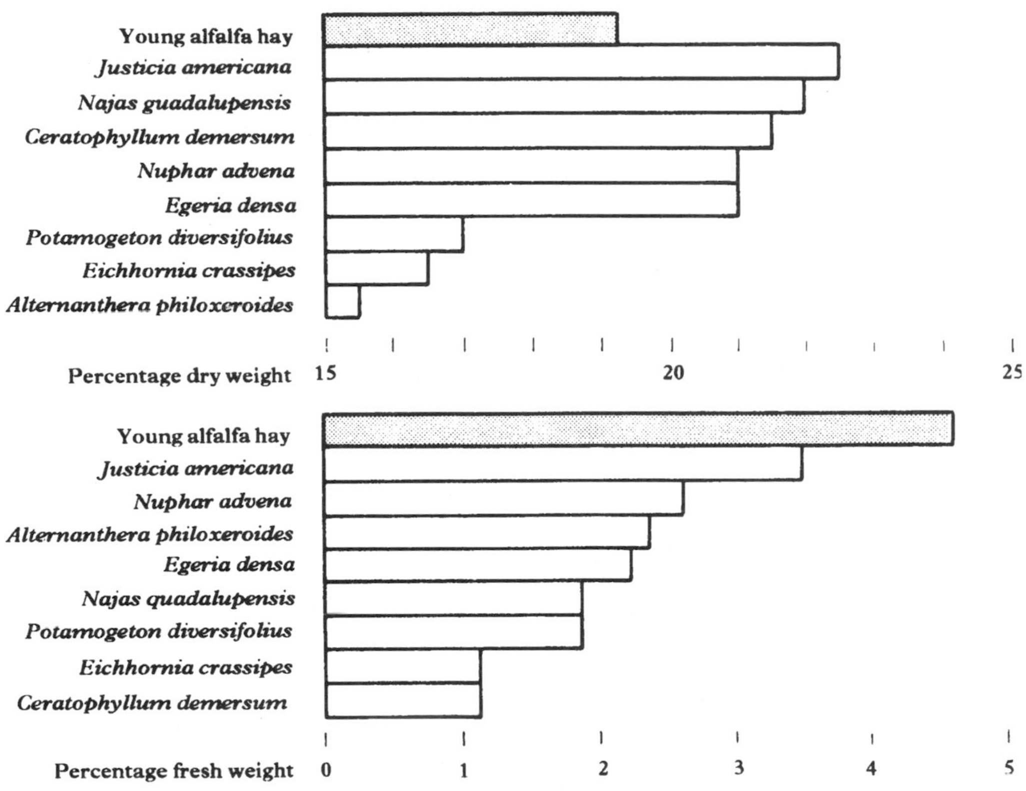

Figure 7.9 Comparison of the percentage crude protein of young alfalfa hay with the crude protein content of eight aquatic weeds (from Boyd, 1974; © UNESCO 1974; reproduced by permission of UNESCO)

the material does not putrefy. But Bagnall *et al.* (1974) found that chopped water hyacinth alone could not be made into silage as it putrefied, and that 50 percent or more of the water had to be pressed from the hyacinth before it could be made into acceptable silage. The aquatic weeds could be wilted in the shade for 48 h, or chopped and pressed to remove some of the water. Also, as silage is bulky, the silos should be located near the animals and supply of plants.

To make silage the aquatic weeds are chopped into small pieces and firmly packed into a silo to produce oxygen-free conditions. Putrefaction is avoided as material is preserved by organic acids such as lactic and acetic acids, which are produced during anaerobic fermentation. The process takes about 20 days, after which the pH falls to about 4. Aquatic plants are often low in fermentable carbohydrates so it is necessary to add either sugar cane, molasses, rice bran, wheat middlings, peanut hulls, cracked corn, or dried citrus pulp, to avoid putrefaction. Silage made from water hyacinth alone is not acceptable to livestock, but the quantity consumed by cattle increases as the level of added carbohydrate is increased, although the addition of sugar cane molasses alone does not improve acceptability. The most acceptable water hyacinth silages to cattle contain 4 percent dried citrus pulp or cracked yellow dent corn (Bagnall *et al.*, 1974).

Polprasert *et al.* (1994) conducted a silaging experiment using a mixture of water hyacinth, molasses, and pig manure at the ratio 85, 10, and 5 percent (wet weight), respectively. No seeding of *Lactobacillus* was used, but the mixture was added with HCl and H_2SO_4 to maintain the acidic pH of 3–4. After about 1 month of fermentation, the silaged product was found to have the following characteristics: dry matter, 17.62 percent; protein, 16.13 percent of dry matter; crude fiber, 13.75 percent of dry matter; crude lipid, 0.75 percent of dry matter; and ash, 17.3 percent of dry matter. It was concluded that this silaged product should be added with rice bran, ground corn or other carbohydrate materials before feeding to animals.

Silage treated with formic acid as a preservative (about 2 L acid/ton pressed water hyacinth) is usually superior to untreated silage as cattle feed. Studies with other organic acid preservatives, e.g. acetic and propionic acids, have also been successful.

Added carbohydrate also functions as an absorbent material which is necessary because of the high water content of the weed. If highly absorbent additives could be found, this may eliminate the need for preliminary dehydration (NAS, 1976).

7.6.3 Human food

Aquatic plants can provide three types of food: foliage for use as a green vegetable; grain or seeds that provide protein, starch or oil; and swollen fleshy roots that provide carbohydrate, mainly starch. There are more than 40 species of aquatic weeds that are edible.

Certain species having potential for more widespread use are Chinese water chestnut (*Eleocharis dulcis*), water spinach (*Ipomoea aquatica*) (Figure 7.10),

Figure 7.10 Water spinach, one aquatic plant that can be used for human consumption

Figure 7.11 Comparing the size of *Spirodella* with the size of *Lemna*

water lilies and taro (*Clocasia esculenta*). Duckweeds, *Spirodela* and *Lemna* (Figure 7.11), which have a high protein content (over 30 percent of dry weight) warrant further study. But before any waste recycling system is started, social acceptance and possible public health hazards have to be investigated.

7.6.4 Feed for other aquatic and amphibious herbivores

The concept of harvesting aquatic weeds *in situ* can be extended to include other herbivorous animals.

Ducks, geese, swans, and other waterfowl forage on vegetation, controlling weeds on the banks of waterways and often clearing aquatic weeds and algae from small lakes, ponds, and canals. In Hawaii, 65 Chinese white goslings were placed in a 1 ha pond which was completely covered with dense paragrass (*Brachiaria mutica*) and cattail (*Typha* spp.) that annually grew 1.8 m above the water. Despite the failure of mechanical and chemical controls to manage the weeds for several previous years, the geese cleared them out in 2.5 years. Also, in 1967, 100 mute swans were added to Nissia Lake near Agras in northern Greece. The lake was being used to produce hydroelectricity, but the turbine inlets were clogged with aquatic vegetation. The swans were placed in the inlet area; they cleared it within a few weeks (NAS, 1976).

These waterfowl can control aquatic weeds noticeably only in small bodies of water such as farm ponds and small lakes. In a larger waterway the number of waterfowl needed to solve the weed problem makes their use impractical. Nonetheless, waterfowl could be used to supplement other weed control efforts such as the use of herbivorous fish or mechanical harvesting.

In addition to being weed-control agents, ducks and geese produce nutritious eggs and highly prized meat. Thus they provide an additional income to farmers, who raise them at little extra cost. However, if not carefully managed ducks and geese can become pests for some crops (especially grains) neighboring their waterway. Where sanitation is poor, salmonellosis sometimes decimates ducks, geese, and swans, and this disease can be transmitted to humans.

Turtles, rodents, and manatees can be used as weed control agents, but they cannot be used for food as they are endangered species and are not usually consumed by people.

7.7 WASTEWATER TREATMENT USING AQUATIC WEEDS

Most aquatic treatment systems consist of one or more shallow ponds in which aquatic weeds are grown to enhance the treatment potential. Typically, each pond will be dominated by one species of plant, but in some cases a variety of plants will be combined for a particular treatment objective.

The presence of aquatic weeds in place of suspended algae is the major physical difference between aquatic treatment systems and waste stabilization ponds. A waste stabilization pond is highly effective if the algal population (often occurring as blooms of extremely high density) can be removed prior to final discharge of the effluent. However, algal biomass is usually carried away with the pond effluent, resulting in an increase in the BOD and the production of inferior-quality effluent. Thus, waste stabilization ponds which sustain good algal growth are often less effective in producing effluents having acceptable levels of suspended solids or BOD. Some of these problems associated with waste stabilization ponds can be overcome by the use of aquatic weeds. Although the latter may have slower growth rates, they can be more readily harvested, especially if they are free-floating, and therefore amenable to a simple mechanized system of continuous harvesting. Moreover, the weed growth tends to shade out the algae and prevents the erratic fluctuations in algal population densities that are a frequent characteristic of algal ponds or waste stabilization ponds.

Contrary to common opinion, the aquatic plants themselves bring about very little actual treatment of the wastewater. Their function is to provide components of the aquatic environment that improve the wastewater treatment capability and/or reliability of that environment. Some specific functions of aquatic plants in aquatic treatment systems are presented in Table 7.5. These functions may be more clearly envisioned and understood by considering them in the light of the morphologies of aquatic plants, some of which are shown in Figures 7.1–7.4.

Aquatic treatment systems are analogous to many more common treatment systems. The principal removal mechanisms are physical sedimentation and bacterial metabolic activity, as is the case in conventional activated sludge and trickling filter systems. Because the plants provide a support medium for bacterial attachment and growth, aquatic treatment systems have thus been equated with a slow-rate, horizontal trickling filter.

Table 7.5 Functions of aquatic plants in aquatic treatment systems

Plant parts	Functions
Roots and/or stems	1. Surfaces on which biofilm bacteria grow in the water column 2. Media for filtration and adsorption of solids
Stems and/or leaves at or above the water surface	1. Attenuate sunlight and thus can prevent the growth of suspended algae 2. Reduce the effects of wind on the water 3. Reduce the transfer of gases between the atmosphere and water 4. Transfer of oxygen from leaves to the root surfaces

From Stowell *et al.*, 1980; reproduced by permission

The fundamental difference between aquatic systems and more conventional technology is the rate of treatment. In conventional systems wastewater is treated rapidly in highly managed environments, whereas in aquatic systems treatment occurs at a relatively slow rate in essentially unmanaged natural environments. The consequences of this difference are:

1. conventional systems require more construction and equipment but less land than aquatic systems;
2. conventional processes are subject to greater operational control and less environmental influence than aquatic process.

Assuming land is available, the comparative economics of conventional and aquatic systems are contrasted in Table 7.6.

Table 7.6 Costs and energy requirements of conventional (Conv.) and aquatic (Aqua.) treatment systems

	Treatment plant size					
	378.5 m^3/day (0.1 mgd)		1892.5 m^3/day (0.5 mgd)		3785 m^3/day (1.0 mgd)	
Item	Conv.[a]	Aqua.[b]	Conv.	Aqua.	Conv.	Aqua.
Capital cost, US$ $\times 10^{-6}$	0.71	0.37	1.23	0.55	1.60	0.90
O&M cost, US$/year $\times 10^{-3}$	35	21	78	48	117	74
Energy, kJ/year $\times 10^{-9}$	0.93	0.53	3.32	1.27	5.06	2.19

[a] Activated sludge + chlorination
[b] Primary clarification + artificial wetlands + chlorination
mgd = million gallons per day; O&M = operation and maintenance; kJ = kilojoules
Adapted from Stowell *et al.*, 1981; reproduced by permission of the American Society of Civil Engineers

Aquatic treatment systems should not be confused with other types of systems that may involve the application of wastewater to wetlands for reasons other than wastewater treatment:

1. aquaculture (growth of organisms having economic value);
2. environmental enhancement (creation of habitat for wildlife);
3. wetlands effluent disposal (nonpoint source disposal of treated wastewater).

The wetlands environment needed for each specific case can be quite varied; therefore a system designed to treat wastewater will be different from that designed for any other purpose.

7.7.1 Wastewater contaminant removal mechanisms

The general mechanisms involved in the removal of contaminants by aquatic weeds are discussed below.

BOD_5 removal

In aquatic systems the BOD_5 associated with settleable solids in a wastewater is removed by sedimentation and anaerobic decay at the pond bottom. The colloidal/soluble BOD_5 remaining in solution is removed as a result of metabolic activity by microorganisms that are:

1. suspended in the water column,
2. attached to the sediments, and
3. attached to the roots and stems of the aquatic plants.

The microbial activity at the roots and stems is the most significant for BOD_5 removal. Reduction of colloidal/soluble BOD_5 will, at least in part, depend on the design of the aquatic system. Direct uptake of BOD_5 by aquatic plants is not significant. Factors affecting the BOD_5 removal rate and efficiency of conventional trickling filters have similar effects on aquatic systems.

The BOD_5 of effluents from aquatic systems will be primarily the result of:

1. extracellular organic compounds produced by plants during the growing season, and
2. organic compounds leached from decaying vegetation during periods of vegetative die-off and dormancy.

These plant-related BOD_5 loads are part of the colloidal/soluble BOD load and should be considered as such in aquatic system design. The release of BOD_5 by plants is species-specific and is affected by environmental factors. The BOD_5 leakage from plants is presently not well quantified in the literature, although releases up to 25 percent of the photosynthetically produced organic matter have been reported (Stowell *et al.*, 1980).

If aquatic systems are not overloaded, effluent BOD_5 concentrations are primarily a function of the plant species grown, the growth phase of the plant, and the wastewater temperature. In such systems, effluent BOD_5 concentrations of 3–10 mg/L during the growing seasons and 5–20 mg/L during periods of dormancy can be expected. An example of the seasonality of aquatic system's effluent BOD_5 concentrations is presented in Figure 7.12. In this example, the magnitude of the winter leakage of BOD is quite high (about 30 mg/L). If the BOD_5 loading rate is either reduced in winter in conjunction with climatically induced reductions in bacterial metabolic rates and in bacterial support structure, or low on a year-round basis so as to avoid overloading the system in winter, then the effluent BOD_5 in winter would be lower than the concentrations reported in Figure 7.12.

Solids removal

Aquatic systems have long hydraulic residence times, generally several days. Consequently, virtually all settleable and floatable solids of wastewater origin are removed.

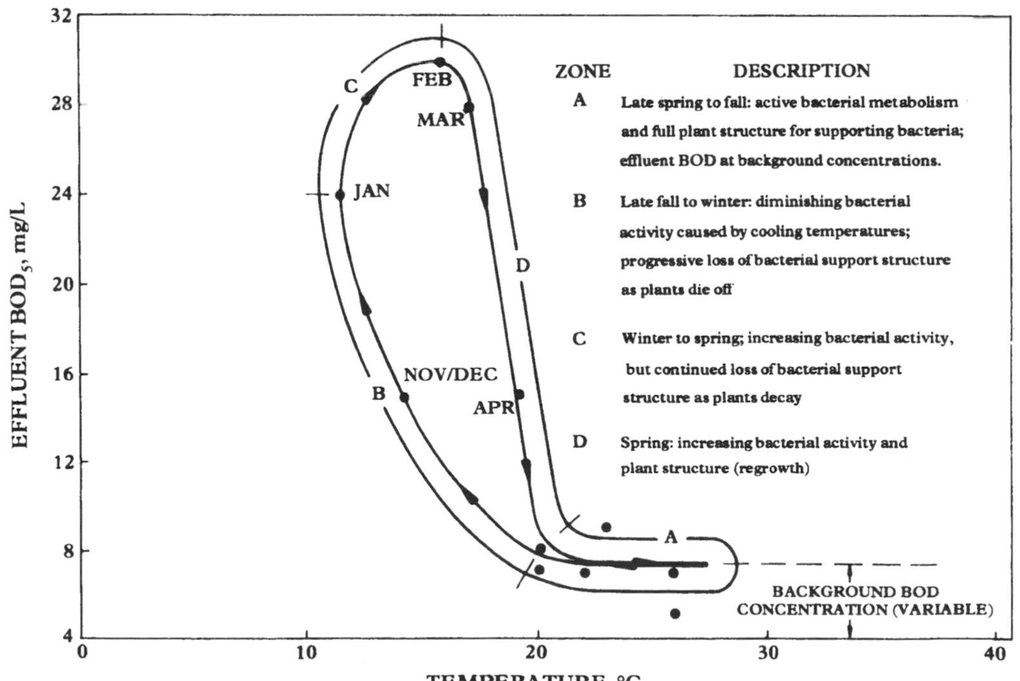

Figure 7.12 An example of the seasonality of effluent BOD_5 concentrations. *Note*: Compiled by Stowell *et al.* (1980), using data from Wolverton and McDonald (1976a). Reproduced by permission

Nonsettling or colloidal solids are removed, at least partially, by a number of mechanisms. Colloidal solids tend to be the foci for bacterial growth. Such growth during the residence time of the water in the aquatic system will result in the settling of some solids and the microbial decay of others. Colloidal solids will also be removed as a result of collisions (inertial and Brownian) with an adsorption to other solids such as plants, pond bottom, and suspended solids.

Ultimate removal of suspended solids will be by bacterial metabolism: anaerobic decay of settled solids and aerobic decay of floating solids entangled in the surfaces of vegetation. The annual build-up of stable residues from these decay processes at the bottom of aquatic systems will be quite less, so that frequent dredging of the pond bottom will not be necessary.

Solids in aquatic system effluents will be decayed or detrital to aquatic plant matter and bacterial flocs. Effluent SS concentrations are a function of water velocity and turbulence in the aquatic system, the type(s) of plant being grown, and the time of the year. An important function of the plants in aquatic systems is attenuation of sufficient light to prevent algae growth. The plants also help reduce the effects of wind (e.g. water turbulence) on aquatic systems. Effluent SS concentrations are normally less than 20 mg/L, typically, and often less than 10 mg/L, particularly during summer and fall.

N removal

N is removed from wastewater by a number of mechanisms:

1. uptake by plants and subsequent harvesting of them,
2. volatilization of ammonia,
3. bacterial nitrification and denitrification.

Of these, bacterial nitrification and denitrification have the greatest N removal potential. To maintain significant populations of the slowly reproducing nitrifying bacteria in an aquatic system, structures to which these organisms can attach are provided by the aquatic plants. For bacterial nitrification (eqs 3.3 and 3.4) to occur the DO concentration of the wastewater must be above 0.6–1.0 mg/L (Metcalf and Eddy, 1991). Thus, the depth of zone below the water surface in which nitrification will occur is a function of BOD loading rate and oxygen flux into the aquatic environment. In estimating the depth of this zone for the purpose of providing a sufficient support structure for the design nitrification rate, the oxygen used in nitrification must be added to that used to metabolize soluble/colloidal BOD. The rate of nitrification also depends on water temperature, and is very slow at temperatures below 10 °C.

Denitrification is an anoxic metabolic pathway used by specific bacterial genera in aquatic environments characterized as having little or no dissolved oxygen, an adequate supply of C for cell synthesis, and neutral pH, as shown below:

$$6NO_3^- + 5CH_3OH \rightarrow 5CO_2 + 3N_2 + 7H_2O + 6OH^- \qquad (7.3)$$

where CH_3OH represents organic C. These criteria are met readily in the bottom sediments and detrital layer of aquatic systems, thus leading to rapid denitrification. The rate of denitrification depends on:

1. the metabolic activity of the bacteria, i.e. environmental factors such as organic C availability, pH, and wastewater temperature;
2. the effective surface area of the bottom sediments; and
3. the potential for the produced N_2 to escape to the atmosphere rather than be fixed in the overlying water of vegetation.

N uptake by plants, and its subsequent removal by harvesting plants, is another mechanism of N removal. The N-uptake is quite high in plants grown in primary sewage effluent, but is very low for plants cultured in secondary sewage effluent. This is due to the following factors:

1. Low concentration of the inorganic N in the effluent.
2. Plant-available N in the secondary sewage effluent is present in NO_3^- (nitrate) form compared with NH_4^+ in the primary sewage effluent. A significant portion of NO_3^- in the secondary sewage effluent is lost through denitrification (Reddy and Sutton, 1984).

Thus harvesting plants, especially those grown in secondary sewage effluents, remove little N directly. The indirect effects of plant harvesting on oxygen transfer rates and the quantity of bacterial support structure in the aerobic zone, however, may affect the rates of bacterial nitrification and denitrification significantly.

P removal

P removal mechanisms in aquatic systems are plant uptake and several chemical adsorption and precipitation reactions (occurring primarily at the sediment/water column interface). In general, chemical adsorption and precipitation are more significant mechanisms of removal (Whigham *et al.*, 1980) because:

1. the potential rates of P removal by these mechanisms are greater than that achievable by plant uptake; and
2. harvesting and disposal of plants are not a necessary part of removing P by chemical means.

However, the removal of P by chemical reactions cannot be predicted accurately because of the many competing and interacting reactions occurring simultaneously in the water column, detrital layer, and sediments. The major factors determining how much, and at what rate, P will be stored in an aquatic system as a result of chemical reactions are pH; redox potential; and the concentrations of iron, aluminum, calcium, and clay minerals. For example, at pH greater than 8, metal phosphates precipitate from solution; while at pH less than 6 and redox potential greater than 200 mV, phosphate absorbs strongly to ferric oxy-hydroxides and similar compounds. Many other phosphate-retaining reactions occur to a greater or lesser extent, depending on the aquatic environment. General conclusions as to what aquatic environmental conditions will maximize P removal have not been reached, and P removals from various aquatic treatment research projects and systems have not been consistent (Wolverton and McDonald, 1976a; Cornwall *et al.*, 1977; Dinges, 1979; Reddy and Sutton, 1984).

Ultimate disposal of P from aqùatic systems is by:

1. harvesting the plants;
2. dredging the sediments; and
3. desorbing and resolubilizing P stored in the sediments and releasing it to the receiving water when it will have the least environmental impact.

Selection of the method will depend on design considerations such as the discharge permit, environmental factors, and the rate of sediment build-up as compared with the rate of P accumulation.

Heavy metals removal

Heavy metals are removed from wastewater during aquatic treatment by:

1. plant uptakes;
2. precipitation as oxides, hydroxides, carbonates, phosphates, and sulfides; and
3. ion exchange with, and adsorption to, sedimented clay and organic compounds.

Although plants can concentrate some metals by over three orders of magnitude above the concentration in water (O'Keefe *et al.*, 1984; Wolverton and McDonald, 1976b), the potential of this removal mechanism is small compared with that of the other removal mechanisms.

The principal means for removing heavy metals during aquatic treatment is by controlling the aquatic environment to optimize the precipitation and ion exchange and absorption removal mechanisms. The rate of removal and storage capacity associated with these mechanisms are functions or redox potential, pH, the presence of clay minerals and insoluble or particulate organic matter, and the concentrations of coprecipitating elements and related compounds such as S, P, iron, aluminum, manganese, and carbonate. Unfortunately, wetland environments are sufficiently complex to preclude accurate theoretical determination of what environmental conditions will result in maximum removal of heavy metals in a given situation. The results of field observations and experiments in this area have not been conclusive.

To some extent heavy metals will be removed in every aquatic system. The extent of removal is affected by aquatic system design and management, as well as wastewater quality (including pretreatment). Resolubilization of sediment-stored heavy metals (intentionally for discharge or accidentally as a result of process upset) can occur. Ultimate disposal of heavy metals from aquatic systems will be by the methods discussed for P in the previous section.

Refractory organic removal

In aquatic systems refractory organic compounds are removed from solution by adsorption to intrasystem surfaces, and are altered chemically by physical, chemical, and biological decay processes. During the residence time of a wastewater in an aquatic system, chemically unstable compounds will disintegrate and the diverse bacterial populations characteristic of aquatic systems will metabolize the more biologically degradable compounds. Refractory organics more resistant to chemical and/or biological decay will be removed, at least partially, by adsorption to the bottom sediments, detrital layer, or 'ooze' layer enveloping the submerged parts of plants. Once adsorbed, decay as a result of bacterial or botanical metabolism and/or physical/chemical processes takes place. Aquatic systems are reported to remove phenolic compounds, chlorinated hydrocarbons, petroleum compounds, and other refractory organics (Dinges, 1979; Seidal, 1976). Quantification of the removal of specific refractory organic compounds from a specific wastewater by means of aquatic treatment methods will require studies (Stowell *et al.*, 1980).

Removal of bacteria and viruses

In aquatic systems concentrations of pathogenic organisms are reduced by prolonged exposure (days) to physical, chemical, and biological factors hostile to these organisms. Physical factors include sedimentation and exposure to ultraviolet (UV) radiation. Chemical factors include oxidation, reduction, and exposure to toxic chemicals, some of which may be excreted by plants (Seidal, 1976). Biological factors include attack by other organisms (particularly bacteria) and natural die-off as a result of being away from a suitable host organism for a long period of time.

However, the extent and reliability of reductions in pathogen concentrations in aquatic systems are unknown. In most wastewater treatment situations some form of disinfection, e.g. chlorination, is necessary to satisfy public health and water quality requirements.

Summary

The principal wastewater contaminant removal mechanisms operative in aquatic treatment systems are summarized in Table 7.7. The effect of each mechanism depends on the design and management of the aquatic system, the quality of the influent wastewater, and climatic environmental factors.

Table 7.7 Contaminant removal mechanisms in aquatic systems

Mechanism	Contaminant affected[a]								Description
	Settleable solids	Colloidal solids	BOD	Nitrogen	Phosphorus	Heavy metals	Refractory organics	Bacteria and viruses	
Physical									
Sedimentation	P	S	I	I	I	I	I	I	Gravitational settling of solids (and constituent contaminants) in pond/marsh settings
Filtration	S	S							Particulates filtered mechanically as water passes through substrate, root masses, or fish
Adsorption		S							Interparticle attractive force (van der Waals force)
Chemical									
Precipitation					P	P			Formation of, or coprecipitation with, insoluble compounds
Adsorption					P	P	S		Adsorption on substrate and plant surfaces
Decomposition							P	P	Decomposition or alteration of less stable compounds by phenomena such as UV irradiation, oxidation, and reduction
Biological									
Bacterial metabolism[b]		P	P	P			P		Removal of colloidal solids and soluble organics by suspended, benthic, and plant-supported bacteria; bacterial nitrification/denitrification
Plant metabolism[b]							S	S	Uptake and metabolism of organics by plants; root excretions may be toxic to organisms of enteric origin
Plant absorption				S	S	S	S		Under proper conditions significant quantities of these contaminants will be taken up by plants
Natural die-off								P	Natural decay of organisms in an unfavorable environment

[a] P = primary effect, S = secondary effect, I = incidental effect (effect occurring incidental to removal of another contaminant)
[b] The term metabolism includes both biosynthesis and catabolic reactions
From Tchobanoglous *et al.*, 1979; reproduced by permission

7.7.2 *Aquatic system design concepts*

Design concepts for aquatic treatment systems differ somewhat from those for conventional systems because aquatic systems have larger surface areas, longer hydraulic detention times, and aquatic plants. The aquatic environment or series of environments necessary to achieve the desired level of wastewater treatment must be envisioned. The plants and equipment to maintain these environments must be selected. The effects of climatic and environmental factors on the aquatic treatment processes must be anticipated and mitigating measures taken as necessary. These concepts and others are part of the rational design of an aquatic treatment system.

Aquatic processing unit (APU)

An APU is an assemblage of plants (and possibly animals) grouped together to achieve specific wastewater treatment objectives. One or more APUs constitute an aquatic system. In this regard APUs are similar to conventional processes making up a conventional system but direct comparisons between APUs and specific types of conventional processes cannot be made in general. However, this does not preclude the use of APUs in conjunction with conventional processes. The conceptual use of APUs to accomplish various treatment objectives is illustrated in Figure 7.13.

The APUs in Figure 7.13 are arranged so that the application is from least to most complex conceptually. In levels (a) and (b) of Figure 7.13 the APUs are used for the removal of nutrients, refractory organics, and/or heavy metals. In contrast to this relatively simple application, the complete treatment of wastewater is envisioned in level (f).

The use of plants and animals in aquatic treatment

The fundamental purpose of plants, animals and the management of these organisms in aquatic systems is to provide components to the aquatic environments that improve the rate and/or reliability of one or more of the contaminant removal mechanisms to be operative by design in the aquatic treatment systems. Plants play a more dominant role than animals in aquatic systems because plants have a greater effect on the aquatic environment and are more adaptable to harsh and/or fluctuating environment.

Aquatic animals are primarily used as biological controls of insect vectors and as bioassay organisms for monitoring the treatment performance of an APU or aquatic system. In certain situations animals can also play a role in plant harvesting and suspended solids removal. In general, however, the direct uptake and/or removal of wastewater contaminants by plants and animals are not significant treatment mechanisms. An observation of interest is that some molecular oxygen produced by the photosynthetic tissue of plants is translocated to the roots and may keep root-tent zone microorganisms metabolizing aerobically, though the surrounding water is anoxic (Tchobanoglous *et al.*, 1979). Reported values of oxygen release rates from the roots of emergent plants are 7.0–8.0 g/day per m^3 of actual root zone and 2.1–5.7 g/day per m^2 of unit surface area of 0.76 m-deep wetland

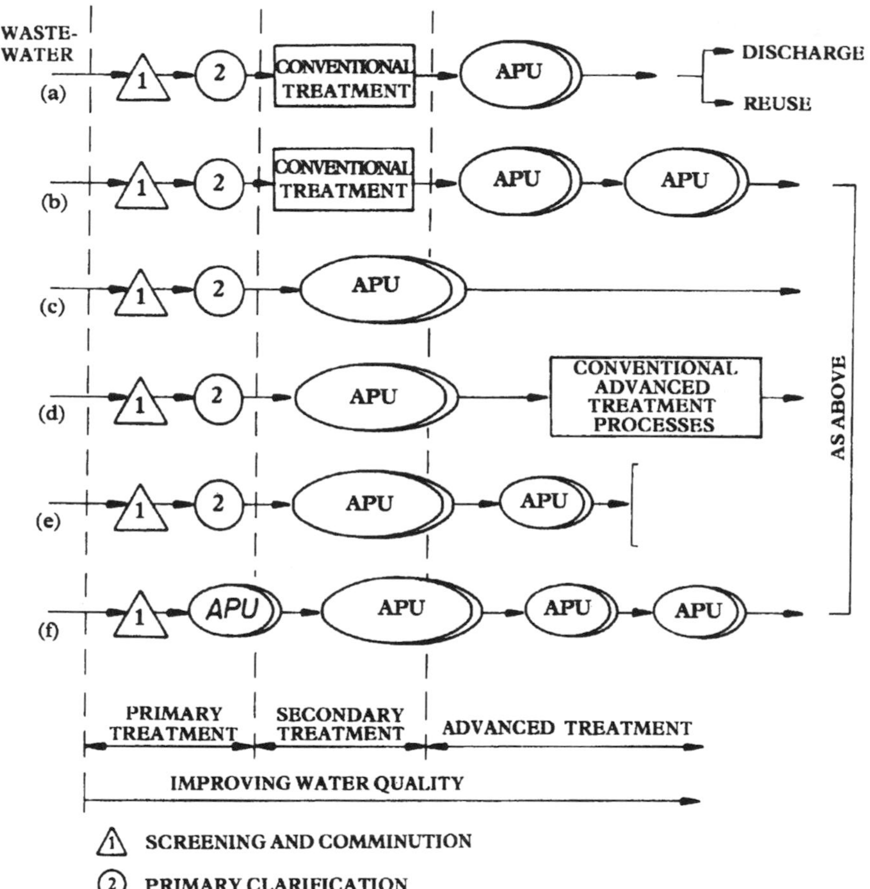

Figure 7.13 Conceptual uses of APUs (aquatic processing units) for the treatment of wastewater (Tchobanoglous *et al.*, 1979)

beds (U.S. EPA, 1993). However, Reed *et al.* (1988) reported the values of oxygen transfer from emergent plants to range from 5 to 45 g O_2/m^2-day, with average being 20 g O_2/m^2-day (Burgoon *et al.*, 1995).

Potential aquatic plants and animals, and their probable role in aquatic systems, are presented in Table 7.9. When selecting organisms for use in an APU the designer must consider not only an organism's effect on the aquatic environment, but also its compatibility with the climate or environment of the design site. Organisms which are incompatible with the climate or environment in which they grow may tend to have unstable populations, resulting in fluctuations in aquatic environmental quality and, ultimately, in diminished APU performance.

The weeds shown in Table 7.9 are by no means the only ones that should be considered for use in an aquatic system. Every aquatic organism has some wastewater treatment potential. Whenever practical, local species should be

Table 7.8 Suggested design criteria for water hyacinth wastewater treatment systems

Parameter	Design value		Expected effluent quality
	Metric	English	
(a) Raw wastewater system			
Hydraulic residence time	> 50 days	> 50 days	$BOD_5 < 30$ mg/L
Hydraulic loading rate	200 m^3/(ha-day)	0.0214 mgad	TSS < 30 mg/L
Maximum depth	< 1.5 m	< 5 ft	
Area of individual basins	0.4 ha	1 acre	
Organic loading rate	< 30 kg of BOD_5/(ha-day)	< 26.7 lb of BOD_5/(acre-day)	
Length to width ratio of hyacinth basin	> 3 : 1	> 3 : 1	
Water temperature	> 10 °C	> 50 °F	
Mosquito control	Essential	Essential	
Diffuser at inlet	Essential	Essential	
Dual systems, each designed to treat total flow	Essential	Essential	
Plant harvesting	Seasonal to annual	Seasonal to annual	
(b) Secondary effluent system			
Hydraulic residence time	> 6 days	> 6 days	$BOD_5 < 10$ mg/L
Hydraulic loading rate	800 m^3/(ha-day)	0.0855 mgad	TSS < 10 mg/L
Maximum depth	0.91 m	3 ft	TP < 5 mg/L
Minimum depth	0.38 m	15 in.	TN < 5 mg/L
Area of individual basins	0.4 ha	1 acre	
Organic loading rate	< 50 kg of BOD_5/(ha-day)	< 44.5 lb of BOD_5/(acre-day)	
Nitrogen loading rate	< 15 kg of TKN/(ha-day)	< 13.4 lb of TKN/(acre-day)	
Length to width ratio of hyacinth basin	> 3 : 1	> 3 : 1	
Water temperature	> 20 °C	> 68 °F	
Mosquito control	Essential	Essential	
Diffuser at inlet	Essential	Essential	
Dual systems, each designed to treat total flow	Essential	Essential	
Plant harvesting	Monthly to weekly	Monthly to weekly	

From O'Brien (1981); reproduced by permission of the American Society of Civil Engineers

Table 7.9 Potential aquatic plants and animals for use in aquatic treatment systems (Tchobanoglous *et al.*, 1979)

Organism	Probable role and remarks
Floating aquatic plants	
Water hyacinth (*Eichhornia* spp.)	Its extensive root system serves as a mechanical filter and a support structure for bacteria. Mats of hyacinth attenuate sufficient light to prevent the growth of algae. Wastewater leaving hyacinth mats is devoid of oxygen, typically. Hyacinths will not winter-over in colder temperate climates. Water hyacinths are potential aquatic pests
Water primrose (*Ludwigia* spp.)	This temperate-climate plant is similar to the water hyacinth, ecologically. The root system is not as extensive as that of the hyacinth, nor is the floating vegetative mat as dense. Water primrose attenuate sufficient light to prevent algae problems. The dissolved oxygen concentration in wastewater leaving primrose mats will depend on BOD loading rate. This plant is a potential nuisance
Duckweed (*Lemna* spp.)	The root system of this plant is very small and will not support an appreciable mass of bacteria. Duckweed grows in dense mats that effectively restrict gas transfer and attenuate light. Ubiquitous in the United States, duckweed is not considered a major aquatic pest. Wind can disrupt duckweed mats. Duckweed can survive throughout the winter in milder temperate climates
Emergent aquatic plants	
Cattails (*Typha* spp.)	The submerged portion of a cattail stand serves as a mechanical filter and a support structure for bacteria. Algae will not grow in dense cattail stands. Water leaving cattail stands is aerobic, typically. Cattails successfully winter-over even in colder climates
Bulrush (*Scirpus* spp.)	Essentially as noted above for cattails, except that stands of bulrush tend to be more open. Bulrushes may be more adaptive than cattails to wastewater environments
Reeds (*Phragmites* spp.)	Reeds are similar to cattails and bulrushes but tend to grow in comparatively open stands. In certain situations algae growth in reed stands could occur
Submerged aquatic plants	
Algae	This broad grouping of very small unicellular plants has very high production rates, but they are difficult and costly to remove from the water once grown. During photosynthesis, molecular oxygen is released into the water at the expense of increasing the BOD of the water. Blue-green algae may increase the nitrogen content of the water by fixing molecular nitrogen. In general, algae are aesthetic and biological nuisances that, if grown, should be removed in subsequent wastewater treatment processes. Various forms of algae will grow throughout the year in open water

Pondweeds (*Potamogeton* spp.)	The value of pondweeds as support structure for bacteria is variable from species to species, as is their potential to compete with, and shade out, algae. Because these plants are for the most part submerged in the wastewater environment, there is greater inherent chance of plant population instability caused by fluctuations in wastewater quality
Other possible aquatic plants	
Water milfoil (*Myriophyllum*) Water velvet (*Azolla*) Coontail (*Ceratophyllum*) Alligator weed (*Alternanthera*) Filamentous green algae	All aquatic species have wastewater treatment potential. The answer to the question 'Which to use?' will depend on the types of aquatic environments necessary to achieve the wastewater treatment objectives, and the potential of the plant species to provide necessary components of these environments reliably. Use of specific plants in specific locales will also depend on whether they will become aquatic pests.
Aquatic animals	
Zooplankton	These organisms feed on algae and other suspended particulates. Their presence and effect are sporadic. The management of zooplankton populations has proven to be difficult
Fish Blackfish Carp Tilapia Catfish White amur Mosquito fish	Fish serve in a role similar to that described for zooplankton. Fish can also be used to reduce the vegetative standing crop, control mosquitoes, or convert plant protein to animal protein. Fish populations are manageable. Fish grown in aquatic systems are a marketable commodity (within the confines of public health regulations)
Bivalves/clams	Clams filter-feed on particulates. To be most effective, clams should be suspended in the water column rather than be placed on the substrate
Crustacea Crayfish Prawn Shrimp	These omnivores would be useful primarily as test and bioassay organisms. They are sensitive to pollutants

From Tchobanoglous *et al.*, 1979; reproduced by permission

used. The fundamental criteria of an aquatic organism's suitability for use in aquatic treatment systems are:

1. the amount and quality of filtration medium/bacterial support structure it provides;
2. the organism's effect on the aquatic environment; and
3. its tolerance to environmental changes.

Up to the present, only floating and emergent aquatic weeds have been employed for wastewater treatment. Owing to their slow growth and low photosynthetic ability, the submerged weeds have not played an important role in the APUs.

7.7.3 Process design darameters for floating aquatic weed system

The different parameters currently being used in the design of aquatic systems include organic loading rate, hydraulic loading rate, hydraulic application rate, hydraulic detention time, and N loading rate. Depending on the treatment objective the major design parameter will be different; for example, organic loading rate is important for organic matter removal. Root depth of the plant is important as it provides the bacterial support and the media for interaction between wastewater and the bacterial mass. Even though there are many potential aquatic plants which can be used for waste treatment they have not yet been investigated to develop rational design criteria. Design criteria for water hyacinth treatment systems are given in Table 7.8, although it is generally known that in tropical areas, water hyacinth treatment systems can be operated at higher loading rates without impairment of effluent quality.

Hydraulic retention time

Hydraulic retention time, typically expressed in days, is the most frequent parameter used to design aquatic systems. The greatest benefit derived from the use of retention time is that a majority of performance data reported in the literature are correlated to retention time. The use of retention time, however, has some serious drawbacks. Depending on system geometry, systems with similar hydraulic retention times may be quite different hydraulically. Also, accurate determination of hydraulic retention time is difficult in aquatic treatment systems because of complex flow patterns and volume displacement attributable to the plants. Generally, hydraulic retention times reported in the literature are based on the assumption of their ideal complete mix or plug-flow hydraulics, neither of which are valid assumptions. In addition, no attempt has been made to account for plant volume displacement in the hydraulic retention times reported in the literature. Very few data are available for hydraulic characterization of existing aquatic systems. In summary, while hydraulic retention time has been used widely as an aquatic system design parameter, little insight into how the hydraulic conditions affect the efficiency of pollutant removal mechanisms can be derived from its usage.

Hydraulic loading rate

Hydraulic loading rate, expressed as $m^3/(m^2\text{-day})$, is the volume of wastewater applied per day per unit surface area of the aquatic system. Most likely, the use of hydraulic loading rate stems from its popular use in land application systems. Typically, aquatic systems are operated in a continuous-flow manner, and as a result hydraulic loading or dosing rate is not a pertinent design parameter.

Hydraulic application rate

Hydraulic application rate refers to the volumetric flow rate of wastewater applied to the aquatic system per unit cross-sectional area of the reactor assuming the wastewater is applied uniformly to the system. The units for this parameter is $m^3/(m^2\text{-day})$, which reduces to an expression of fluid velocity. Hydraulic application rate has not been used widely, but offers a much better unit of comparison for system performance data than the aforementioned parameters, for several reasons. One reason hydraulic application rate should be a better design parameter is because hydraulic application rate is a gage of the fluid velocity which is thought to have a significant effect on removal mechanisms operative in aquatic systems. In addition, hydraulic application rate can be compared directly from system to system.

Organic loading rate

Organic loading rate, expressed as $kg/(m^2\text{-day})$, is the mass of applied organic material per unit surface area of the system per unit time. It is a function of flow rate and concentration of organic matter. Theoretically, organic loading rates are dictated by a balance between the applied C and available oxygen on the oxygen/C stoichiometry of bacterial conversion. In practice, organic loading rates, based on experience, are dependent on effective distribution of wastewater to the system. To avoid odor problems due to uneven organic loading, the wastewater should be distributed evenly over the entire pond system.

A relationship between average BOD_5 loading and removal rates for aquatic systems obtained from 24 different studies is shown in Figure 7.14. These data suggest that aquatic systems using either emergent or floating plants could be designed at BOD_5 loading rates up to approximately 110 kg/(ha-day) with 80 percent or more of the applied BOD_5 being removed. It is apparent that a low effluent BOD_5 concentration from aquatic systems can be achieved with a reduced BOD_5 loading rate. BOD_5 removal rates naturally decrease during the winter period if the influent wastewater has a low temperature.

N loading rate

N loading rate, expressed as $kg/(m^2\text{-day})$, is defined as the mass of applied N to the system per unit surface area per unit time. If plants are being harvested on a regular basis the N requirement for new plant reproduction can be matched with the system N loading rate. Early in the development of aquatic systems, N removal by plant uptake and subsequent harvesting was suggested as the principal removal mechanism. More recently the potential N removal through nitrification and

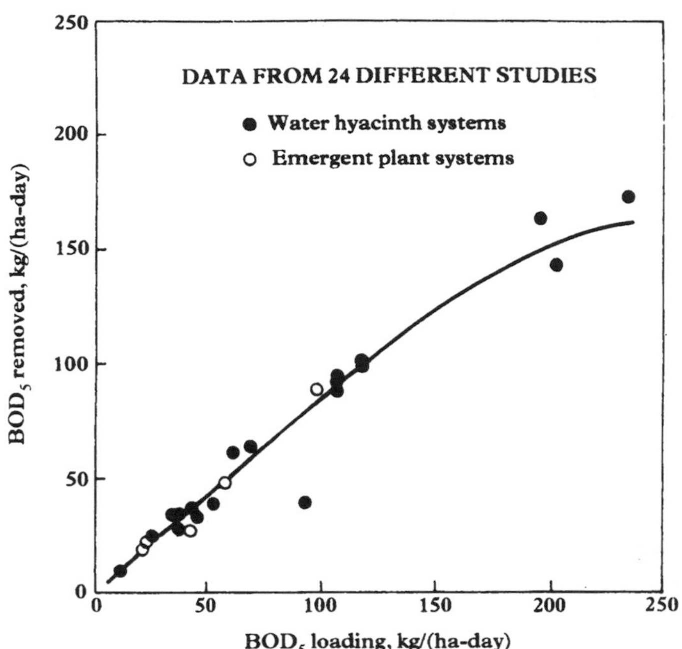

Figure 7.14 BOD removal as a function of BOD loading in aquatic systems (after Stowell *et al.*, 1981; reproduced by permission of the American Society of Civil Engineers)

denitrification has been identified. If nitrification is the major step for total N removal, then N loading rate is not a valid design parameter. System environmental conditions favoring nitrification are different from those that would enhance denitrification. As a result, a loading rate based on a lumping of the various forms of influent N into a single unit is not fundamentally sound.

From their experimental study, Weber and Tchobanoglous (1985) found ammonium N removal rate in an aquatic (water hyacinth) system to be a function of the hydraulic application rate and reactor length. The ammonium N conversion rate was independent of the ammonium loading rate; and hydraulic retention time was not recommended as a design parameter. As shown in Figure 7.15, ammonium removal rates were observed to be inversely proportional to the hydraulic application rates, and Figure 7.16 shows the concentrations of ammonium N removal to be proportional to the reactor length within the experimental conditions employed.

Weber and Tchobanoglous (1986) developed an equation for ammonium N removal in water hyacinth reactors as follows:

$$N_r = \frac{k \cdot l}{q} \tag{7.4}$$

where N_r = ammonium N removal, mg/L;
k = rate coefficient, mg/(L-day);
q = hydraulic application rate, $m^3/(m^2\text{-day})$;
l = reactor distance, m.

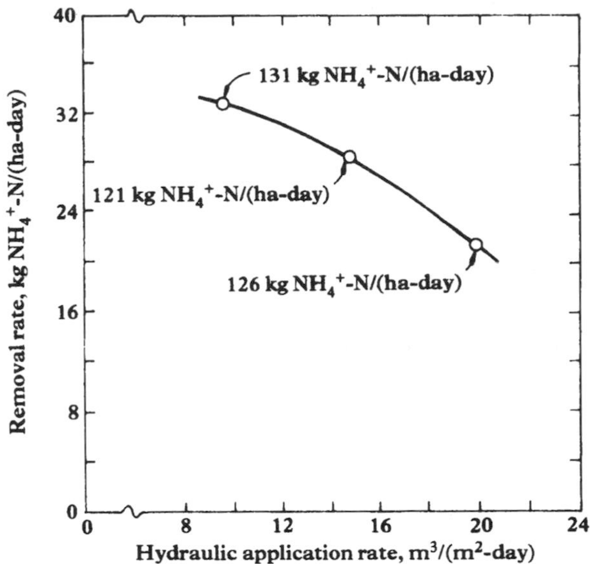

Figure 7.15 NH_4^+–N removal as a function of hydraulic application rate for ammonium-nitrogen loading rate study (corresponding loading rates for NH_4^+–N are as shown) (after Weber and Tchobanoglous, 1985; reproduced by permission of the Water Pollution Control Federation, U.S.A.)

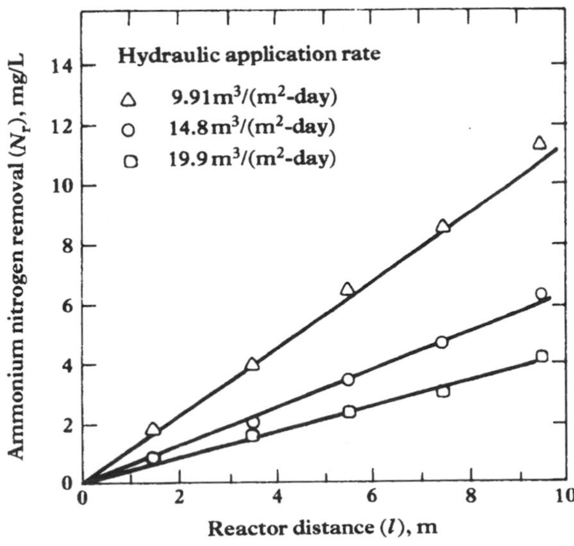

Figure 7.16 Ammonium-nitrogen removal (N_r) as a function of reactor distance for three hydraulic application rates (after Weber and Tchobanoglous, 1986; reproduced by permission of the Water Pollution Control Federation, U.S.A.)

Table 7.10 Calculated value of the rate coefficient, k, as a function of hydraulic application rate

Hydraulic application rate, $m^3/(m^2\text{-day})$	k, mg/(L-day)
9.9	11.9
14.8	9.5
19.9	8.8

k = Hydraulic application rate multiplied by the slopes, from Figure 7.16
From Weber and Tchobanoglous (1986); reproduced by permission of the Water Pollution Control Federation, U.S.A.

Based on the data shown in Figure 7.16 the values of k as a function of q are given in Table 7.10.

Climatic influences

Surface area loading rates are much lower in aquatic systems than in conventional systems; typically more than two orders of magnitude lower. A consequence of these lower loading rates is greater exposure to climatic factors such as temperature, rain, and wind. Any of these factors can disrupt treatment by altering the aquatic environment or damaging the plants. In general, the use of aquatic plants native to the climatic regime at the aquatic system site is encouraged, but in certain cases this may not be desirable or possible.

Temperature Wastewater in aquatic systems will be warm in summer and cool in winter, because of the large surface areas and long hydraulic residence times. A 10 °C rise or drop in temperature is generally known to double or halve the metabolic rate of bacteria and plants. Temperature has a similar effect on the reproductive rates of these organisms.

Summer high temperatures may upset treatment by causing damage to some plant species and/or by increasing bacterial metabolic activity, resulting in altered balances between BOD_5 reduction rates and oxygen flux from the atmosphere. Another potential problem is that the water may become sufficiently warm to result in violations of discharge permit requirements on water temperature or other factors affected by water temperature such as un-ionized ammonia concentration.

Winter low temperatures will reduce the metabolic rate of bacteria and may kill many plant species. A plant kill may, but will not necessarily, result in a loss of bacterial structure and/or leakage of plant-contained compounds. Research in this area is necessary. At water temperatures approaching 0 °C metabolic activity virtually cases; thus wastewater treatment would result only from physical and chemical mechanisms. This reduced level of treatment may be inadequate in some situations. In these cases aquatic treatment would have to be considered a seasonal treatment process. Wastewater storage or some other form of treatment would be necessary during the periods when the aquatic system is not functional.

Rain Rain storms could overload aquatic systems hydraulically. Water inflow will increase from I/I (inflow/infiltration) and direct rainfall into the APUs. Direct rainfall alone from a typical storm could more than double the hydraulic load to an aquatic system. The nature, magnitude, and duration of effects resulting from hydraulic overloading are unknown. Probable effects of hydraulic overloading would include wash-through of BOD_5, resuspension of sediment, and addition of dissolved oxygen to APUs designed to be anaerobic.

Wind Wind can alter the aquatic environment directly or indirectly, the latter by damaging the plants. Wind increases the rate of gas transfer between the wastewater and atmosphere. Wind also increases water turbulence, which may result in resuspension of sediment. Plants broken or otherwise damaged by wind may not function as intended, which would upset the wastewater treatment process until the plants have recovered. Floating plants, particularly duckweeds, are more sensitive to wind than rooted plants because the former tend to be blown around, and may result in openwater surfaces which are vulnerable to the direct effects of wind and other environmental factors. Though more sensitive to winds, floating plants, by virtue of their mobility, are less likely to be physically damaged by wind than rooted plants.

Environmental factors

Environmental impacts that could result from the presence of an aquatic treatment system include odors, fog generation, increased vector-organism population, and introduction of nuisance organisms to the local environment. The potential for significant environmental impact from any of these or other factors must be evaluated, and mitigating measures taken as necessary.

The potential of the local environment to have significant impact on the organisms in the aquatic system must also be evaluated. Of particular concern in this regard are:

1. local and/or migratory animals that may eat the APU plants, damage APU levees by burrowing, or upset APU hydraulics by either channelizing or blocking flow; and
2. local plant diseases or plant-consuming insects.

As an example, water hyacinths could not be used for wastewater treatment in an area where biological control of water hyacinths has been established.

Wastewater characteristics

Contaminant concentrations in domestic wastewaters should not be toxic to aquatic plants in properly designed APUs. When an appreciable proportion of the wastewater is of industrial origin, small-scale pilot studies should be conducted to evaluate the toxicity of the wastewater. If toxicity from shock loadings occurs, mitigating measures such as flow equalization and storage facilities will be necessary. If the wastewater is consistently toxic, source control, pretreatment, or processes other than aquatic treatment will be necessary.

Process reliability, upsets, and recovery

Little is known about the reliability of aquatic systems. The greater influence of climatic and other environmental factors on aquatic processes should reduce their reliability in comparison to conventional processes. However, the slower rate of treatment, reduced reliance on equipment, and greater bacterial diversity of aquatic processes should improve their reliability, comparatively. Currently operating aquatic processes have not been as reliable as conventional processes; but considering how little is known about the design, monitoring, and management of aquatic processes, significant improvement in reliability should be achievable.

Causes of aquatic process upsets and the nature, magnitude, and duration of disruption of treatment have not been studied. Once rational aquatic treatment design criteria have been established, most process upsets will result from:

1. unusual climatic and environmental events insufficiently accounted for in design, and
2. improper APU management.

The nature, magnitude, and duration of treatment process disruption will depend, primarily, on the severity of the event causing the upset, which APUs are affected, and the extent of damage or alteration to the aquatic environment.

Essentially nothing is known about the recovery times of aquatic processes from various types and magnitudes of upsets. Recovery times are expected to be longer in aquatic processes than in conventional processes, because environmental conditions in the former are less conductive to rapid bacterial growth rates. If a necessary function of the APU plants is damaged, the recovery time could be several weeks or more if a new group of plants must be grown.

O'Brien (1981) summarized the design and performance characteristics of such systems in the U.S., as shown in Tables 7.11 and 7.12. Most of these systems are in the hotter parts of the U.S. and the climate there closely resembles the tropical climates of developing countries. Suggestions by Dinges (1979) to supplement the design criteria in Table 7.8 include the use of:

1. perforated pipes to achieve uniform influent flow distribution;
2. filters around the effluent structure to prevent the escape of plants; and
3. circular galvanized-wire mesh enclosures in ponds to improve production of the fish used for controlling mosquitoes.

The data in Table 7.12 suggest that advanced waste treatment standards can be met if intensive management techniques are practiced. Ultimate disposal of plants as surface mulch, compost, animal feed, or for generation of biogas (see sections 7.5 and 7.6) ensures that resource recovery is also practiced. But in most wastewater treatment installations these processes will not be sufficiently profitable to offset the cost of solids disposal.

Example 7.1 Design a water hyacinth treatment system to treat the following wastewater.

Wastewater characteristics:

BOD_5	$= 200$ mg/L
COD	$= 300$ mg/L
Flow rate	$= 100\ m^3/day$

Choose organic loading	$= 25$ kg BOD_5/(ha-day) (from Table 7.9)
Area required	$= 100 \times 200 \times 10^{-3}/(25)$
	$= 0.8$ ha

If two basins are to be constructed,

Area of individual basin	$= 0.4$ ha
Length/width	$= 4:1$
Width	$= (0.4 \times 10^4/4)^{0.5}$
	$= 31.6$ m
Choose width	$= 32$ m
Length	$= 125$ m
Choose depth	$= 1.2$ m (root depth $= 0.9$ m; bottom layer $= 0.3$ m)

Check with data in Table 7.8

Hydraulic loading rate	$= 100/(2 \times 0.4)$
	$= 125\ m^3/(ha\text{-}day) < 200\ m^3/(ha\text{-}day)$
Hydraulic retention time	$= 4000 \times 1.2 \times 2/(100)$
	$= 96$ days > 50 days

The design is acceptable.

Example 7.2 Estimate the yield of duckweeds from an aquatic treatment system treating domestic sewage.

Area of ponds	$= 1$ ha (or $10^4\ m^2$)
Doubling time of duckweed	$= 10$ days
Choose initial duckweed density	$= 400\ g/m^2$ (wet weight)

Notes:

1. Duckweed density reported in the literature varies from $200\ g/m^2$ to $700\ g/m^2$.
2. Lower initial duckweed density may cause algal blooms suppressing the growth of duckweed.
3. Harvesting interval has to be chosen such that the growth of duckweed is not reduced due to overcrowding.

Table 7.11 Summary of design and performance characteristics of aquatic weed wastewater treatment systems

Location	Type of pretreatment	Surface area of hyacinth pond (ha)	Depth of hyacinth pond (m)	Hydraulic loading rate (m^3/ha per day)	Hydraulic residence time (days)	Organic loading rate (kg/ha per day)	Hyacinth cover (%)	Nutrient content of plants, dry weight, (%)	
								Total Kjeldahl nitrogen	Total phosphorus (TP)
National Space Technology Lab:	None	2	1.22	240	54	26		2.73	0.45
Before hyacinths									
After hyacinths							100		
Lucedale, Miss.:	None	3.6	1.73	260[a]	≈67[a]	44		3.56	0.89
Before hyacinths									
After hyacinths							100		
Orange Grove, Miss.:	Two aerated, Three facultative lagoons in experiment conducted in parallel systems	0.28	1.83	3,570	6.8	179		3.75	0.85
Without hyacinths									
With hyacinths							100		
Cedar Lake, Miss. (duckweed)	One-aerated and one facultative lagoon	0.07	1.5	700	22	31	100 (duckweed)	—	—
Williamson Creek, Tex.:									
Phase I	Plant A: aeration basin, clarifier, three lagoons in series	0.0585	1.0	1,860	5.3	43	100	—	—
Phase II	Plant B: two-aeration basins in parallel, three lagoons		0.85	1,860	4.5	89	100	—	—
Full-scale	Plant A: aeration basin, clarifier, two lagoons in series (1.2 ha each)	1.2	0.7–1.3	1,100–1,400	6–9	56–72	100	—	—
Austin-Hornsby Bend, Tex.	Excess activated sludge lagoons overflow to hyacinth pond	1.4	1.23	430	≈3	—	100		

San Juan, Tex.	Aerated lagoon, followed by stabilization ponds	1.0 (two ponds, each 1.0 ha; only one is in operation)	0.61 0.91 0.91 1.37	870 850 1,550 1,860	≈7 ≈11 ≈6 ≈7	—	100
Alamo, Tex.: Before hyacinths	Imhoff tank, trickling filter, aeration	Two basins: 1.35; 1.05					
After hyacinths	basin, two 4.04-ha stabilization ponds in parallel						25
San Benito, Tex.: Before hyacinths After hyacinths	Stabilization pond, four ponds in series	Three in series: 0.8; 0.8; 2.0		3,100; 3,100; 1,200 (overall system: 0.068) 680			First in series: 0; Second in series: 50; Third in series: 100
Rio Hondo, Tex.	None	Three in series: 0.41; 0.41; 0.41		1,110; 1,110 1,110	197	Basin 1:	
Lakeland, Fla.	Secondary treatment	Three in series, 0.40; 0.40; 0.40					
Walt Disney World, Fla.	Primary treatment	Three experi-mental basins: 8.84 × 109.8 m	Variations: 0.38; 0.61; 0.91	650–7,850			
Coral Springs, Fla.	Activated sludge effluent	Five in series: 0.18; 0.08; 0.08; 0.08; 0.08	0.38	870	6 days total: 2; 1; 1; 1; 1	Basin 1: 31	100
University of Florida, Gainesville, Fla.	Trickling filter and activated sludge with polishing pond	0.76	1.4	1,220	0.63	5.2	100

[a] Based on effluent flow rate

From O'Brien (1981); reproduced by permission of the American Society of Civil Engineers

Table 7.12 Summary of influent/effluent characteristics of aquatic weed wastewater treatment systems

Location	BOD_5 (mg/L)		Suspended solids (mg/L)		TN (mg/L)		TP (mg/L)		Dissolved oxygen, (mg/L)	Effluent pH	Sampling period and comments
	Influent	Effluent	Influent	Effluent	Influent	Effluent	Influent	Effluent			
National Space Technology Lab:											
Before hyacinths	91	17	70	49	9.8	5.2	2.9	2.1	6.9	9.3	May 1974–May 1976
After hyacinths	110	7	97	10	12.0	3.4	3.7	1.6	2.3	7.1	June 1976–September 1977
Lucedale, Miss.:											
Before hyacinths	127	57	140	77					0		Odors at night
After hyacinths	161	23	125	6							
Orange Grove, Miss.:											
Without hyacinths	50	37	49	53					2.0		
With hyacinths	50	14	49	15							
Cedar Lake, Miss. (duckweed)	44	18	188	11					0.5		
Williamson Creek, Tex.:											
Phase I	23	5	43	7	8.2	2.5					
Phase II	46	6	117	8	9.9	3.6					
Full-scale	42	12	40	9							October 1977–August 1979

Austin-Hornsby Bend, Tex.										No change at loading rates used
San Juan, Tex.	—	20	30							April 1978–March 1979
		9	11							
		35	30							
		31	32							
Alamo, Tex.:										
Before hyacinths		33	86							April 1978–March 1979
After hyacinths		38–50	68–82							May 1979–June 1979
San Benito, Tex.:										
Before hyacinths		17	35							April 1978–March 1979
After hyacinths		17	20							June 1979
Rio Hondo, Tex.			24							July 1978–May 1979
Lakeland, Fla.					[a]	[a]	[a]	[a]	—	
Walt Disney World, Fla.					[b]	[b]	[b]	[b]	—	
Coral Springs, Fla.	13	3	3	22.4	1.0	11.0	3.6			March 1979–May 1979
University of Florida, Gainesville, Fla.	4.2	3.8	11.8	9.4	15.7	13.2	5.5	4.3		April 1972–January 1973

[a] Only partial cover and limited preliminary data
[b] Experimental units not fully operational
From O'Brien (1981); reproduced by permission of the American Society of Civil Engineers

Choose the harvesting interval to be once every 10 days.

Amount of duckweeds to be harvested each time

$= 400 \times 10^4$ g
$= 4$ tons

Monthly harvest $= 4 \times 30/(10)$
$= 12$ tons wet mass

If duckweeds contains 95 percent water, dry mass

$= 12 \times 0.05$

Amount of duckweed

$= 0.6$ ton dry mass

The yield of duckweed is 0.6 ton dry mass/(ha-month)

Example 7.3 Wastewater from a food processing factory has the following characteristics:

Flow rate $= 100\ m^3/day$
COD $= 300$ mg/L
BOD_5 $= 200$ mg/L
NH_4^+–N $= 10$ mg/L
PO_4^{3-}–P $= 5$ mg/L

The factory plans to convert this wastewater to produce protein biomass using the existing pond system which has an area of 1.2 ha. The available alternatives are to grow either water hyacinth or duckweeds. Determine a suitable weed option which will give more financial return annually. The following information is given:

Water hyacinth

Water hyacinth density $= 2\ kg/m^2$ (at the time of harvesting)
Doubling time $= 10$ days
Protein content $= 17$ percent by weight (dry)
Harvesting is done manually (no additional cost)
Market price of protein $= 1$ Baht/kg

Duckweeds

Duckweed density in pond $= 0.7\ kg/m^2$ (at the time of harvesting)
Area doubling time $= 15$ days
Protein content $= 20$ percent by weight (dry)
Harvesting cost $= 100$ Baht/ton dry weight
Market price of protein $= 3$ Baht/kg

(a) *Water hyacinth*

$$\text{Organic loading of the pond} = 100\left(\frac{m^3}{day}\right) \times 200\left(\frac{mg}{L}\right) \times \frac{10^{-6}\ kg/mg}{10^{-3}\ m^3/L} \times \frac{1}{1.2\ ha}$$
$$= 16.67\ kg\ BOD_5/(ha\text{-}day)$$
$$< 30\ kg\ BOD_5/(ha\text{-}day)$$

Hydraulic loading

$$= 100\left(\frac{m^3}{day}\right) \times \frac{1}{1.2\ ha} = 83.3 < 200\ m^3/(ha\text{-}day)$$

Because the doubling time is 10 days, the number of water hyacinth harvestings is 36 per year. The amount of water hyacinth harvested is half of the total pond area.

Amount of water hyacinth harvested/year

$$= 36 \times \tfrac{1}{2} \times 1.2 \times 10^4 \times 2.0\ kg = 432 \times 10^3\ kg$$

Assuming solid content $= 5$ percent
Total protein obtained $= 0.05 \times 0.17 \times 432 \times 10^3\ kg = 3{,}672\ kg/year$
Income from protein $= 1.0 \times 3{,}672 = 3{,}672$ Baht/year
Annual income $= 3{,}672$ Baht

(b) *Duckweed*

Because the doubling time is 15 days, the number of duckweed harvestings is 24 per year. The amount of duckweed harvested is half of the total pond area.

Amount of duckweed harvested per year $= \tfrac{1}{2} \times 24 \times 0.7 \times 1.2 \times 10^4\ kg$
$= 100.8 \times 10^3\ kg$
Solid content in duckweed $= 20$ percent
Total protein obtained $= 0.2 \times 0.2 \times 100.8 \times 10^3 = 4{,}032\ kg/year$
Cost for harvesting $= 100.8 \times 10^3 \times \dfrac{0.2 \times 100}{1{,}000} = 2{,}016$ Baht
Income from protein $= 4{,}032 \times 3.0$ Baht/year
$= 12{,}096$ Baht/year
Net income $= (12{,}096 - 2{,}016)$ Baht/year
$= 10{,}080$ Baht/year

Duckweed culture will give a better financial return annually.

7.7.4 *Emergent aquatic weeds and constructed wetlands*

Wetland is an area which is inundated or saturated by surface or groundwater at a frequency or duration sufficient to maintain saturated conditions and growth of related vegetation. Emerged aquatic weeds, such as cattails (*Typha*), bulrushes (*Scirpus*) and reeds (*Phragmites*) are the major and typical component of the wetland systems (Figure 7.3). Wetland is a natural system where complex physical, chemical, and biological reactions essential for wastewater treatment, listed in Table 7.7, exist. To avoid interference with the natural ecosystems, constructed wetlands in which the hydraulic regime is controlled have been used for treatment of a variety of wastewaters (Hammer, 1989; WPCF, 1990). Constructed wetlands can range from creation of a marshland to intensive construction involving earth moving, grading, impermeable barriers, or erection of tanks or trenches (U.S. EPA, 1988).

Two types of constructed wetlands have been developed for wastewater treatment, namely, free water surface (FWS) and subsurface flow (SF). An FWS system consists of parallel basins or channels with relatively impermeable bottom and soil and rock layers to support the emergent vegetation, and the water depth is maintained at 0.1–0.6 m above the soil surface. An SF system, also called 'root zone', 'rock-bed filter' or 'reed beds', consists of channels or trenches with impermeable bottom and soil and rock layers to support the emergent vegetation, but the water depth is maintained at or below the soil surface (Figure 7.17). To reduce short-circuiting, the length to width ratios of constructed wetland units should be more than 2 : 1. Although it might appear that SF constructed wetlands could be subjected to frequent clogging problems, performance data reported so far have shown them to function satisfactorily with a high degree of removal efficiencies (Reed and Brown, 1992, 1995).

Table 7.13 shows removal efficiencies of some constructed wetlands located in the U.S.A. which, in general, are comparable with other constructed wetlands in operation worldwide. Although more than 80 percent of N and P removal can be expected in constructed wetlands, not much information about the removal of fecal microorganisms is available (see section 7.1.3). Based on the performance data, a summary of wetland design considerations is given in Table 7.14, which should serve as guidelines in the design and operation of natural and constructed wetlands for wastewater treatment.

Owing to the complex reactions occurring in wetland beds, it is very difficult to develop a comprehensive model that would adequately describe or predict the treatment efficiency. Nevertheless, Reed and Brown (1995) recommended a first-order reaction for BOD_5 removal in constructed wetlands:

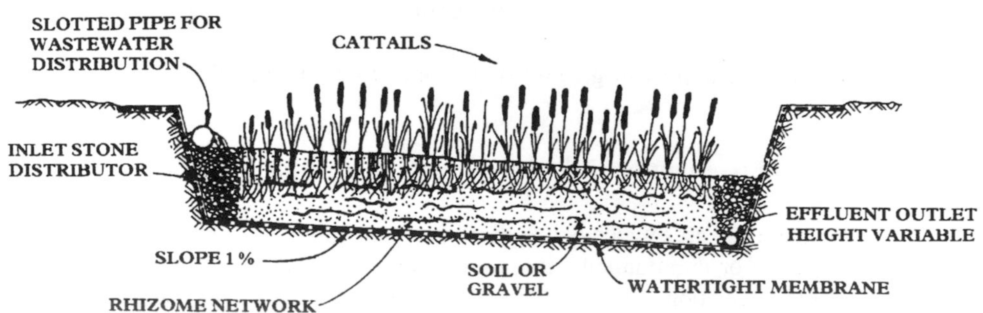

Figure 7.17 Typical cross-section of SF constructed wetlands (U.S. EPA, 1988)

Table 7.13 Summary of BOD_5 and SS removal from constructed wetlands (U.S. EPA, 1988)

Project	Flow, m^3/day	Wetland type	BOD_5 (mg/L)		SS (mg/L)		Percent reduction		Hydraulic surface loading rate, m^3/ha-day
			Influent	Effluent	Influent	Effluent	BOD_5	SS	
Listowel, Ontario	17	FWS[a]	56	10	111	8	82	93	—
Santee, CA	—	SFS[b]	118	30	57	5.5	75	90	—
Sydney, Australia	240	SFS	33	4.6	57	4.5	86	92	—
Arcata, CA	11,350	FWS	36	13	43	31	64	28	907
Emmitsburg, MD	132	SFS	62	18	30	8.3	71	73	1,543
Gustine, CA	3,785	FWS	150	24	140	19	84	86	412

[a] Free water surface system
[b] Subsurface flow system

Table 7.14 Summary of wetland design considerations (WPCF, 1990). Reproduced by permission of the Water Environment Federation

Design consideration	Constructed: Free water surface	Constructed: Subsurface flow	Natural
Minimum size requirement (ha/1,000 m^3-day)	3–4	1.2–1.7	5–10
Maximum water depth (cm)	50	Water level below ground surface	50; depends on native vegetation
Bed depth (cm)	Not applicable	30–90	Not applicable
Minimum aspect ratio (length/width)	2 : 1	Not applicable	Prefer 2 : 1
Minimum hydraulic residence time (days)	5–10	5–10	14
Maximum hydraulic loading rate (cm/day)	2.5–5	6–8	1–2
Minimum pretreatment	Primary; secondary is optional	Primary	Primary; secondary; nitrification, TP reduction
Configuration	Multiple cells in parallel and series	Multiple beds in parallel	Multiple discharge sites
Distribution	Swale; perforated pipe	Inlet zone (> 0.5 m wide) of large gravel	Swale; perforated pipe
Maximum loading (kg/ha-day)			
BOD	100–110	80–120	4
TN	60	60	3
Additional considerations	Mosquito control with mosquitofish; remove vegetation once each 1–5 years	Allow flooding capability for weed control	Natural hydroperiod should be > 50%; no vegetation harvest

$$\frac{C_e}{C_0} = e^{-K_T \cdot t} \qquad (7.5)$$

where C_e = effluent BOD_5 concentration, mg/L,
C_0 = influent BOD_5 concentration, mg/L,
K_T = first-order BOD_5 removal rate constant, day^{-1},
$K_T = K_{20}(1.06)^{T-20}$,
T = liquid temperature, °C,
$K_{20} = 1.104\ day^{-1}$, for SF constructed wetlands,
$K_{20} = 0.678\ day^{-1}$, for FWS constructed wetlands,
t = hydraulic retention time in the wetland beds, days*.

*In calculating the t value, the wetland bed porosity and water loss by evapotranspiration must be taken into account.

It is apparent from the K_{20} values that the SF constructed wetlands are more efficient in the removal of BOD_5 than the FWS constructed wetlands. This superior efficiency is obviously due to better contact between the liquid wastewater and the microorganisms (mainly bacteria) present in the SF wetland beds responsible for BOD_5 biodegradation.

Because of current interests in the utilization of constructed wetlands in wastewater treatment/recycling, more comprehensive design models with respect to BOD_5, nutrient and fecal coliform removal that will encompass, not only temperature but other important parameters, should be developed.

From their surveys of more than 20 constructed wetland sites in the U.S.A., Reed and Brown (1992) did not observe the need for annual harvesting of the common emergent vegetation, but recommended that the vegetative detritus present in the wetland beds be cleaned out on some extended schedule. However, because vegetative plants obviously grow faster in the tropics than in temperate climates, it would be useful to study the need and frequency of plant harvesting for constructed wetlands located in tropical areas; the results of this study are essential in determining the amount of nutrients and/or toxic compounds that could be removed from the wetland beds by harvesting.

7.8 HEALTH HAZARDS RELATING TO AQUATIC WEEDS

The cultivation of aquatic weeds may cause health problems by:

1. contamination of the people who work in the aquatic pond operation with pathogens in the water;
2. contamination of the plants with pathogens and toxic materials such as heavy metals and pesticides; and
3. providing habitats (pond and plants) for various vectors.

Concerning the first and second health problems, the risk of contacting pathogens by workers and contamination of plants with pathogens can be minimized if the waste is adequately treated to eliminate pathogens before applying to aquatic plants. This step is necessary as pathogen removal in an aquatic system is not very effective. Direct fertilization of aquatic weed ponds with nightsoil should be avoided, as the risk is high and it reduces the acceptability of the product.

The third health problem involves the possibility of disease transmission through infection by Metacercariae (infective stage) which has its intermediate host in the pond, such as snails (in case of schistosomiasis) or attached to the plant leaves and stems. According to Feachem *et al.* (1983), the parasitic fluke *Fasciolopsis buski* is important in some parts of Asia. The worm has a life cycle that moves from human (or pig or dog) to snail to water plant and then to human again. Animals or people become infected by eating the encysted Metacercariae on water plants, such as seed-pods of water caltrop, bulbs of water chestnut, and roots of lotus, water bamboo, etc. In areas where those diseases associated with intermediate hosts in the pond are endemic and their respective hosts are present, waste recycling using aquatic plants is not advisable.

Aquatic vegetation enhances the production of mosquito by protecting the larvae from wave action, providing a habitat for breeding and interfering with mosquito control procedures. The two major vectors are *Anopheles*, which transmits malaria, and *Mansonia*, which carries filariasis and encephalitis.

The eggs of *Mansonia* are laid on the undersides of leaves of aquatic weeds just above the surface of the water. The mosquito larva inserts its respiratory siphon into the air-containing tissues of the plant and need not come to the water surface for air. The air is obtained from the submerged portions of the plants, especially from the roots. Different *Mansonia* species have a preference for certain plants but water lettuce seems to be the most common host, followed by water hyacinth, then *Azolla* and duckweeds.

An effective way to prevent breeding of mosquito in the pond is to stock fish that feed on mosquito larvae. Successful control was reported by *Sarotherodon mossambicus* in water spinach ponds and *Gambusia affinis* in water hyacinth ponds (Wolverton *et al.*, 1975; Edwards, 1980).

Biological accumulation in aquatic weeds will occur when wastewater containing pesticides and heavy metals is treated. This reduces the weeds' potential usage as food and for agricultural purposes. So wastewaters containing toxic materials should be pretreated to remove these specific contaminants, or other alternatives for treatment and recovery have to be considered.

REFERENCES

Bagnall, L. O. (1979). Resource recovery from wastewater aquaculture. In *Aquaculture Systems for Wastewater Treatment*—Seminar Proceedings and Engineering Assessment, Report No. EPA 430/9–80–006, pp. 421–39. U.S. Environmental Protection Agency, Washington, D.C.

Bagnall, L. O., Furman, T. D. S., Hentges, J. F., Nolan, W. J., and Shirley, R. L. (1974). *Feed and Fiber from Effluent-Grown Water Hyacinth*. Environmental Protection Series: EPA-660/2–74–041. Washington, D.C.

Boyd, C. E. (1970). Utilization of aquatic plants. In *Aquatic Vegetation and Its Use and Control* (ed. D. S. Mitchell), pp. 107–15. UNESCO, Paris.

Burgoon, P. S., Reddy, K. R. and DeBusk, T. A. (1995). Performance of subsurface flow wetlands with batch-load and continuous-flow conditions. *J. Water Environ. Res.*, **67**, 855–62.

Cooper, A. (1976). *Nutrient Film Technique of Growing Crops*. Grower Books, London.

Cornwell, D. A., Zoltek, J., Jr, Patrinely, C. D., Furman, T. D. S. and Kim, J. I. (1977). Nutrient removal by water hyacinths. *J. Water Pollut. Control Fed.*, **49**, 57–65.

Dassanayake, M. D. (1976). Noxious aquatic vegetation control in Sri Lanka. In *Aquatic Weeds in South East Asia* (eds. C. K. Varshney and J. Rzoska), pp. 59–61. Dr. W. Junk B.V., The Hague.

DeBusk, T. A., Williams, L. D., and Ryther, J. H. (1977). Growth and yields of the freshwater weeds *Eichhornia crassipes* (water hyacinth), *Lemna minor* (duckweed), and *Hydrilla verticillata*. In *Cultivation of Macroscopic Marine Algae and Freshwater Aquatic Weeds* (ed. J. H. Ryther), pp. 275–95. Contract No. E(11–1)-2948. U.S. Department of Energy, Washington, D.C.

Dinges, R. (1979). Development of water hyacinth wastewater treatment systems in Texas. In *Agriculture Systems for Wastewater Treatment*—Seminar Proceedings and Engineering Assessment. Report No. EPA 430/9–80–006, pp. 193–226. U.S. Environmental Protection Agency, Washington, D.C.

Edwards, P. (1980). *Food Potential of Aquatic Macrophytes*. ICLARM Studies and Reviews No. 5, International Center for Living Aquatic Resources Management, Manila.

Edwards, P., Polprasert, C., Rajput, V. S., and Pacharaprakiti, C. (1988). Integrated biogas technology in the tropics. 2. Use of slurry for fish culture. *Waste Manage. Res.*, **6**, 51–61.

Feachem, R. G., Bradley, D. J., Garelick, H., and Mara, D. D. (1983). *Sanitation and Disease—Health Aspects of Excreta and Wastewater Management*. Wiley, Chichester.

Frank, P. A. (1976). Distribution and utilization-research on tropical and subtropical aquatic weeds in the United States. In *Aquatic Weeds in South East Asia* (eds. C. K. Varshney and J. Rzoska), pp. 353–9. Dr. W. Junk B.V., The Hague.

Gopal, B. and Sharma, K. P. (1981). *Water Hyacinth—The Most Troublesome Weed of the World*. Hindasia Publishers, Delhi.

Gupta, O. P. and Lamba, P. S. (1976). Some aspects of utilization of aquatic weeds. In *Aquatic Weeds in South East Asia* (eds. C. K. Varshney and J. Rzoska), pp. 361–7. Dr. W. Junk B.V., The Hague.

Hammer, D. A. (1989). *Constructed Wetlands for Wastewater Treatment: Municipal, Industrial and Agricultural*. Lewis Publisher, Inc., Boca Raton, FL.

Jewell, W. J., Madras, J. J., Clarkson, W. W., Delancey-Pompe, H., and Kabrick, R. M. (1982). *Wastewater Treatment with Plant in Nutrient Films*. U.S. EPA Report No. PB 83–247–494. Springfield, Virginia.

Little, E. C. S. (1968). *Handbook of Utilization of Aquatic Plants: a Compilation of the World's Publications*. Food and Agriculture Organization, Rome.

Mehta, I., Sharma, R. K., and Tuank, A. P. (1976). The aquatic weed problem in the Chambal irrigated area and its control by grass carp. In *Aquatic Weeds in South East Asia* (eds. C. K. Varshney and J. Rzoska), pp. 307–14. Dr. W. Junk B.V., The Hague.

Metcalf and Eddy, Inc. (1991). *Wastewater Engineering: Treatment, Disposal, Reuse*, 3rd Edition, McGraw-Hill, New York.

Mitchell, D. S. (1974). *Aquatic Vegetation and Its Use and Control*. UNESCO, Paris.

Mitchell, D. S. (1976). The growth and management of *Eichhornia crassipes* and *Salvinia* spp. in their native environment and in alien situations. In *Aquatic Weeds in South East Asia* (eds. C. K. Varshney and J. Rzoska), pp. 167–76. Dr. W. Junk B.V., The Hague.

NAS (1976). *Making Aquatic Weeds Useful: Some Perspectives for Developing Countries*. National Academy of Sciences, Washington, D.C.

Ngo, H. N. (1984). Waste recycling through bio-fertilization. *Enfo* (Environmental Sanitation Information Center, Asian Institute of Technology, Bangkok), **2**, 11–13.

O'Brien, W. J. (1981). Use of aquatic macrophytes for wastewater treatment. *J. Env. Eng. Div.*—ASCE, **107**, 681–98.

O'Keefe, D. H., Hardy, J. K., and Rao, R. A. (1984). Cadmium uptake by the water hyacinth: effects of solution factors. *Env. Pollut.*, **34**, 133–47.

Oron, G., Porath, D., and Wildschut, L. R. (1986). Wastewater treatment and renovation by different duckweed species. *J. Env. Eng. ASCE*, **112**, 247–63.

Pirie, N. W. (1978). Freshwater weeds are a resource. *Appropriate Technol.*, **4**, 15–17.

Polprasert, C. (1984). Utilization of composted nightsoil in fish production. *Conservation and Recycling*, **7**, 199–206.

Polprasert, C., Edwards, P., Pacharaprakiti, C., Rajput, V. S., and Suthirawut, S. (1982). Recycling rural and urban nightsoil in Thailand. AIT Research Report No. 143, submitted to International Development Research Centre, Ottawa, Canada. Asian Institute of Technology, Bangkok, Thailand.

Polprasert, C., Yang, P. Y., Kongscricharoern, N., and Kanjanaprapin, W. (1994). Productive utilization of pig farm waste: a case study for developing countries. *Resources Conservation Recycling*, **11**, 245–59.

Reddy, K. R. and Sutton, D. L. (1984). Water hyacinths for water quality improvement and biomass production. *J. Env. Qual.*, **13**, 1–8.

Reed, S. C. and Brown, D. S. (1992). Constructed wetland design—the first generation. *Water Environ. Res.*, **64**, 776–81.

Reed, S. C. and Brown, D. S. (1995). Subsurface flow wetlands—a performance evaluation. *Water Environ. Res.*, **67**, 244–8.

Reed, S. C., Middlebrooks, E. J., and Crites, R. W. (1988). *Natural Systems for Waste Management and Treatment*. McGraw-Hill, New York.

Reimer, D. N. (1984). *Introduction to Freshwater Vegetation*. AVI Publishing Co., Inc., Westport, CT.

Seidal, K. (1976). Macrophytes and water purification. In *Biological Control of Water Pollution* (eds. J. Tourbier and R. W. Pierson), pp. 109–121. University of Pennsylvania Press, Philadelphia.

Stowell, R., Ludwig, R., Colt, J., and Tchobanoglous, G. (1980). Towards the rational design of aquatic treatment systems. Paper presented at the ASCE Convention, 14–18 April, Portland, Oregon. Department of Civil Engineering, University of California, Davis, California.

Stowell, R., Ludwig, R., Colt, J., and Tchobanoglous, G. (1981). Concepts in aquatic treatment system design. *J. Env. Eng. Div.*—ASCE, **107**, 919–40.

Taylor, J. S. and Stewart, E. A. (1978). Hyacinths. In *Advances in Water and Wastewater Treatment—Biological Nutrient Removal* (eds. M. P. Wanielista and W. W. Eckenfelder, Jr), pp. 143–180, Ann Arbor Science, Ann Arbor, Michigan.

Tchobanoglous, G., Stowell, R., and Ludwig, R. (1979). The use of aquatic plants and animals for the treatment of wastewater: an overview. In *Aquaculture Systems for Wastewater Treatment*—Seminar Proceedings and Engineering Assessment, pp. 35–55. Report No. EPA 430/9–80–006. U.S. Environmental Protection Agency, Washington, D.C.

U.S. EPA (1988). *Design Manual—Constructed Wetlands and Aquatic Plant Systems for Municipal Wastewater Treatment*. EPA/625/1-88/022, United States Environmental Protection Agency, Cincinnati, OH.

U.S. EPA (1993). *Subsurface Flow Constructed Wetlands for Wastewater Treatment—A Technology Assessment*, EPA 832-R-93-001, Office of Water, Washington, D.C.

Weber, A. S. and Tchobanoglous, G. (1985). Rational design parameters for ammonia conversion in water hyacinth treatment systems. *J. Water Pollut. Control Fed.*, **57**, 316–23.

Weber, A. S. and Tchobanoglous, G. (1986). Prediction of nitrification in water hyacinth treatment systems. *J. Water Pollut. Control Fed.*, **58**, 376–80.

Westlake, D. F. (1963). Comparisons of plant productivity. *Biol. Rev.*, **38**, 385–425.

Whigham, D. F., Simpson, R. L., and Lee, K. (1980). *The Effect of Sewage Effluent on the Structure and Function of a Freshwater Tidal Marsh Ecosystem*. New Jersey Water Resources Research Institute, New Jersey.

Wolverton, B. C. and McDonald, R. C. (1975). *Water Hyacinths for Upgrading Sewage Lagoons to Meet Advanced Wastewater Treatment Standards, Part I.* NASA Technical Memorandum, TM-X-72729. National Aeronautics and Space Administration, Washington, D.C.

Wolverton, B. C. and McDonald, R. C. (1976a). *Water Hyacinths for Upgrading Sewage Lagoons to Meet Advanced Wastewater Treatment Standards, Part II.* NASA Technical Memorandum, TM-X-72730. National Aeronautics and Space Administration, Washington, D.C.

Wolverton, B. C. and McDonald, R. C. (1976b). *Water Hyacinths (Eichhornia crassipes) for Removing Chemical and Photographic Pollutants from Laboratory Wastewater.* NASA Technical Memorandum, TM-X-72731. National Aeronautics and Space Administration, Washington, D.C.

Wolverton, B. C. and McDonald, R. C. (1978). Water hyacinth (*Eichhornia crassipes*) productivity and harvesting studies. *Econ. Bot.*, **33**, 1–10.

Wolverton, B. C. and McDonald, R. C. (1979). The water hyacinth: from prolific pest to potential provider. *Ambio*, **8**, 2–9.

Wolverton, B. C., McDonald, R. C., and Gordon, J. (1975). *Bio-conversion of Water Hyacinth into Methane Gas: Part I.* NASA Technical Memorandum, TM-X-72725. National Aeronautics and Space Administration, Washington, D.C.

WPCF (1990). *Natural Systems for Wastewater Treatment*, Manual of Practice No. FD-16, Water Pollution Control Federation, Alexandria, VA.
Yount, J. L. and Crossman, R. A., Jr (1970). Eutrophication control by plant harvesting. *J. Water Pollut. Control Fed.*, **42**, 173–83.

EXERCISES

7.1 From eqs 7.1 and 7.2, derive the following equation:

$$\text{Doubling time} = \frac{\ln 2}{\ln x}$$

where $x =$ daily incremental factor.

7.2 It is estimated that water hyacinths have a doubling time of 10 days in natural freshwater. Determine how much time is needed for the initial 10 water hyacinth plants to cover 0.4 ha of a natural freshwater surface. The average density of water hyacinths in water bodies is 150 plants/m^2.

7.3 Design a constructed wetland system to treat a raw wastewater with the following characteristics:

$$\text{Flow rate} = 500\ \text{m}^3/\text{day}$$
$$\text{BOD}_5 = 200\ \text{mg/L}$$
$$\text{COD} = 300\ \text{mg/L}$$

The effluent BOD_5 concentration after treatment should be less than 10 mg/L.
Give reasons about the types of constructed wetlands you plan to choose, and draw schematic diagram of the constructed wetlands including the hydraulic profiles.

7.4 A rural household plans to build a 200 m^2 fish pond to raise tilapia. It intends to grow duckweeds as fish feed. To produce a fish yield of 110 kg/pond per year, a duckweed loading of 4 kg (dry weight) per pond per year is needed. If the doubling time of water hyacinth is 10 days, determine the required area of the water hyacinth pond. Assume an average duckweed density of 8 kg (wet weight)/m^2 of pond area and a moisture content of 94 percent for duckweeds.

7.5 You are to visit and evaluate a floating aquatic weed pond or a constructed wetland unit being used for wastewater treatment, and determine the first-order K_T values, as shown in eq. 7.5. Compare the results and discuss whether the K_T values of these two systems should be the same or different.

7.6 If you are to modify eq. 7.5 to encompass other parameters responsible for wastewater treatment in constructed wetlands or floating aquatic weed ponds, what are the parameters to be included and how is eq. 7.5 to be modified?

8

Land treatment of wastewater

Land treatment is defined as the controlled application of wastes to the land surface to achieve a specified degree of treatment through natural physical, chemical, and biological processes within the plant–soil–water matrix.

Land treatment of wastewater has been practiced for a long time, and is receiving considerable attention at present as an alternative treatment to other existing treatment processes. In developing countries where land is plentiful, land treatment is considered as an inexpensive treatment method.

8.1 OBJECTIVES, BENEFITS, AND LIMITATIONS

In the past the objective of land treatment was simply to dispose of the wastes, but the current trend includes utilization of nutrients in wastewater and sludge for crop production and tertiary (advanced) treatment of wastewater, or to recharge groundwater.

Wastewater disposal on land will be a viable treatment method only if the protection of groundwater from possible degradation is held as a prime objective. With the treatment of wastewater, other benefits such as economic return from marketable agricultural crops, exchange of wastewater for irrigation purposes in arid climates to achieve overall water conservation, and development and preservation of open space and green belts, can also be obtained from the land treatment process.

Land treatment systems are less energy-intensive than conventional systems. (Examples of conventional systems include activated sludge, trickling filters, aerated lagoons, etc.) In land treatment, energy is needed for transportation and application of wastewater to the land. But in conventional treatment systems, such as activated sludge, energy is needed for transportation of wastewater, mixing and aeration of wastewater and sludge, return of sludge, effluent recirculation, and transportation of digested sludge.

As less mechanical equipment is needed for the land treatment process when compared with other conventional treatment processes, the maintenance of the land treatment system is easy and inexpensive.

However, land area, soil condition, and climate are the main limiting factors of this treatment process. Large areas of land are normally required for application of wastewater; for some cities these land treatment sites may be too expensive or too far away from the wastewater sources. Soil condition is important for the removal mechanisms of wastewater constituents, and a soil having too coarse or too fine a texture is not appropriate for land treatment. Climatic conditions where the magnitude of evaporation and evapotranspiration is greater than that of precipitation are generally preferred for land treatment of wastewater and sludge.

8.2 WASTEWATER RENOVATION PROCESSES

Depending on the rate of water movement and the flow path within the process, the treatment of wastewater by land is classified as follows: (i) slow rate (or irrigation) process (SR); (ii) rapid infiltration process (RI); and (iii) overland flow process (OF).

Selection of process depends on the required objectives, as well as the soil condition of the land. Figure 8.1 gives the relationship between loading rate and soil type for the above three processes. A description of soil textural classes is given in Table 8.1.

A comparison of design features for alternative land treatment processes is shown in Table 8.2 and the expected quality of treated wastewater from land treatment processes is given in Table 8.3. The site characteristics for the land treatment processes are shown comparatively in Table 8.4. Excellent effluent quality is usually obtained with land treatment systems (Table 8.3) because the

Table 8.1 Soil textural classes and general terminology used in soil descriptions (U.S. EPA, 1981)

General terms		Basic soil textural class names
Common name	Texture	
Sandy soils	Coarse	Sand
		Loamy sand
	Moderately coarse	Sandy loam
		Fine sandy loam
Loamy soils	Medium	Very fine sandy loam
		Loam
		Silt loam
		Silt
	Moderately fine	Clay loam
		Sandy clay loam
		Silty clay loam
Clayey soils	Fine	Sandy clay
		Silty clay
		Clay

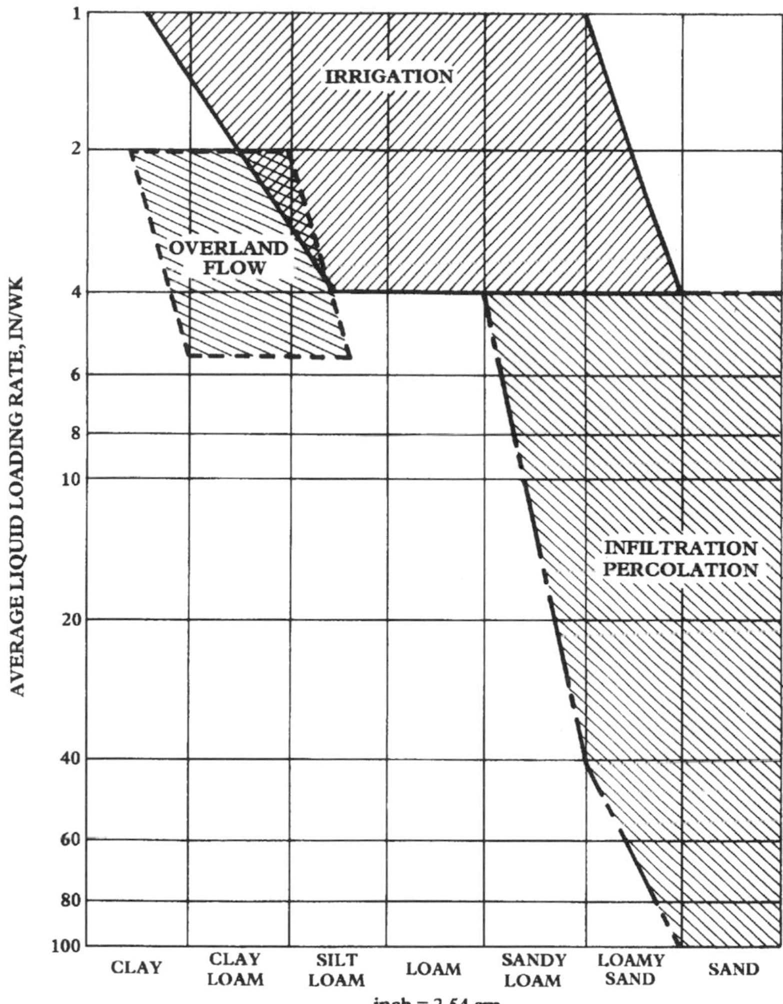

Figure 8.1 Soil type vs liquid loading rates for different land application approaches (U.S. EPA, 1976)

BOD_5 loadings applied (Table 8.2) are much lower than the BOD_5 removal capabilities. Design of a land treatment system is generally based on hydraulic application rates and N loadings. However, BOD_5 loading to a land treatment site needs to be carefully evaluated so that anaerobic conditions do not occur. The occurrence of anaerobic conditions is usually harmful to the root system of crops, and affects the availability of nutrients to the crop; it may also enhance mobility of toxic materials previously precipitated there.

8.2.1 Slow rate process (SR)

SR is the controlled application of wastewater to a vegetated land at a rate of a few centimeters of liquid per week. The flow path depends on infiltration, and usually

Table 8.2 Comparison of typical design features for land treatment processes (U.S. EPA, 1981)

Feature	Slow rate	Rapid infiltration	Overland flow
Application techniques	Sprinkler or surface[a]	Usually surface	Sprinkler or surface
Annual loading rate, m	0.5–6	6–125	3–20
Field area required, ha[b]	23–280	3–23	6.5–44
Typical weekly loading rate, cm	1.3–10	10–240	6–40[c]
Minimum preapplication treatment provided in the United States	Primary sedimentation[d]	Primary sedimentation[e]	Grit removal and comminution[e]
Disposition of applied wastewater	Evapotranspiration and percolation	Mainly percolation	Surface runoff and evapotranspiration with some percolation
Need for vegetation	Required	Optional	Required
BOD_5 loading[f], kg/ (ha-year)	370–1,830	8,000–46,000	2,000–7,500

[a] Includes ridge-and-furrow and border strip
[b] Field area in hectares not including buffer area, roads, or ditches for 3,785 m^3/day (1 million gallons/day) flow
[c] Range includes raw wastewater to secondary effluent, higher rates for higher level of preapplication treatment
[d] With restricted public access; crops not for direct human consumption
[e] With restricted public access
[f] Range for municipal wastewater

Table 8.3 Expected quality of treated water from land treatment processes[a] (U.S. EPA, 1981)

	Slow rate[b]		Rapid infiltration[c]		Overland flow[d]	
Constituent	Average	Upper range	Average	Upper range	Average	Upper range
BOD_5	< 2	< 5	5	< 10	10	< 15
Suspended solids	< 1	< 5	2	< 5	10	< 20
Ammonia nitrogen as N	< 0.5	< 2	0.5	< 2	< 4	< 8
Total nitrogen as N	3[e]	< 8[e]	10	< 20	5[f]	< 10[f]
Total phosphorus as P	< 0.1	< 0.3	1	< 5	4	< 6
Fecal coliforms, no./100 mL	0	< 10	10	< 200	200	< 2,000

Units in mg/L unless otherwise noted
[a] Quality expected with loading rates at the mid to low end of the range shown in Table 8.2
[b] Percolation of primary or secondary effluent through 1.5 m (5 ft) of unsaturated soil
[c] Percolation of primary or secondary effluent through 4.5 m (15 ft) of unsaturated soil; phosphorus and fecal coliform removals increase with distance of wastewater flow
[d] Treating comminuted, screened wastewater using a slope length of 30–36 m (100–120 ft)
[e] Concentration depends on loading rate and crop
[f] Higher values expected when operating through a moderately cold winter or when using secondary effluent at high rates

Table 8.4 Comparison of site characteristics for land treatment processes (U.S. EPA, 1981)

	Slow rate	Rapid infiltration	Overland flow
Grade	Less than 20% on cultivated land; less than 40% on noncultivated land	Not critical; excessive grades require much earthwork	Finish slopes 2–8%[a]
Soil permeability	Moderately slow to moderately rapid (clay loam to sand loam), > 0.15 cm/h	Rapid (sands, sandy loams), > 5.0 cm/h	Slow (clays, silts, and soils with impermeable barriers), < 0.5 cm/h
Depth to groundwater	0.6–1 m (minimum)[b]	1 m during flood cycle;[b] 1.5–3 m during drying cycle	Not critical[c]
Climatic restrictions	Storage often needed for cold weather and during heavy precipitation	None (possibly modify operation in cold weather)	Storage usually needed for cold weather

[a] Steeper grades might be feasible at reduced hydraulic loadings
[b] Underdrains can be used to maintain this level at sites with high groundwater table
[c] Impact on groundwater should be considered for more permeable soils

on lateral flow within the treatment site. Treatment occurs by means of physical, chemical, and biological processes at the surface and as the wastewater flows through the plant–soil matrix. A portion of the flow may reach the groundwater, some is used by the vegetation, but off-site runoff of the applied wastewater is avoided by the proper system design. Typical hydraulic pathways for SR treatment are shown in Figure 8.2a in which surface vegetation responsible for evapotranspiration and soil percolation are essential components in this treatment process. The percolated water can be collected through underdrains placed under the vegetation soil or from the recovery wells constructed within the vicinity.

The objectives of the SR process are: treatment of the applied wastewater; conservation of water through irrigating landscape areas; and economic gain from the use of wastewater and nutrients to produce marketable crops.

The principal limitations to the practice of SR are: the considerable land area required, its relatively high operating cost, and the treatment site is usually a long way from the wastewater sources. As shown in Figure 8.1 and Table 8.2, wastewater application rates for SR are smaller than for the other two land treatment processes. The SR application rates vary from 2.5 to 10 cm/week, depending on the types of crops and soil. In some cases certain wastewater characteristics, such as high salt and boron concentrations, may preclude irrigation of many crops, especially in arid regions.

Adequately disinfected wastewater should pose no danger to health when it is used for irrigation. Adequate disinfection, which can be very costly, requires complete and rapid mixing and a specified contact time of the disinfectant in the effluent. Any aerosolizing of inadequately disinfected water produces possible health risks to humans, and these risks should be minimized. Before harvesting the

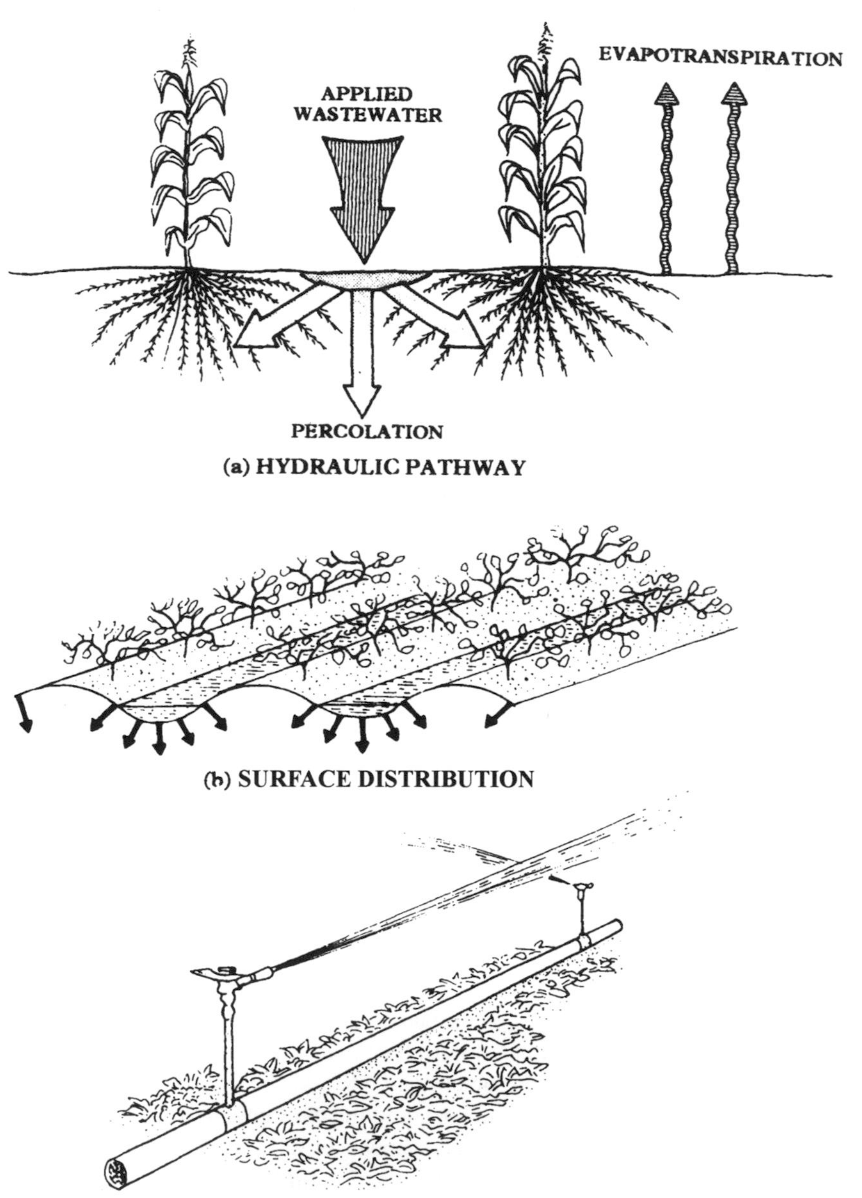

Figure 8.2 Slow-rate land treatment (U.S. EPA, 1981)

irrigated crops, wastewater application must be stopped to allow for drying of the soil and die-off of the pathogens that may be present on the crops.

In general maximum cost-effectiveness for both municipal and industrial wastewater systems will be achieved by applying the maximum possible amount of wastewater to the smallest possible land area. This will in turn limit the choice of suitable vegetation, and possibly the market value of the harvested crop. Optimization of a system for wastewater treatment usually results in the selection of perennial grasses because of the longer application season, higher hydraulic loadings, and greater N uptake, compared with other agricultural crops. Annual planting and cultivation can also be avoided with perennial grasses.

Forest systems also offer the advantage of a longer application season and higher hydraulic loadings than typical agricultural crops, but may be less efficient than perennial grasses for N removal, depending on the type of tree, its stage of growth and the general site conditions.

METHODS OF IRRIGATION

Spray, ridge and furrow, and flood are the wastewater irrigation methods commonly adopted in this process (Figure 8.2). Spray irrigation may be accomplished by using a variety of systems from portable to solid-set sprinklers (Figure 8.2c). Ridge and furrow irrigation consists of applying water by gravity flow into furrows. The relatively flat land is groomed into alternating ridges and furrows with crops grown on the ridges (Figure 8.2b). Flood irrigation is accomplished by inundation of land with water. The type of irrigation system to be used depends upon soil drainability, the crop, the topography, process objective and the economics. Table 8.5 gives the methods of surface irrigation (flood and ridge and furrow) and their conditions of use applicable for wastewater and other irrigation water.

RELIABILITY

As shown in Table 8.3, a high-quality effluent can be expected from SR process. For a well-operated SR system, treatment efficiency is in the order of 99 percent of BOD_5, suspended solids, and fecal coliforms. As irrigation soils are loamy (see Figure 8.1), considerable amounts of organic matter, heavy metals, P, and microorganisms are retained in the soil by adsorption and other mechanisms. It should be noted that plant uptake and/or adsorption of heavy metals and pathogens are a potential problem in the reuse of these irrigated plants. N is taken up by plant growth, and if the crop is harvested the removal rate can be in the order of 90 percent.

SITE SELECTION

Soils ranging from clay loams to sandy loams are suitable for irrigation (Figure 8.1). Soil depth should be at least 0.3 m of homogeneous material and preferably 1.5–2 m throughout the site. The depth is needed for extensive root development of some plants and for wastewater renovation. The minimum depth to groundwater should be 0.6–1.0 m to avoid groundwater contamination (Table 8.4). If the site drainage is poor, control procedures such as underdrains or wells may be required.

Table 8.5 Surface irrigation methods and conditions of use

Irrigation method	Suitabilities and conditions of use				Remarks
	Crops	Topography	Water quantity	Soils	
Flooding					
Small rectangular basins	Grain, field crops, orchards, rice	Relatively flat land; area within each basin should be leveled	Can be adapted to streams of various sizes	Suitable for soils of high or low intake rates; should not be used on soils that tend to puddle	High installation costs. Considerable labor required for irrigating. When used for close-spaced crops a high percentage of land is used for levées and distribution ditches. High efficiencies of water use possible
Large rectangular basins	Grain, field crops, rice	Flat land; must be graded to uniform plane	Large flows of water	Soils of fine texture with low intake rates	Lower installation costs and less labor required for irrigation than small basins. Substantial levées needed
Contour checks	Orchards, grain, rice, forage crops	Irregular land, slopes less than 2%	Flows greater than 1 ft^3/s	Soils of medium to heavy texture that do not crack on drying	Little land grading required. Checks can be continuously flooded (rice), water ponded (orchards), or intermittently flooded (pastures)
Narrow borders up to 16 ft wide	Pasture, grain, alfalfa, vineyards, orchards	Uniform slopes less than 7%	Moderately large flows	Soils of medium to heavy texture	Borders should be in direction of maximum slope. Accurate cross-leveling required between guide levées
Wide borders up to 100 ft wide	Grain, alfalfa, orchards	Land graded to uniform plane with maximum slope less than 0.5%	Large flows up to 20 ft^3/s	Deep soils of medium to fine texture	Very careful land grading necessary. Minimum of labor required for irrigation. Little interference with use of farm machinery

Benched terraces	Grain, field crops	Slope up to 20%	Streams of small to medium size	Soils must be sufficiently deep that grading operations will not impair crop growth	Care must be taken in constructing benches and providing adequate drainage channel for excess water. Irrigation water must be properly managed. Misuse of water can result in serious soil erosion
Furrow					
Straight furrows	Vegetables, row crops, orchards, vineyards	Uniform slopes not exceeding 2% for cultivated crops	Flows up to 12 ft^3/s	Can be used on all soils if length of furrows is adjusted to type of soil	Best suited for crops that cannot be flooded. High irrigation efficiency possible. Well adapted to mechanized farming
Graded contour furrows	Vegetables, field crops, orchards, vineyards	Undulating land with slopes up to 8%	Flows up to 3 ft^3/s	Soils of medium to fine texture that do not crack on drying	Rodent control is essential. Erosion hazard from heavy rains or water breaking out of furrows. High labor requirement for irrigation
Corrugations	Close-spaced crops such as grain, pasture, alfalfa	Uniform slopes of up to 10%	Flows up to 1 ft^3/s	Best on soils of medium to fine texture	High water losses possible from deep percolation or surface runoff. Care must be used in limiting size of flow in corrugations to reduce soil erosion. Little land grading required
Basin furrows	Vegetables, cotton, maize, and other row crops	Relatively flat land	Flows up to 5 ft^3/s	Can be used with most soil types	Similar to small rectangular basins, except crops planted on ridges
Zigzag furrows	Vineyards, bush berries, orchards	Land graded to uniform slopes of less than 1%	Flows required are usually less than for straight furrows	Used on soils with low intake rates	This method is used to slow the flow of water in furrows to increase water penetration into soil

1 $ft^3/s = 0.028\ m^3/s$; 1 ft = 0.305 m. From Booher (1974); reproduced by permission of the Food and Agriculture Organization of the United Nations

For crop irrigation, slopes should be limited to 20 percent or less, depending upon the type of farm equipment to be used. Forested hillsides and noncultivated land up to 40 percent in slope have been spray-irrigated successfully (Table 8.4).

8.2.2 Rapid infiltration process (RI)

RI is the controlled application of wastewater to earthen basins in rapidly permeable soils (e.g. sandy loam, loamy sand, and sand) at a high rate (Figure 8.1). Treatment is accomplished by biological, chemical, and physical interactions in the soil matrix with the near-surface layer being the most active zone. The design flow path requires infiltration and typically lateral flow away from the application site. A cyclic application is the typical mode of operation with the flooding period followed by the drying period. This allows aerobic restoration of the infiltration surface and drainage of the applied percolate. The geohydrological aspects of the RI site are more critical than for other processes. Proper subsurface conditions and the local groundwater system are essential for design. Schematic views of the typical hydraulic pathways for RI process are shown in Figure 8.3. A much greater portion of the applied wastewater percolates to the groundwater than with slow-rate land treatment. The percolated water is collected for reuse by underdrains or recovery wells (Figure 8.3b,c). In some cases the percolated water can move through underground aquifers into surface streams nearby.

The main objective of a rapid infiltration system is wastewater treatment, and the system design and operating criteria are developed to achieve this goal. However, there are other objectives with respect to the utilization or final disposal of the treated water. They are:

1. groundwater recharge;
2. recovery of treated water for subsequent reuse or discharge;
3. recharge of surface streams;
4. seasonal storage of treated water beneath the site with recovery and agricultural reuse during the growing season.

The wastewater is applied to rapidly permeable soils by spreading in basins or by sprinkling, and is treated as it travels through the soil matrix. Vegetation is not usually used, but there are some exceptions.

Advantages of this process are: it is a treatment system with nearly complete recovery of renovated water; and it is a method of repelling saltwater intrusion into the aquifers. Where groundwater quality is being degraded by salinity intrusion, groundwater recharge can be used to reverse the hydraulic gradient and protect the groundwater.

The limitations of this process are in connection with groundwater effects: influent N may be converted to the nitrate form, which is leached to the groundwater; if the zone becomes anaerobic or anoxic conversion of sulfates to hydrogen sulfide may be a problem; and P retention in the soil matrix may be neither complete nor of long duration.

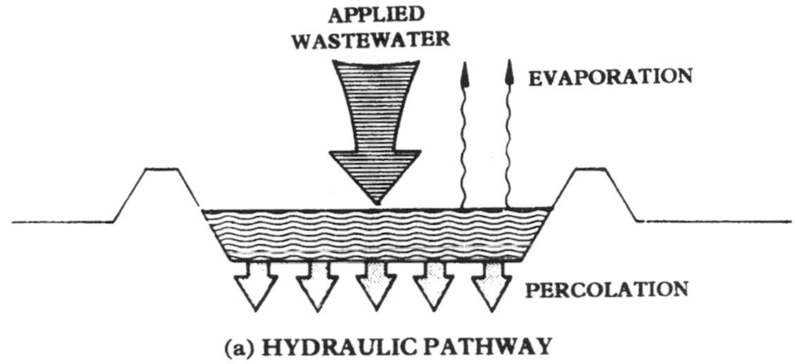

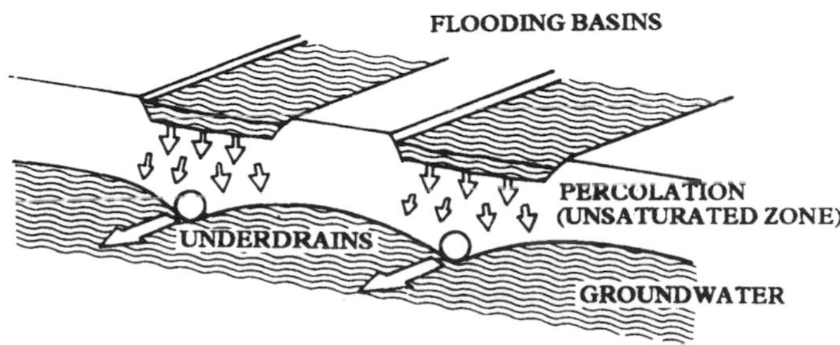

(b) RECOVERY OF RENOVATED WATER BY UNDERDRAINS

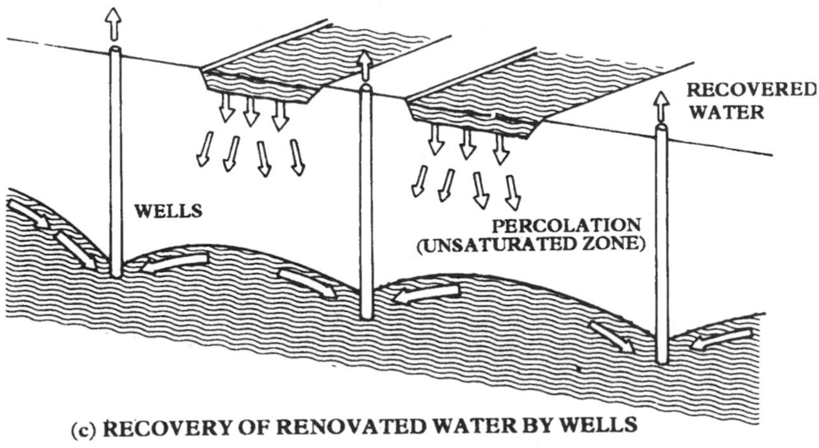

(c) RECOVERY OF RENOVATED WATER BY WELLS

Figure 8.3 Rapid infiltration (U.S. EPA, 1981)

RELIABILITY

Removal of constituents by the filtering and straining action of the soil is effective. Suspended solids, fecal coliforms, and BOD_5 are almost completely removed in most cases. N removal is generally about 50 percent, unless specific operating procedures are established to maximize denitrification. P removals range from 70 to 95 percent, depending on the physical and chemical characteristics of the soil that influence retention of P.

SITE SELECTION

Acceptable soil types include sand, sandy loams, and loamy sand (Figure 8.1). Very coarse sand and gravel are not ideal because they allow wastewater to pass too rapidly through the upper few centimeters of soil where biological and chemical actions take place. Other factors include percolation rate, depth, movement and quality of groundwater, topography, and underlying geological formations.

8.2.3 *Overland flow process (OF)*

A schematic view of overland flow treatment, including a pictorial view of a typical OF sprinkler system, are shown in Figure 8.4.

OF is a treatment method in which wastewater is applied over the upper reaches of sloped terraces and allowed to flow across the vegetated surface to runoff collection ditches. The objectives of OF process are: (i) to treat the wastewater to a degree comparable with that of secondary to tertiary treatment, and (ii) to produce forage grasses or preserving greenbelts and open spaces. OF is subject to the same type of limitations as SR, but it can be applied to relatively impermeable soil and a gently sloping terrain. The technique has considerable potential for treatment of municipal wastewater. The effluent is of a quality approaching that from tertiary treatment. In addition to a low construction cost the system produces little or no sludge, which is convenient for the system's operation and maintenance.

Operating costs are considerably lower than the conventional and advanced waste treatment because of the relative simplicity of operation. It has the advantages of avoiding groundwater degradation, providing economic return through the growth and sale of hay, and providing a high-quality effluent suitable for industrial or agricultural reuse. Sodium effects (discussed in section 8.3) on overland flow systems are less critical because the infiltration rates of the soils are low.

RELIABILITY

Overland flow systems at various places using wastewater have been monitored to determine removal efficiencies. The expected ranges based on results at these sites

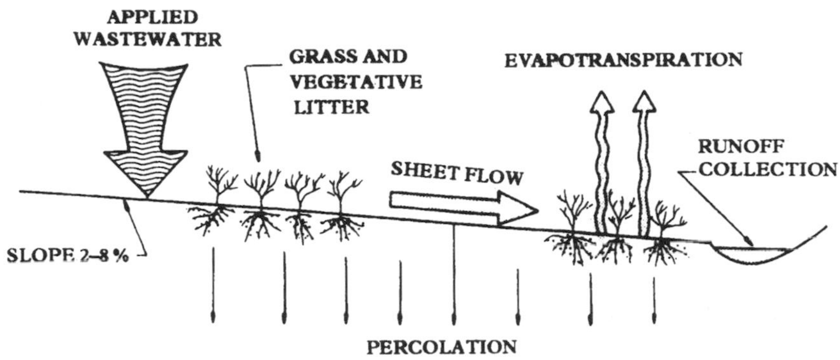

(a) HYDRAULIC PATHWAY

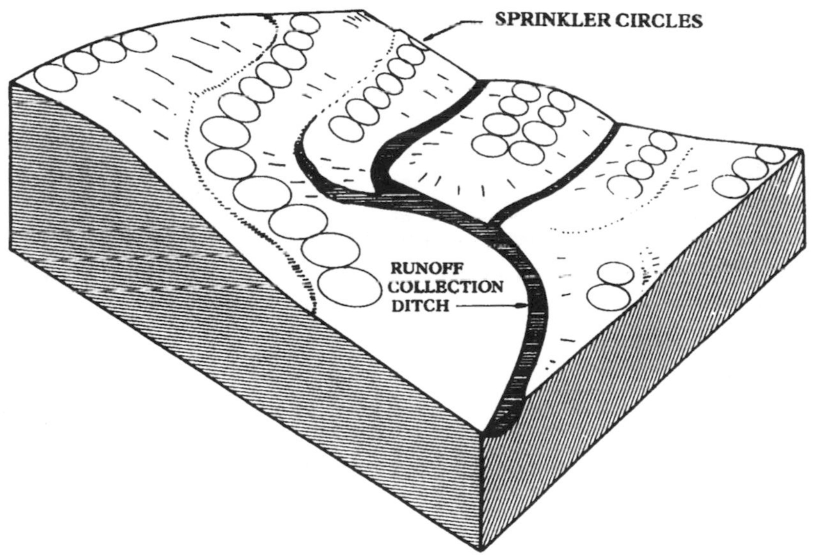

(b) PICTORIAL VIEW OF SPRINKLER APPLICATION

Figure 8.4 Overland flow (U.S. EPA, 1981)

are BOD_5 and SS removal of 95–99 percent, N removal of 70–90 percent, and P removal of 50–60 percent (see Table 8.3). Biological oxidation, sedimentation, and grass filtration are the primary removal mechanisms for organics and suspended solids. N removal mechanisms include crop uptake, biological uptake, denitrification, and fixation in soil. P is removed by adsorption and precipitation.

SITE SELECTION

Soils with limited drainability, such as clay and clay loams, are suitable for OF systems (Figure 8.1). The land may have a slope between 2 and 8 percent and a very smooth surface so that the wastewater will flow in a sheet over the ground surface. Grass is planted to provide a habitat for the bacteria to stabilize the wastewater. Because groundwater is not likely to be affected, it is of little concern in site selection.

8.2.4 Combined processes

Rapid infiltration can be used after overland flow to further reduce concentrations of BOD_5, suspended solids, and P. Because of the increased reliability and overall treatment capability, the application rates for the overland flow process could be higher than normal. This is shown schematically in Figure 8.5(a).

In another scheme, Figure 8.5(b), the rapid infiltration process precedes slow-rate treatment. The recovered renovated water should meet even the most restrictive requirements for use on food crops. The unsaturated zone can be used for storage of renovated water to be withdrawn on a schedule consistent with crop needs.

8.2.5 Groundwater recharge

Recharging underground aquifers with treated wastewater is one of the generally accepted forms of water reuse. The water disappears from the site and is usually

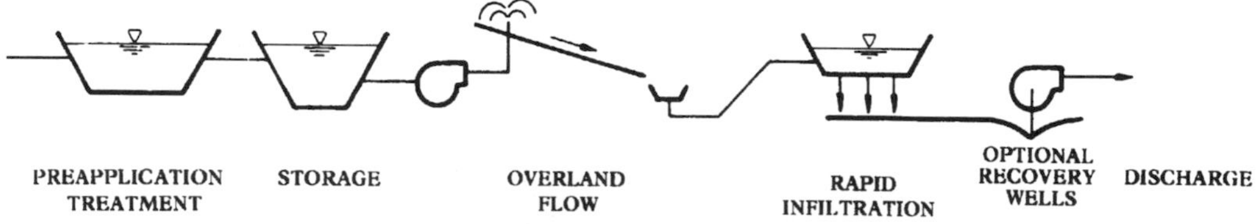

(a) OVERLAND FLOW FOLLOWED BY RAPID INFILTRATION

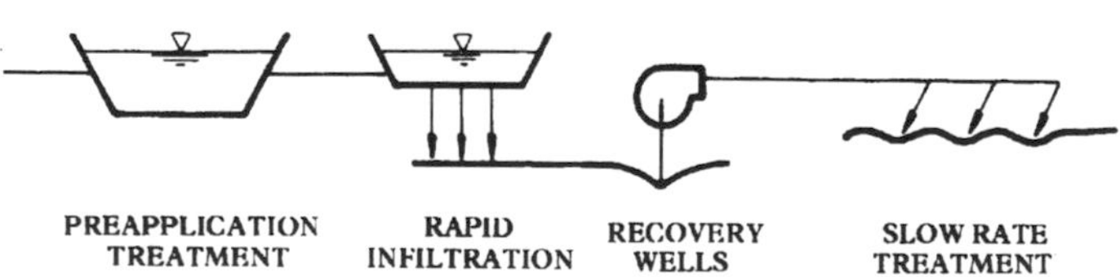

(b) RAPID INFILTRATION FOLLOWED BY SLOW RATE TREATMENT

Figure 8.5 Examples of combined systems (U.S. EPA, 1981)

diluted by other sources of water. Furthermore, a long time ordinarily elapses before recharged water is extracted, so most microbial contaminants will die off. Basically, there are two different methods for recharging an aquifer. Either water may be allowed to enter the aquifer from the surface through spreading beds or spray irrigation, or it may be pumped under pressure through to down-wells which will penetrate a deep aquifer.

Water applied to the soil surface will be filtered and biologically oxidized, and many chemical constituents will be adsorbed or precipitated on the soil particles. The treatment given by surface infiltration can be equivalent to, or better than, conventional biological treatment followed by filtration and disinfection.

8.3 WASTEWATER RENOVATION MECHANISMS

Any wastewater treatment system, including a land treatment system, is designed to convert the wastewater into an acceptable effluent and to dispose of the solids removed and produced in the system. The basic approach is to determine the characteristics of the untreated waste and to utilize the capabilities of various treatment processes to achieve the desired effluent quality.

The effectiveness of a land treatment system is related to the characteristics of the soil and the resultant pollutant removal mechanisms. When wastewater is applied to the soil some constituents may pass through the soil to the groundwater, some are utilized by growing plants, some are metabolized by the soil micro-organisms, and others are retained within the soil. The design of a wastewater land treatment system and the need for preapplication (or pretreatment) methods must relate the quantity and types of pollutants in the wastewater to the pollutant removal mechanisms in the soil. These can be categorized as physical, such as gravitational settling, filtration and dilution; chemical, such as adsorption and precipitation; and biological, such as microbial transformations and plant uptake. The need for preapplication can be determined by evaluating the possibility of overloading the removal mechanisms.

The waste management relationship that occurs with land treatment of waste-water is shown in Figure 8.6.

8.3.1 Physical removal mechanisms

As wastewater moves through the soil pores, suspended solids are removed by filtration. The depth at which removal occurs varies with the size of particles, soil texture, and rate of water movement. The higher the hydraulic application rate and the coarser the soil, the greater the distance the particles will move. However, at the wastewater application rates used with the slow-rate process, large suspended solids are removed in the surface soil and smaller particulates, including bacteria, are removed in the upper few centimeters of most soils, except the very coarse soils.

Constituents in the applied wastewater can be diluted by rain and, for cold climates, snow-melt. Chemical and biological transformations and removals in the

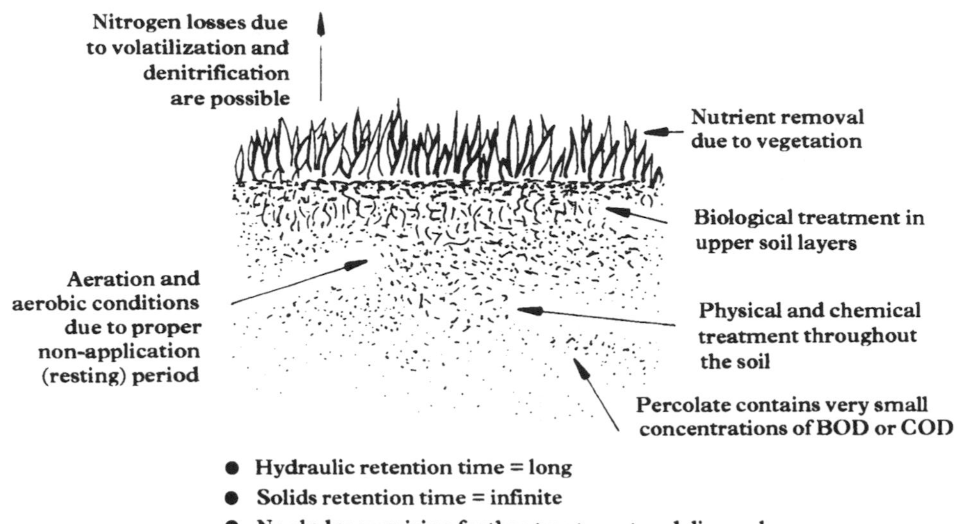

Figure 8.6 Waste management relationships that occur with land treatment (from Loehr and Overcash, 1985; reproduced by permission of the American Society of Civil Engineers)

soil can also reduce concentrations of specific constituents. Where evaporation losses are high, as in arid climates, increases in the concentration of conservative constituents such as salts can occur.

Excessive suspended solids can clog the soil pores as well as the wastewater distribution systems. Clogging of the soil will reduce the soil infiltration rate. Natural decomposition of the organic solids during non-application of wastewater or 'resting' periods will allow the infiltration rate to recover.

The design hydraulic application rate for a land treatment system should be less than the infiltration rate of the soil. At this application rate, soil clogging due to suspended solids will not be a significant problem. Thus preapplication methods for suspended solids removal should be limited to methods such as screening or primary sedimentation that will avoid clogging of the irrigation distribution equipment or avoid excessive wear of pumps and piping.

8.3.2 Chemical removal mechanisms

Chemical reactions in the soil affect the mobility of dissolved ions or compounds, with the result that some constituents are retained within the soil profile for extended periods of time while the movement of others may be only temporarily restricted. Liquid residence times for normal wastewater irrigation rates are of the order of weeks, while some constituents may be retained in the soil for a much longer period.

Adsorption and chemical precipitation are the most important chemical reactions governing the movement of constituents in the irrigated wastewater, with cation exchange being the most important adsorption phenomenon. The cation exchange capacity (CEC) of soils can range from 2 to 60 meq/100 g of soil, with most soils having a CEC value between 10 and 30. The difference occurs because soils vary widely in their humus and clay content, the components that have the highest CEC.

Typical soils have considerable capacity to adsorb many of the cations in wastewater, including many of the metals which may adversely affect the health of humans and animals eating the crops grown on the irrigated fields.

Cation exchange of ammonium N is a possible control mechanism for N. However, the ammonium ion is biologically oxidized to nitrate in aerobic soils. Nitrate is an anion and will move with the soil water.

Phosphate is the only anion appreciably retained in soil. The primary mechanism is the formation of insoluble or slowly soluble precipitates.

In arid regions wastewater irrigation rates may not be enough to avoid the accumulation of sodium ions in the soil. Such accumulations can lead to a degradation of soil structure and reduction in water percolation rates.

The relationship between the principal cations in wastewater (calcium, magnesium, and K) is of importance. When the ratio of sodium to the other cations, especially calcium and magnesium, becomes too high, the sodium tends to replace the calcium and magnesium ions on clay particles. The predominance of sodium ions on clay particles has the effect of dispersing the soil particles and decreasing the soil permeability. To determine the sodium hazard, the SAR (sodium adsorption ratio) had been developed. It is defined as follows.

$$SAR = \frac{Na}{[(Ca + Mg)/2]^{0.5}} \tag{8.1}$$

where Na, Ca, and Mg are concentrations of the respective ions in milliequivalents per liter of water (meq/L).

When industrial wastes are included in the wastewater to be irrigated, adjustment of the pH and SAR may be needed. The wastewater to be applied should have a pH within the range of 6.0–9.5 to avoid adverse effects to site vegetation. Wastewater with high SAR must be pretreated to remove Na, or accompanied by special soil management procedures to compensate for the effect of the sodium. The SAR of wastewaters used for irrigation should be no more than 8–10, otherwise these wastewaters should not be used in irrigation.

Some preapplication methods that may be used to control the chemical reactions in the soil could include:

1. pretreatment controls, such as industrial source control or chemical precipitation, if the amounts of potentially toxic elements and chemicals in the wastewater are likely to exceed the chemical removal mechanisms in the soil;
2. adjustment of wastewater pH and SAR to acceptable levels.

The relationship of potential problems to concentrations of major inorganic constituents in irrigation waters is shown in Table 8.6.

Table 8.6 Relationship of potential problems to concentration of major inorganic constituents in irrigation waters for arid and semi-arid climates

Problem and related constituent	No problem	Increasing problems	Severe problems
Salinity[a]			
EC of irrigation water, mmhos/cm	< 0.75	0.75–3.0	> 3.0
Permeability			
EC of irrigation water, mmhos/cm	> 0.5	< 0.5	< 0.2
SAR (sodium adsorption ratio)[b]	< 6.0	6.0–9.0	> 9.0
Specific ion toxicity[c]			
From root absorption			
Sodium (evaluate by SAR)	< 3	3.0–9.0	> 9.0
Chloride, meq/L	< 4	4.0–10	> 10
Chloride, mg/L	< 142	142–355	> 355
Boron, mg/L	< 0.5	0.5–2.0	2.0–10.0
From foliar absorption (sprinklers)[d]			
Sodium, meq/L	< 3.0	> 3.0	—
Sodium, mg/L	< 69	> 69	—
Chloride, meq/L	< 3.0	> 3.0	—
Chloride, mg/L	< 106	> 106	—
Miscellaneous[e]			
HCO_3^-, meq/L	< 1.5	1.5–8.5	> 8.5
HCO_3^-, mg/L	< 90	90–250	> 520
pH	Normal range = 6.5–8.4		

Note: Interpretations are based on possible effects of constituents on crops and/or soils. Suggested values are flexible and should be modified when warranted by local experience or special conditions of crop, soil, and method of irrigation

[a] Assuming water for crop plus water needed for leaching requirement will be applied. Crops vary in tolerance to salinity. Electrical conductivity (EC) mmhos/cm × 640 = approximate total dissolved solids (TDS) in mg/L or ppm

[b] $SAR = Na/\sqrt{(Ca + Mg)/2}$
where Na = sodium; Ca = calcium; Mg = magnesium, all in meq/L

[c] Most tree crops and woody ornamentals are sensitive to sodium and chloride (use values shown). Most annual crops are not sensitive

[d] Leaf areas wet by sprinklers (rotating heads) may show a leaf burn due to sodium or chloride absorption under low-humidity, high-evaporation conditions. (Evaporation increases ion concentration in water films on leaves between rotations of sprinkler heads)

[e] HCO_3^- with overhead sprinkler irrigation may cause a white carbonate deposit to form on fruit and leaves

From Ayers and Westcot (1976); reproduced by permission of the Food and Agriculture Organization of the United Nations

8.3.3 Biological removal mechanisms

The biological transformations that occur in the soil include organic matter decomposition and nutrient assimilation by plants. These transformations occur in the biologically active upper few centimeters of the soil, i.e. the rooting zone. The numbers of bacteria are large, ranging from one to three billion per gram of soil. The great diversity of native organisms enhances the capability of a soil to degrade the variety of natural and man-made organic compounds in the applied wastewater.

The presence or absence of oxygen in the soil has a significant effect on the rate and end-products of degradation. The oxygen status of the soil is a function of soil porosity. Soil properties that allow for rapid infiltration and transmission of the applied wastewater also yield good oxygen movement. Low and intermittent wastewater application rates used with the SR system normally result in aerobic conditions, rapid organic matter decomposition, and oxidized end-products.

Oxygen at the soil surface must diffuse into the soil layer or waste–soil matrix, depending on the nature of the wastewater application and soil–water migration of liquid and waste organics (Figure 8.7). The transfer mechanism is oxygen diffusion. Because soil pores are usually smaller and have solids deposits, the diffusion of oxygen into soils may be the rate-limiting step in satisfying the waste oxygen demand and maintaining aerobic soil conditions. The phenomenon of the transfer of photosynthetic oxygen from the crop's leaves to the root zone, similar to that of aquatic weed (see Chapter 7), is not yet well understood.

As a result of organic matter decomposition, elements such as N, P, and S are converted from organic to inorganic forms. Many of these mineralized constituents can be assimilated by plants. Crops are an essential part of the SR and OF processes (see Table 8.2).

The biological nitrification processes in the soil produce nitrate from ammonia and organic N under aerobic conditions (eq. 3.3). However, nitrate compounds can be reduced into N gas under anoxic conditions as a result of denitrification. It is possible to consider both gaseous N losses (volatilization and denitrification) and N removal by plant uptake as control mechanisms for the N in the applied wastewater.

Crop selection and management are important components of the wastewater irrigation system. Plant uptake of N is in the range of 100–400 kg/ha per growing season, depending upon specific crop and management techniques. Data on nutrient (N, P, K) uptake rates for selected crops are given in Table 8.7. It appears from this table that forage crops have the ability to take up nutrients better than do field crops. However, nutrient contents of the soil and soil characteristics have influence on the nutrient uptake rates, and detailed investigation at the land treatment site would be useful.

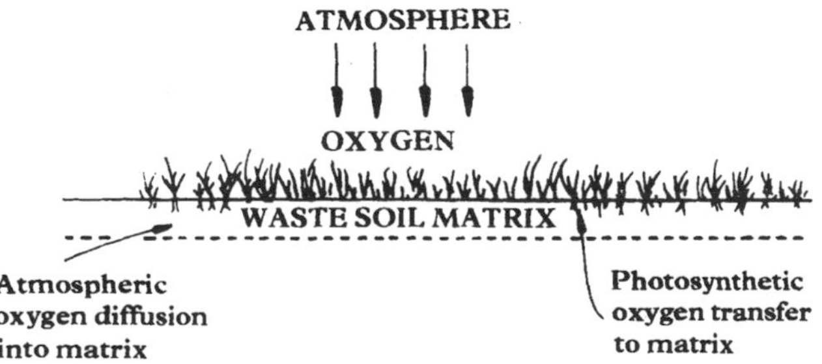

Figure 8.7 Schematic of oxygen transfer to stabilize organic compounds applied to land

Table 8.7 Nutrient uptake rates for selected crops, kg/(ha-year) (U.S. EPA, 1981)

	Nitrogen	Phosphorus	Potassium
Forage crops			
Alfalfa[a]	225–540	22–35	175–225
Bromegrass	130–225	40–55	245
Coastal bermudagrass	400–675	35–45	225
Kentucky bluegrass	200–270	45	200
Quackgrass	235–280	30–45	275
Reed carnarygrass	335–450	40–45	315
Ryegrass	200–280	60–85	270–325
Sweet clover[a]	175	20	100
Tall fescue	150–325	30	300
Orchardgrass	250–350	20–50	225–315
Field crops			
Barley	125	15	20
Corn	175–200	20–30	110
Cotton	75–110	15	40
Grain sorghum	135	15	70
Potatoes	230	20	245–325
Soya beans[a]	250	10–20	30–55
Wheat	160	15	20–45

[a] Legumes will also take nitrogen from the atmosphere

The N application rate should be determined from an N balance on the system. The important processes involved in N removal from wastewater applied to the land are ammonia volatilization, crop uptake and removal, soil adsorption of ammonia, incorporation into the soil organic fraction, and denitrification. With the slow-rate process, N management is principally due to crop uptake with some denitrification. The proper application rate will be that which, when crop uptake and denitrification are considered, maintains the N concentration in the percolating water below allowable limits, generally less than 10 mg/L nitrate N.

The factors that favor denitrification in the soil are: high organic matter, fine-textured soil, frequent wetting or high groundwater table, neutral to slightly alkaline pH, vegetative cover, and warm temperature.

Denitrification losses of N can range up to 50 percent, depending upon how the land disposal site is managed. A conservative estimate would be to assume denitrification and volatilization losses to be 20–30 percent of the applied N. Consideration of plant uptake and these losses as N control mechanisms will reduce the risk of excessive N in the percolate (Broadbent *et al.*, 1977).

Preapplication approaches related to biological mechanisms could include N removal prior to land application where the N application rate is the limiting factor and the required land area is excessive. Such removal could include nitrification followed by denitrification in preliminary wastewater storage ponds.

8.4 SYSTEM DESIGN AND OPERATION

Three methods can be used to determine wastewater application rates in land treatment systems: field measurements; comparison with application rates from similar existing projects; and combination of the above two methods together with previous experience and judgement. In view of the present state of the art in land treatment technology, the third method is recommended, so as to maximize land treatment efficiency and minimize operation and maintenance costs.

In SR or irrigation systems, application rates are determined by using water balance, N balance, and measured or estimated percolation rates. In rapid infiltration systems field measurements are necessary, coupled with a detailed knowledge of the subsurface hydrology and comparisons with existing systems. In OF systems, application rates depend mostly on required wastewater treatment and are currently determined by comparison with existing treatment systems or by calculation from kinetic equations.

8.4.1 Irrigation or SR system

WASTEWATER APPLICATION RATE

Typical wastewater application rates for irrigation or slow-rate systems range from about 1.5 to 10 cm/week (Figure 8.1 and Table 8.2). The choice depends on the climate, soil permeability, crop type and management practices, and required quality of the treated water. Information on climate and soil permeability can be obtained from meteorological data and site investigation, respectively; this information is used in the water balance calculation as shown in the next section on 'Hydraulic loading rate'. The required quality of the treated water is usually evaluated by using N balance. When crop production is the priority, the irrigation requirements of the crop may limit the application rate.

Example 8.1 For a clay-loam soil which is irrigated periodically, the moisture content before irrigation is 19 percent by weight. If the effective infiltrated irrigation water is 1,000 m^3/hectare, estimate the soil moisture content after irrigation. Assume that the soil's bulk density is 1.35, and the depth to which the soil is wetted is 90 cm. What should be the period of irrigation if the evapotranspiration rate is 250 mm/month?

Infiltrated water, $I = 1{,}000\ m^3/10{,}000\ m^2 = 100$ mm

The effect of irrigation on the soil moisture content can be estimated from the following equation.

$$I = (P'' - P')SD/100 \qquad (8.2)$$

where I = effective infiltrated irrigated water, mm;
P'' = final moisture content, percentage by weight;
P' = initial moisture content, percentage by weight;
S = soil's bulk density;
D = depth of soil to the root zone to be wetted, mm

Hence,

$$100 \text{ mm} = (P'' - 19 \text{ percent})1.35 \times 900/100$$

Thus, $P'' = 27.3$ percent.

Therefore, soil moisture content after irrigation = 27.3 percent.

$$\text{Required irrigation period} = \frac{100 \text{ mm}}{250 \text{ mm/month}} = 12 \text{ days.}$$

Select 5 days for irrigation period and 7 days for resting or drying period. Wastewater application rate is 100 mm/5 days or 20 mm/day during the irrigation period.

HYDRAULIC LOADING RATE

In the water balance the inputs are wastewater application and precipitation, and the outputs are evapotranspiration, percolation, and runoff. Precipitation and evapotranspiration values should be determined for a design year which is wetter than the normal climatic condition. The percolation rate can be either estimated from the soil permeability or measured in the field. Recommended field measurement techniques for soil percolation rates are basin flooding or the sprinkler infiltrometer, the details of which can be found in U.S. EPA (1981) or other handbooks on groundwater and hydrology.

Using water balance, the following is obtained:

$$L_h + P_p = ET + W + R \tag{8.3}$$

where L_h = wastewater hydraulic loading rate, cm/week;
P_p = design precipitation, cm/week;
ET = evapotranspiration, cm/week;
W = percolation, cm/week;
R = net runoff, cm/week.

For slow-rate systems the design is such that there is no surface runoff. Therefore net runoff, R, is usually negligible in the calculation.

N LOADING RATE

Because of the ability of N to move with the percolating water, N is often the limiting water quality parameter in SR and RI systems. In slow-rate systems the input N is balanced against crop uptake, denitrification, and the N that percolates through the root zone. The climate can also influence the N balance. For example,

Table 8.8 Comparison of nitrogen loadings in humid and arid climates (Olsen and Kemper, 1968)

Parameter	Humid climate	Arid climate
Applied nitrogen, mg/L	25	25
Precipitation minus evapotranspiration, m/year	0.5	– 0.5
Crop uptake, kg/ha-year	336	336
Denitrification, percentage of applied nitrogen	20	20
Hydraulic loading, m/year	4.0	2.9
Wastewater nitrogen loading, kg/ha-year	980	720

in a humid climate the water from precipitation (in excess of evapotranspiration) can dilute the percolating N concentration, as shown in Example 8.2 and Table 8.8, for a percolate containing a maximum of 10 mg/L N. Thus, under a humid climate condition (as shown in Table 8.8), 36 percent more N can be applied than in the arid climate.

The annual N balance is given by:

$$L'_n = U + D + 0.1WC_n \tag{8.4}$$

where L'_n = wastewater loading, kg/(ha-year);
U = crop N uptake, kg/(ha-year);
W = percolation rate, cm/year;
C_n = percolate N concentration, mg/L;
D = denitrification rate, kg/(ha-year).

The term $0.1WC_n$ on the right-hand side of eq. 8.4 is the percolate N loading or P'_n, whose unit is kg/(ha-year).

Example 8.2 Determine the N balance for a designed percolate N concentration (C_n) of 10 mg/L for the following conditions:

	Humid	*Arid*
1. Applied N concentration, C'_n, mg/L	25	25
2. Crop N uptake, U, kg/(ha-year)	336	336
3. Denitrification as percentage of applied N	20	20
4. Precipitation (P_p) – evapotranspiration (ET), cm/year	50	–50

Assuming denitrification (D) is 20% of the wastewater N loading (L'_n).

From eq. 8.3 the annual water balance is:

$$\begin{aligned} L_h + P_p &= ET + W + R \\ \text{or} \quad W &= L_h + P_p - ET - 0 \text{ (assuming } R \text{ is negligible)} \\ W &= L_h + 50 \text{ (humid)} \\ W &= L_h - 50 \text{ (arid)} \end{aligned}$$

The amount of percolating water, W, resulting from the applied wastewater, L_h, has a significant effect on the allowable N loading, L_n.

From eq. 8.4 the annual N balance is:

$$L'_n = U + D + 0.1WC_n$$
$$\text{for humid} \quad L'_n = 336 + 0.2L'_n + (0.1)(L_h + 50)(10)$$
$$\text{for arid} \quad L'_n = 336 + 0.2L'_n + (0.1)(L_h - 50)(10)$$

The relationship between the N and the hydraulic loading is: $L'_n = 0.1C'_nL_h$ (SI units). Where L'_n = wastewater N loading, kg/(ha-year); C'_n = applied N concentration, mg/L; L_h = wastewater hydraulic loading, cm/year.

$$L'_n = 0.1 \times 25L_h = 2.5L_h$$
$$\text{or } L_h = 0.4L'_n$$

for humid climate,

$$L'_n = 336 + 0.2L'_n + 0.1(0.4L'_n + 50)10$$
$$= 965 \text{ kg/(ha-year)}$$
$$L_h = 965 \times 0.4 = 386 \text{ cm/year}$$

for arid climate,

$$L'_n = 336 + 0.2L'_n + 0.1(0.4L'_n + 50)10$$
$$= 715 \text{ kg/(ha-year)}$$
$$L_h = 715 \times 0.4 = 286 \text{ cm/year}$$

Summary of solutions

	Humid	*Arid*
1. Wastewater N loading (L'_n), kg/(ha-year)	965	715
2. Wastewater hydraulic loading (L_h), cm/year	386	286
3. Percolating water (W), cm/year	436	236
4. Denitrification (D), kg/(ha-year)	193	143
5. Percolate N loading (P'_n), kg/(ha-year)	436	236
($P'_n = 0.1C_nW = L'_n - U - D$)		

APPLICATION SCHEDULE

The application schedule for irrigation depends on the soil permeability, type of crop, application technique, and climate. Operator convenience should also be considered. For permeable soils the schedule should be to irrigate once a week or more frequently. For less permeable soils using surface irrigation the application schedule should include a longer period of drying.

8.4.2 *Rapid infiltration of RI system*

WASTEWATER APPLICATION RATE

In rapid infiltration, evaporation and vegetation are relatively unimportant, but the soil permeability is critical. It is therefore important to concentrate the planning and design on determining the optimal infiltration rate to ensure that the system will work hydraulically while providing the necessary wastewater treatment. As given in Figure 8.1 and Table 8.2, typical application rates for RI systems range from 10 to 250 cm/week.

HYDRAULIC LOADING RATE

Although the soil permeability can be related to the infiltration rate, as previously described, it is recommended that the soil profile be evaluated and that field measurements of infiltration rates be conducted. The preferred method of determining the infiltration rate depends on the nature of the soil profile. If the profile is generally homogeneous a surface flooding basin 2 m or more in diameter can be used. The basin is filled with clean water until the soil is saturated, and then the rate if infiltration is measured. Clean water is generally used unless the actual wastewater is available.

TREATMENT PERFORMANCE

In rapid infiltration systems the required treatment performance is of primary importance in determining the application rate. Lance and Gerba (1977) showed that decreasing the application rate from the hydraulic limit can result in increased removals of constituents, especially N. Because the chief mechanism of N removal in rapid infiltration systems is denitrification, the following requirements of biological denitrification must be met: adequate detention time, anoxic conditions (or at least anaerobic micro-sites), and adequate C to drive the reaction. The reduction in application rate increases detection time and increases the potential for denitrification. The effect of infiltration rate on N removal for a rapid infiltration site in Phoenix, Arizona, is shown in Figure 8.8.

APPLICATION SCHEDULE

Treatment efficiency in rapid infiltration systems responds to variations in application cycle. Short and frequent application schedules such as 0.5–3 days on and 1–5 days drying maximize nitrification, probably because the soil is exposed more to air and aerobic conditions, but minimize denitrification or N removal. As the drying time during the cycle increases, the potential for N removal increases. For example, the application schedule at Hollister, California, is 1–2 days on and 14–21 days off. At Phoenix, Arizona, maximum N removal occurred when 10 days on was followed by 10–20 days off.

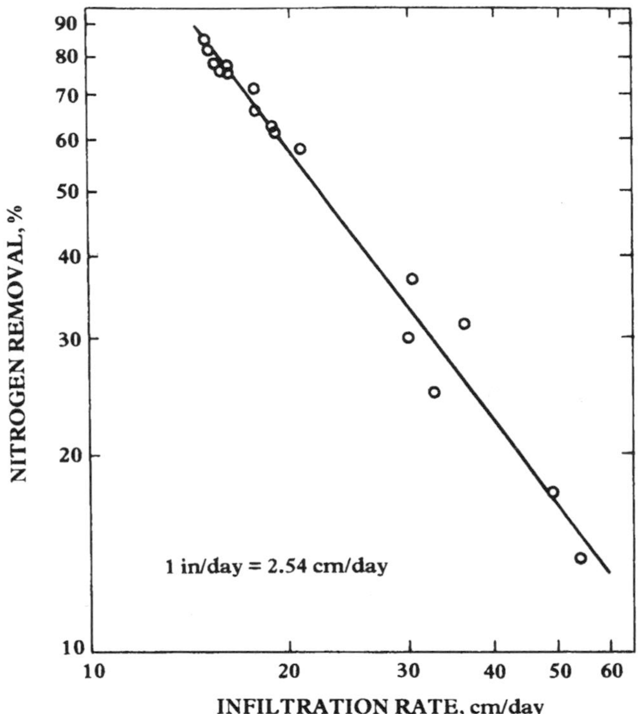

Figure 8.8 Effect of infiltration rate on nitrogen removal for rapid infiltration, Phoenix, Arizona (from Lance and Gerba, 1977; reproduced by permission of Pergamon Books Ltd)

8.4.3 Overland flow or OF system

WASTEWATER APPLICATION RATE

OF is a form of a fixed-film biological reactor. The application rate should therefore be predictable on the basis of the kinetics of treatment and the required treatment levels. For example, high-quality effluent has been produced in research projects using these application rates: raw wastewater, 10 cm/week; primary effluent, 15–20 cm/week; and secondary effluent, 25–40 cm/week. With all three application rates the runoff contained less than 10 mg/L BOD_5 on the average. Slopes of the land were 2–4 percent and 36 m long (Tucker and Vivado, 1980).

Wastewater application rates for OF systems are usually determined from comparisons with existing systems and research projects (see also Figure 8.1 and Table 8.2). For raw wastewater 7.5–10 cm/week should be considered. For primary effluent, 10–20 cm/week could be used, depending on the level of overall treatment required. For secondary treatment facilities (e.g. activated sludge or trickling filters), the need is either for polishing (further reductions for BOD_5 and suspended solids) or for nutrient removal. For polishing, 20–40 cm/week could be considered. For nutrient removal only 10–20 cm/week may be possible because adequate detention time is necessary for denitrification.

APPLICATION SCHEDULE

Experience with existing systems suggests application schedules to be 6–8 h on and 16–18 h off over 5–6 days a week. At Melbourne, Australia, however, the application has been continuous rather than intermittent. The optimum will depend on the climate and the BOD_5 loading. Other considerations include the harvesting of grass and the potential for propagation of insects.

8.4.4 Other design considerations

There are many factors that determine the area required for a wastewater land treatment system. These factors are related to the characteristics of the soil, wastewater, climate, crop, and preapplication treatment, and should be evaluated using site-specific information.

In determining the required land area, the land area for each potentially limiting parameter (such as hydraulic, N, toxic compounds, and salt loadings) should be evaluated. The parameter which requires the largest land area to avoid environmental problems becomes the limiting parameter. When the land area determined for the limiting parameter is used for the design of a land treatment system, there is an additional degree of safety in terms of the application rates of the other constituents or parameters of concern.

It should be possible to determine application rates for land treatment systems on a case-by-case basis using data on soil infiltration rates, climate, wastewater characteristics, and required treatment performance. Unfortunately, neither the relationship between the infiltration rate and loading rate nor the relationship between the loading rate and treatment performance has been developed to the point where they can be used in design without requiring comparisons with other land treatment systems.

PREAPPLICATION TREATMENT

Preapplication treatment of wastewater to be used for land treatment is necessary to:

1. inactivate the enteropathogenic organisms to an acceptable level, so that public health risks to the workers, grazing animals or human beings who consume those irrigated crops are minimized (see section 8.7);
2. minimize the operational problems in distribution systems such as clogging of the spray nozzles and distribution pipes, etc.;

3. lengthen the working life of land treatment sites as clogging of the soil pores will be less when pretreated wastewater is applied;
4. prevent nuisance conditions during wastewater storage prior to being applied to land.

Screening, screening with comminution, aeration, ponding, and disinfection are the common types of pretreatment facilities used. In case of industrial wastewaters containing some toxic compounds, other preapplication treatment such as chemical precipitation and flocculation or sedimentation might be necessary. Where concern is on nitrate contamination to groundwater aquifers, preapplication treatment may include ammonia stripping, water hyacinth pond treatment, or storage ponds having a long detention time, to reduce the N content.

The level of preapplication treatment provided should be the minimum necessary to achieve objectives stated above. Any additional treatment, in most cases, will only increase cost and energy use.

CROP SELECTION

Crop selection is an important factor in SR and OF processes. Presence of suitable crops on the application area will reduce the erosion, provide support media for microorganisms, remove the wastewater nutrients by plant uptake, and produce revenue where markets for these crops exist. Nutrient uptake capacity (see Table 8.7), tolerance to high soil moisture conditions, consumptive use of water and irrigation requirements, and revenue potential should be considered when selecting a crop for SR and OF systems. In SR systems presence of crop will maintain or increase the soil infiltration rate. In OF systems a mixture of grasses is generally preferred over a single species. Corn and paddy are also suitable crops to be used but public health aspects should be studied in detail, as these crops are consumed by the public.

STORAGE

Storage of wastewater is required for any process of land treatment in order to equalize the incoming flows, to maintain constant application rates, and in the case of repairs or operational problems in the distribution system. Even though SR and RI systems are capable of operating during adverse climatic conditions storage may be required. In the case of OF systems, storage is necessary to accommodate the wastewater during rainy and winter seasons due to reduced hydraulic loading or complete shutdown, and also to accommodate stormwater runoff. Generally 5–10 days of storage is sufficient to overcome the problems.

DISTRIBUTION SYSTEM

Distribution system design consists of two important components: (i) selection of the type of distribution system; and (ii) details of system components.

Method of wastewater distribution varies with the type of land treatment system. For SR systems, either a surface distribution system in which wastewater is applied to the land at one end and allowed to spread over the field by gravity (Figure 8.2b) or a sprinkler distribution system where the wastewater is spread over the land by sprinklers (Figure 8.2c) is suitable. In the case of an RI system, wastewater distribution is usually by surface spreading (Figure 8.3). This method employs gravity flow from piping systems or ditches to flood the application area. Wastewater distribution onto OF slopes can be achieved by surface methods, low-pressure sprays and high-pressure impact sprinklers. Choice of the appropriate system should be based on minimization of operational difficulties such as uneven distribution on to the slope, and also the operational cost of the system. If the surface distribution method is adopted, the land should be perfectly leveled to avoid ponding. Extreme care should be taken when a sprinkler system is used for the treatment areas situated near residential areas, to avoid spreading pathogens through aerosols and wind action.

8.5 LAND TREATMENT—DESIGN EQUATIONS

Mathematical models for the three land treatment processes have not been extensively developed. This is due partly to the complex nature of land treatment involving physical, chemical, and biological reactions. Efficiency of a land treatment system also depends on climate and soil conditions at the specific site. An SR process is normally designed based on practical experience. Some empirical and rational models for the RI and OF processes have been developed.

8.5.1 RI process

The treatment performance of RI systems with respect to N and P removal is given below.

N REMOVAL

Application rates up to 30 cm/day with 20 mg/L of ammonia will result in a nitrified effluent. As wastewater temperatures drop, the rate of nitrification will also decrease. For nitrification the loading cycle should consist of short application periods (1 day or so) and relatively long drying periods (5–10 days) to allow for aeration and the occurrence of aerobic conditions in the soil favorable for the growth of nitrifying bacteria.

N removal by denitrification requires adequate organic C and adequate detention time. N removal can be expressed by the following equation (Reed and Crites, 1984):

$$N_{\mathrm{t}} = \frac{\mathrm{TOC} - 5}{2} \tag{8.5}$$

where N_{t} = total N removal, mg/L;
TOC = total organic C in the applied wastewater, mg/L.

The 5 mg/L of residual TOC is typical for municipal wastewater passage through about 5 ft (1.5 m) of soil. The coefficient 2 in the denominator is based on experimental data where 2 g of wastewater C were required to denitrify 1 g of wastewater N.

N removal is also related to infiltration rate, and an application rate of 15 cm/day is recommended as a maximum where 80 percent N removal is needed with secondary effluent. When primary effluent is used the maximum application rate is recommended not to exceed 20 cm/day.

P REMOVAL

There is no crop uptake in RI systems, and the nature of the soils and high hydraulic rates used require greater travel distances in the soil for effective P removal.

An equation to predict P removal at RI sites has been developed from data collected at a number of operating systems (Reed and Crites, 1984).

$$P_{\mathrm{x}} = P_0 \cdot \exp(-kt) \tag{8.6}$$

where P_{x} = total P concentration at a distance x along the percolate flow path, mg/L;
P_0 = total P concentration in the applied wastewater, mg/L;
k = instantaneous rate constant, 0.002/h at neutral pH;
t = detention time $= x\theta/I$, h;
x = distance along the flow path, inch;
θ = volumetric water content, inch^3/inch^3, usually 0.4;
I = infiltration rate, inch/h.

8.5.2 OF process

Based on a pilot study conducted by Smith and Shroeder (1985), it was found that the removal of organic material from primary effluent is accomplished in two stages as a function of slope length (z) and application rate (q).

In the first stage a rapid removal of organic matter takes place within the first 6–10 m of slope length, and in the second stage a slower rate of organic removal takes place in the remainder of the slope.

From a practical design and operation standpoint only the second stage would be the basis for design and operation.

The probable reason for the observation of the two distinct rates of organic removal over the length of the overland flow slope is that settleable matter is removed readily by sedimentation in the first few meters, leaving colloidal and soluble organic material to be removed at a slower rate by adsorption and, ultimately, bacterial assimilation over the remaining length of the slope.

The mathematical model was developed from the studies made on lands with a down-slope grade of 2 percent and a cross-slope grade of 0.2 percent with a mixture of perennial rye grass, annual rye grass, orchard grass, and tall fescue grass. The tall fescue grass was predominant.

BOD_5 AND TOC REMOVAL

Removal of BOD_5 and TOC from wastewater and primary effluent after the first 9 m of travel down the overland flow slope followed the first-order removal model:

$$C_z/C_0 = A \cdot \exp(-Kz) \tag{8.7}$$

where z = distance down-slope, in m ($z > 9$ m);
C_z = organic concentration (BOD_5 or TOC) at a distance (z) down-slope, mg/L;
C_0 = initial organic concentration (BOD_5 or TOC), mg/L;
K = empirically determined overall rate constant;
A = empirically determined coefficient.

It has been shown that BOD_5 concentrations in the runoff from overland flow slopes approach a non-zero value in the range of 3–5 mg/L. Effluent BOD_5 and TOC concentrations approach average values of 5 and 7 mg/L, respectively, after 64 m of down-slope travel. Hence the removal models were modified to account for this as follows:

BOD_5 removal model:

$$(C - 5)/C_0 = A \exp(-Kz) \tag{8.8}$$

TOC removal model:

$$(C - 7)/C_0 = A' \exp(-K'z) \tag{8.9}$$

where C = required effluent BOD_5 or TOC concentrations at distance z;
K and K' = empirical rate constants for BOD_5 and TOC removal, respectively;
A and A' = empirical coefficients for BOD_5 and TOC removal, respectively.

The overall rate constants K and K' in eqs 8.8 and 8.9 vary as a function of application rate q, according to the following equation:

$$K \text{ or } K' = k/q^n \tag{8.10}$$

Table 8.9 Summary of coefficients for BOD_5 and TOC removal models

Applied wastewater	Coefficient value	
	k	n
Primary effluent		
BOD_5	0.043	0.136
TOC	0.038	0.170
Screened raw wastewater		
BOD_5	0.030	0.402
TOC	0.032	0.350

From Smith and Schroeder (1985); reproduced by permission of the Water Pollution Control Federation, U.S.A.

where k and n are empirical coefficients and q is wastewater application rate whose unit is m^3/(h-m of slope width). Then:

$$BOD_5 \text{ model}: (C-5)/C_0 = A \cdot \exp(-kz/q^n) \tag{8.11}$$

$$\text{TOC model}: (C-7)/C_0 = A' \cdot \exp(-kz/q^n) \tag{8.12}$$

The values of coefficients k and n, obtained for q values ranging from 0.1 to 0.37 m^3/(h-m) are summarized in Table 8.9.

A family of curves that describe BOD_5 and TOC removal as functions of slope length (up to 64 m) and application rates (0.1, 0.16, 0.25, and 0.37 m^3/(h-m)) for raw wastewaters and primary effluent are shown in Figures 8.9–8.12 (application periods were up to 16 h/day). Either application rate or slope length may be used as the independent design variable. The variation of A and A' with q was found to be inconsistent. Therefore, for design purposes, it would be satisfactory to use directly the family of curves given in Figures 8.9–8.12.

When designing a system the required or desired level of treatment is known. Thus a value for $(C-c)/C_0$, according to eq. 8.8 or eq. 8.9 can be established. If slope length is used as the independent variable then a design value for slope length, z, is selected. The required value of application rate, q, is then determined by entering the appropriate curve with the known values of z and $(C-c)/C_0$. Alternatively, if application rate q is the independent design variable, then a value for q is selected and the required value of slope length, z, is taken from the curves. The independent design parameter used depends on the individual case.

A later study by Witherow and Bledsoe (1986) confirmed the removal of BOD_5, TOC, total suspended solids (TSS) and NH_4–N to follow the models given in eqs 8.11 and 8.12. By setting n at 0.5 and using field data from a 4 ha OF site in Oklahoma, U.S.A., the intercept constant A and rate constant k of eq. 8.13 for the four pollutants are given in Table 8.10.

$$C_z/C_0 = A \cdot \exp(-kz/q^{0.5}) \tag{8.13}$$

The total area required for slopes must be determined to complete the system design. Total area required may be computed by the following equation:

$$\text{Area} = (Qz)/(qP_d) \tag{8.14}$$

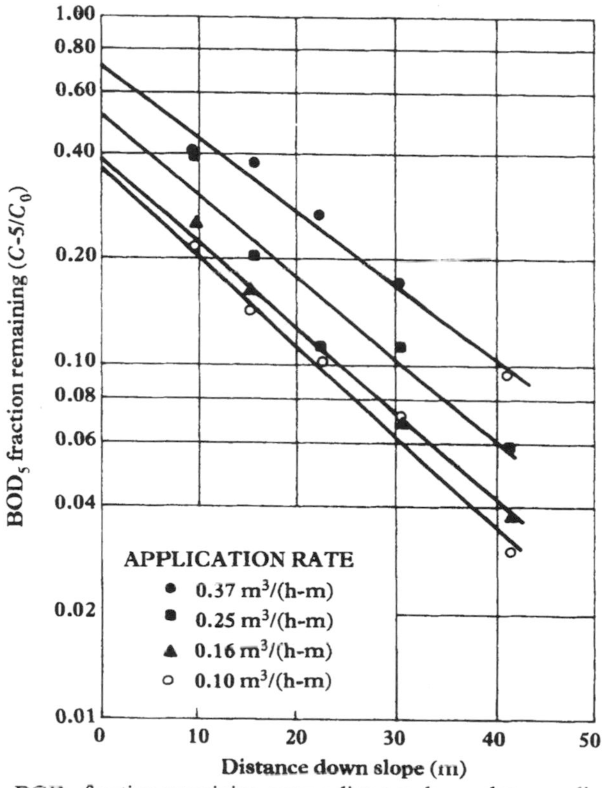

Figure 8.9 BOD_5 fraction remaining versus distance down-slope; application rate study with screened raw wastewater (from Smith and Schroeder, 1985; reproduced by permission of the Water Pollution Control Federation, U.S.A.)

where area = total area of OF site, m^2;
Q = design daily flow, m^3/day;
z = slope length, m;
q = application rate, $m^3/(h\text{-}m)$;
P_d = design application period, h/day.

Design values for Q, P_d, and q should be selected to allow operating flexibility during periods when parts of the system must be shut down for harvesting or repairing.

Table 8.10 Intercepts (A) and rate constants (k) of eq. 8.13 (adapted from Witherow and Bledsoe, 1986)

Pollutants	A	k (m/h)
TSS	0.44–0.96	0.024–0.056
BOD_5	0.58–0.88	0.030–0.046
TOC	0.57–0.88	0.016–0.039
NH_4–N	0.99–1.12	0.016–0.054

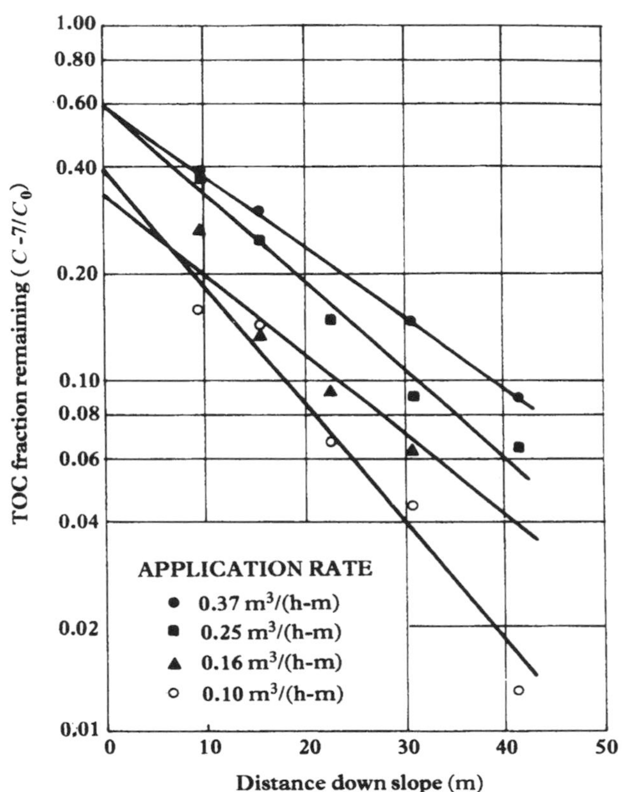

Figure 8.10 TOC fraction remaining versus distance down-slope; application rate study with screened raw wastewater (from Smith and Schroeder, 1985; reproduced by permission of the Water Pollution Control Federation, U.S.A.)

Design daily flow (Q)

It is preferable to provide equalization or offline storage facilities to allow for constant application rates and short-term storage during power outage or extremely heavy precipitation. The design daily flow should be based on peak seasonal average daily flow. For systems without such storage facilities the design daily flow Q should be based on the average flow during selected periods, depending on the flow conditions.

Design application rate (q)

The design q value for an OF system should generally match the infiltration rate of the soil or vegetated surface of the soil, to prevent excessive runoff and impairment of effluent quality. The value of q in eq. 8.13 should be less than the value used to determine z, to provide a safety factor in process performance and to allow operating flexibility by permitting a range of application rates to be used. A factor of 1.5 is suggested for design. Thus, if a design value for q of 0.37 $m^3/(h\text{-}m)$ is

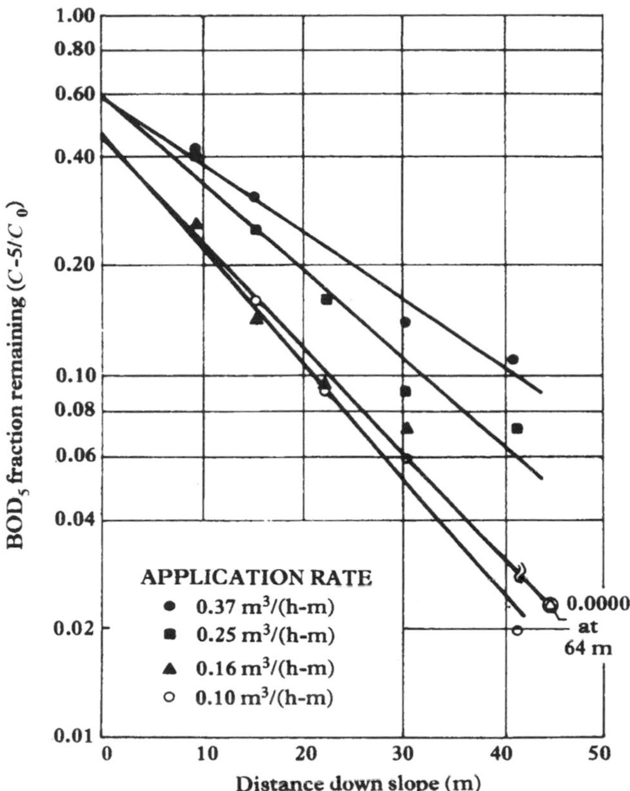

Figure 8.11 BOD_5 fraction remaining vs distance down-slope; application rate study with primary effluent (from Smith and Schroeder, 1985; reproduced by permission of the Water Pollution Control Federation, U.S.A.)

used to determine z, then a design value for q of 0.25 $m^3/(h\text{-}m)$ should be used in eq. 8.14 to calculate the required area.

Design application period (P_d)

Design application periods of 8–12 h/day are recommended. For small systems it may be more convenient or cost-effective to operate only during one working shift. In this case, the entire area would receive the design daily flow during 8 h application period. Storage facilities would be required to hold wastewater during the 16 h non-operating period.

Example 8.3 Determine the area required for land treatment of wastewater using the overland flow method.

The following information is given: screened raw wastewater is to be used; the flow (Q) is 3,000 m^3/day; the influent BOD_5 (C_0) is 200 mg/L; and the required effluent BOD_5 (C) is 20 mg/L.

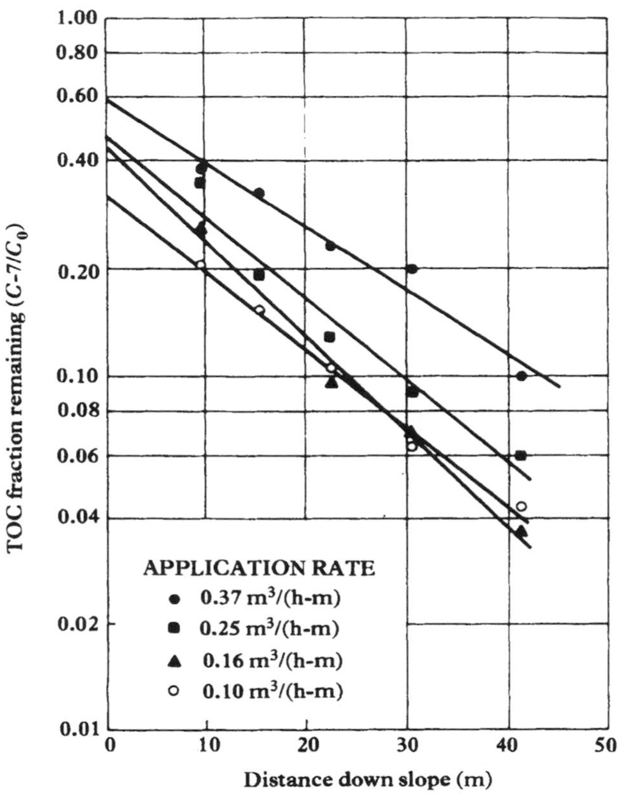

Figure 8.12 TOC fraction remaining versus distance down-slope; application rate study with primary effluent (from Smith and Schroeder, 1985; reproduced by permission of the Water Pollution Control Federation, U.S.A.)

The necessary design calculations are as follows:

Compute the required removal ratio, $(C - 5)/C_0$, from eq. 8.8.

$$(C - 5)/C_0 = (20 - 5)/200 = 0.07$$

Select an application rate, q, in valid range of the model.

$$q = 0.25 \text{ m}^3/(\text{h-m})$$

Determine required value of slope length, z, using Figure 8.9.

$$z = 37 \text{ m}$$

Select application period, P_d.

$$P_d = 8 \text{ h/day}$$

Compute q for area calculation

$$q = (0.25 \text{ m}^3/(\text{h-m}))/1.5$$
$$q = 0.17 \text{ m}^3/(\text{h-m})$$

Compute required total area from eq. 8.14. Assume 7 days/week application frequency.

$$\text{Area} = (Qz)/(qP_d)$$
$$\text{Area} = (3{,}000\ \text{m}^3/\text{day})(37\ \text{m})/(0.17\ \text{m}^3/(\text{h-m}))(8\ \text{h/day})$$
$$= 81{,}620\ \text{m}^2 \approx 8.2\ \text{ha}$$

Comparison of current design based on hydraulic loading

Assume the hydraulic loading = 3 cm/day (Figure 8.1)
Land area required is given by:

$$\text{Area} = 3.65Q/L_h \tag{8.15}$$

where area = wetted land area, ha;
Q = design flow, m^3/day;
L_h = hydraulic loading, cm/year.

Therefore,

$$\text{Area} = (4.65 \times 3{,}000\ \text{m}^3/\text{day})/(3\ \text{cm/day} \times 365\ \text{days/year})$$
$$= 10\ \text{ha}$$

From the required area values obtained using both design criteria, it can be seen that the mathematical model presented based on application rate, slope length and application period (eq. 8.14), gives an economical value of 8.2 ha compared with an area of 10 ha with a 3 cm/day hydraulic loading.

Example 8.4 A wastewater has the following characteristics:

Flow rate	= 100 m^3/day
BOD_5	= 250 mg/L
Fecal coliforms	= 6×10^5/100 mL
Design precipitation	= 1,000 mm/year
Percolation	= 400 mm/year
Evapotranspiration	= 1,200 mm/year

(a) If the soil type is clay loam, which type of land treatment could be suitable?
(b) Determine whether there will be runoff from this land treatment or not, if so calculate the runoff in m^3/day.
(c) From the experimental results it was found that the wastewater application cycle to this land should be 3 days of application and 11 days of resting. Application area of the site is 14 ha. Determine the approximate wastewater application rate to be applied which will not produce runoff.

Solution

(a) As the soil type is clay loam, the suitable type of land treatment will be the OF system (according to Figure 8.1).

(b) From Figure 8.1, choose a liquid loading rate
$= 4$ in./week
$= 4 \times 25.4 \times 52$ mm/year
$= 5{,}283$ mm/year.

General mass balance equation for land treatment of wastewater is given in eq. 8.3:
Design precipitation + wastewater applied
= percolation + evapotranspiration + runoff
1,000 + 5,283 = 400 + 1,200 + runoff
runoff = 4,683 mm/year
Application rate $= 4 \times 2.54$ cm/week.
Assuming that the wastewater is applied continuously

$$\text{Liquid loading rate} = \frac{4 \times 2.54 \times 10^{-2}}{7} \text{m/day} = 0.0145 \text{ m/day}$$

$$\text{Area required} = \frac{100 \text{ m}^3/\text{day}}{0.0145} = 6,889 \text{ m}^2$$

Choose the land area required = 7,000 m^2

$$\text{Total runoff from site} = \frac{7 \times 10^3 \times 4.683}{365} \text{m}^3/\text{day}$$

$$= 89.8 \text{ m}^3/\text{day}$$

(c) If there should not be any runoff from the land then, from eq. 8.3, wastewater applied = 400 + 1,200 − 1,000 = 600 mm/year, which is less than the liquid loading rate in (b).

No. of days in one cycle = 14 days
No. of cycles in year = 26

$$\text{Application rate} = \frac{600}{26} \text{mm/cycle}$$

No. of days of application/cycle = 3 days

$$\text{Application rate} = \frac{600}{3 \times 26} \text{mm/day} = 7.7 \text{ mm/day during the 3-day application period}$$

If the application area is 14 ha, wastewater flow rate to be applied during the application period is $0.0077 \times 14 \times 10^4 = 1{,}078 \text{ m}^3/\text{day}$, which is greater than the available flow rate of 100 m^3/day.

Example 8.5 A wastewater has the same characteristics as in Example 8.4. Design an OF system to treat this wastewater so that the treated effluent contains BOD_5 and fecal coliforms less than 50 mg/L and 100/100 mL, respectively. The application rate is 0.1 m^3/(h-m width).

The following information is given:

For BOD_5 removal $\dfrac{C-5}{C_0} = 0.72\ \exp\left(-\dfrac{0.02z}{q^{0.5}}\right)$

For fecal coliform removal $\dfrac{N}{N_0} = \exp(-K_f t)$

where C and C_0 = effluent and influent BOD_5 concentrations, mg/L

N and N_0 = effluent and influent fecal coliform concentrations, no./100 mL;

K_f = fecal coliform removal rate in OF, 0.2/min;

t = time of wastewater flow in OF, min;

$v = \dfrac{1}{n} R^{2/3} S^{1/2}$;

where v = flow velocity of wastewater during OF treatment, m/s;

R = hydraulic radius of flow, m;

S = slope of soil = 0.02;

n = coefficient of roughness = 0.38.

For BOD_5 removal

$$\frac{50-5}{250} = 0.72\exp\left(-\frac{0.02z}{0.1^{0.5}}\right)$$

Slope length, $z = 22$ m for BOD_5 removal

For fecal coliform removal

Assuming that, during wastewater application, the depth of wastewater is 5 cm throughout the entire down-slope distance

$$\text{Hydraulic radius, } R = \frac{\text{wetted area}}{\text{wetted perimeter}}$$

$$\text{Width of the applied area} = \frac{Q}{q} = \frac{100\,\text{m}^3/\text{day}}{0.1\,\text{m}^3/\text{h-m}} = 41.7\,\text{m}$$
$$\approx 42\text{m}$$

$$R = \frac{42\,(0.05)}{42 + 2(0.05)} = 0.05$$

$$v = \frac{1}{0.38}(0.05)^{2/3}(0.02)^{1/2} = 0.05\,\text{m/s}$$

$$\text{Fecal coliform reduction } \frac{100}{6 \times 10^5} = \exp(-0.2t)$$

$$t = 43.5\,\text{min}$$

$$\text{Slope length for fecal coliform removal} = 0.05 \times 43.5 \times 60$$
$$= 130.5\,\text{m}$$

Choose slope length of the OF area of 130.5 m to provide the treated effluent quality meeting the required BOD_5 and fecal coliform concentrations. Width of the applied area is 42 m.

8.6 SYSTEM MONITORING

Monitoring of land treatment systems involves the observation of significant changes resulting from the application of wastewater. The monitoring data are used to confirm environmental assessment and to determine if any corrective action is necessary to protect the environment or maintain the renovation capacity of the system. The components of the environment that need to be observed include wastewater, groundwater, and soils upon which wastewater is applied and, in some cases, vegetation growing in soils that are receiving wastewater.

8.6.1 Water quality

Monitoring of water quality for land application systems is generally more involved than for conventional treatment systems because non-point discharges of system effluent into the environment are involved. Monitoring of water quality at several stages of a land treatment process may be needed for process control. These stages may be at: the applied wastewater, the renovated water, and the receiving waters—surface water or groundwater.

The water quality parameters and the frequency of analysis will vary from site to site depending on the nature of the applied wastewater. The measured parameters may include:

1. those adversely affecting receiving water quality, either as drinking water supply or as irrigation water supply;
2. those required by regulatory agencies; and
3. those necessary for system control.

An example of a suggested water quality monitoring program for a large-scale SR system is presented in Table 8.11. Renovated water may be recovered as runoff in an overland flow system, or as drainage from underdrains or groundwater from recovery wells in SR and RI systems.

Table 8.11 Monitoring program for a large slow-rate system (U.S. EPA, 1973)

Parameter	Frequency of analysis					
				Groundwater		
	Applied water	Soil	Plants	Onsite wells	Perimeter wells	Background wells
Flow	C	—	—	—	—	—
BOD_5 or TOC	W	—	—	Q	Q	Q
COD	W	—	—	Q	Q	Q
Suspended solids	W	—	—	—	—	—
Nitrogen, total	W	2A	A	Q	Q	Q
Nitrogen, nitrate	—	—	—	Q	Q	Q
Phosphorus, total	M	2A	A	Q	Q	Q
Coliforms, total	W	—	—	Q	Q	Q
pH	D	Q	—	Q	Q	Q
Total dissolved solids	M	—	—	Q	Q	Q
Alkalinity	M	—	—	Q	Q	Q
SAR	M	Q	—	Q	Q	Q
Static water level	—	—	—	M	M	M

Legend: C = Continuously, 2A = Two samples per year
D = Daily, A = Annually
Q = Quarterly, M = Monthly
W = Weekly, SAR = Sodium adsorption ratio

Note: Wastewater applied and groundwater should be tested initially and periodically thereafter, as appropriate, for heavy metals, trace organics, or other constituents of environmental concern

8.6.2 Groundwater

In groundwater, travel time of constituents is slow and mixing is not significant compared with surface water. Surface inputs near a sampling well will move vertically and arrive at the well much sooner than inputs several hundred meters away from the well. Thus the groundwater sample represents contributions from all parts of the surface area with each contribution arriving at the well at a different time. A groundwater sample may reflect surface inputs, especially inorganic constituents such as heavy metals, from several years before sampling, and have no association with the land application system. Consequently, it is imperative to obtain adequate background quality data and to locate the sampling well so that response times are minimized.

If possible, existing background data should be obtained from wells in the same aquifer, both beyond and within the anticipated area of influence of the land application system.

Wells with the longest history of data are preferable. Monitoring of background wells should continue after the system is in operation, to provide a base for comparison.

In addition to quality, the depth to groundwater should be measured at the sampling wells to determine if the hydraulic response of the aquifer is consistent with what was anticipated. For SR systems a rise in water-table levels to the root zone would necessitate corrective actions such as reduced hydraulic loading or adding underdrainage. The appearance of seeps or perched groundwater tables might also indicate the need for corrective actions.

8.6.3 Soil

In some cases application of wastewater to the land will result in changes in soil properties. Results of soil sampling and testing will serve as the basis for deciding whether the soil properties should be adjusted by the application of chemical and organic amendments. Soil properties that are important to land treatment of wastewater include: pH, exchangeable sodium percentage or SAR, salinity, nutrient status, and heavy metals.

Soil pH below 5.5 or above 8.5 is generally harmful to most plants. Below pH 6.5 the capacity of soils to retain metals is reduced significantly; the soil with pH above 8.5 generally indicates a high sodium content and possible permeability problems.

The level at which salinity becomes harmful to plant growth depends on the type of crop. Salinity in the root zone is controlled by leaching soluble salts to the subsoil or drainage system.

8.6.4 Crop tissue

Plant tissue analysis is probably more revealing than soil analysis with regard to deficient or toxic levels of elements. All the environmental factors that affect the uptake of an element are integrated by the plant, thus eliminating much of the complexity associated with interpretation of soil test results. If a regular plant tissue monitoring program is established, deficiencies and toxicities can be determined and corrective actions can be taken.

8.6.5 Case studies

The following are successful case studies for the three land treatment processes operating at full-scale in the U.S.A.

SLOW-RATE PROCESS

The city of San Angelo in Texas, U.S.A., owns and operates a wastewater irrigation system with an area of 259 ha and a capacity of 1.9×10^4 m^3/day. The surface soils are clay and clay loam. A hay crop is grown, which is baled and sold, and an average of 500 cattle are grazed on the pasture land (Crites and Pound, 1976). Primary treatment of the wastewater precedes surface irrigation, and about 1 week of storage capacity is available when irrigation is not needed. Operating and performance data of this system are shown in Table 8.12.

Table 8.12 Operating and performance data of an irrigation system in San Angelo, Texas (U.S. EPA, 1981)

Annual wastewater loading rate, cm/year	290
BOD in applied wastewater, mg/L	89
BOD in percolate, mg/L	0.7
BOD removal, %	99
Total N in applied wastewater, mg/L as N	35.4
Total N in percolate, mg/L as N	6.1
N removal, %	83

RAPID INFILTRATION PROCESS

At Lake George, New York, a rapid filtration system has been in operation since 1933. The soil type is sand. Wastewater that has received secondary treatment from trickling filters is spread at a rate of about 76 cm/week in the summer. In the winter the application rate is reduced to 18 cm/week because of reduced flows (the city's population fluctuates seasonally). Spreading of the wastewater is continuous, and when ice forms on the surface it is not removed but merely floated by the next wastewater application (Crites and Pound, 1976). The operating and performance data are shown in Table 8.13.

OVERLAND FLOW PROCESS

The Campbell Soup Company in Paris, Texas, U.S.A., operates a complete line of heat-processed soups as well as beans and spaghetti products. Its 364-ha overland flow treatment site is planted with water tolerant grasses, such as reed canary grass, perennial rye grass, bermuda grass, tall fescue, and native vegetation (Loehr *et al.*,

Table 8.13 Operating and performance data of a rapid infiltration system in Lake George, New York (U.S. EPA, 1981)

Hydraulic loading rate, cm/year	58,000
Summer loading rate, cm/week	76
Winter loading rate, cm/week	18
BOD/N ratio	2:1
Flooding to drying time ratio	1:4
Average BOD loading rate, kg/ha-day	53
BOD in applied wastewater, mg/L	60
BOD in percolate water, mg/L	1.2
BOD removal, %	98
Total N in applied wastewater, mg/L as N	11.5 (summer), 12.0 (winter)
Total N in percolate, mg/L as N	7.70 (summer), 7.50 (winter)
Removal of total N, %	33 (summer), 38 (winter)
Soluble phosphate in applied wastewater, mg/L as P	2.1
Soluble phosphate in percolate, mg/L as P	< 1 (summer), 0.014 (winter)
Removal of soluble phosphate, %	> 52 (summer), 99 (winter)
Fecal coliforms in applied wastewater, MPN/100 mL	359,000
Fecal coliforms in percolate, MPN/100 mL	72 (summer), 0 (winter)

Table 8.14 Design, operation and performance data of the Campbell Soup System, Paris, Texas (Loehr *et al.*, 1979; Hinrichs *et al.*, 1980)

Length of slope	45–60 m
Slope	2–6%
Size of sprinkler nozzles	6.5 (50 L/min)–8 (80 L/min) mm
Distance between sprinklers	25 m
Operating pressure at sprinkler heads	340–480 kPa
Pretreatment	Screen and degrease
Volume of wastewater treated	22,700 m^3/day
Number of months working in a year	12 months/year
Wastewater application rate	5–7.6 (cm/week)
Wastewater application schedule	6–8 h/day on, 16–18 h/day off

	Mean concentration (mg/L)		Removal (%)	
Parameter	Influent	Effluent	Concentration	Mass (based on runoff)
TSS	263	16	93.5	98.2
BOD	616	9	98.5	99.1
N	17.4	2.8	83.9	91.5
P	7.6	4.3	42.5	61.5

1979). The soils at the site are sandy loam, clay, and clay loam. Study of the fate of the applied wastewater indicates that about 20 percent percolates down through the soil, 10–30 percent is lost by evapotranspiration, and the remaining 60 percent returns to the surface stream as runoff (Loehr *et al.*, 1979). A sprayed effluent is allowed to flow down 45–60 m slopes ranging from 2 to 6 percent. Grasses are harvested two or three times a year as cash hay crops. The purified water (or runoff) is discharged to the receiving stream via prepared waterways. Table 8.14 shows the design, operation and performance data of this system. This overland flow system operates as well as in other seasons. Although the removal of P was about 42 percent, a later change in the operating procedure to provide a longer rest period between applications, with no change in the total wastewater volume, has increased P removal to nearly 90 percent without affecting the BOD or N removals.

Chemical analyses of the soil indicate that while the concentration of sodium salts is increasing, it has not reached a level injurious to plants and is not expected to do so in the future. Similarly, nitrates percolating into the groundwater reserve are not expected to build up to the point of concern.

8.7 PUBLIC HEALTH ASPECTS AND PUBLIC ACCEPTANCE

Land treatment of wastewater provides many potential benefits, but carries with it the risk of contamination of human and animal food. Land treatment can produce an equal or better quality of effluent than a conventional treatment process, but the public is not aware of it. For a successful method of wastewater treatment and reuse, public acceptance of land treatment as a safe and suitable method is very important.

The major health concern is possible pollution by N, heavy metals, toxic organic compounds, and pathogens.

8.7.1 Nitrogen

Excess concentration of nitrate (greater than 10 mg/L NO_3^-–N) in drinking water will be the major health concern for infants under 6 months of age. The major pathway of concern in land treatment is conversion of the wastewater N to nitrate, and percolation to drinking water aquifers. All three land treatment methods are quite efficient in nitrification of the wastewater N. As OF is a surface discharge system it is only the SR and RI systems that are of concern for groundwater impacts. N is often the limiting factor for design of an SR system achieving reasonable amounts of N in the percolate by selecting proper crops and application rates. The very high loading rates inherent in RI systems result in the greatest potential for nitrate contamination of drinking water in underground aquifers. The extent of nitrification and nitrate contamination depends on temperature, C source, hydraulic loading rate, loading period, and physical properties of the soil. To overcome this problem the treatment site can be located far away from drinking water sources, or located above non-drinking water aquifers. Alternatively, the percolate can be removed using wells or underdrains, for reuse elsewhere. An alternative that has been proposed is to use a two-step land treatment consisting of an OF system followed by an RI system; in this way most N could be recovered during the OF process through biomass uptake, nitrification/denitrification, and volatilization.

8.7.2 Heavy metals and other toxic organic compounds

The pathways of potential concern are movement of metals and other toxic organic compounds to drinking water sources, and translocation of these compounds through the food chain to humans. The plants grown on a land treatment site can accumulate toxic compounds in their leaves, stems, and fruits. Removal of metals and toxic compounds is generally effective in all three land treatment methods, with that effectiveness retained for considerable periods of time. As the problem of heavy metal and toxic compound pollution is more serious with respect to sludge application to land, this matter will be discussed more in Chapter 9.

8.7.3 Pathogens

The pathogens of concern in land treatment systems are bacteria, parasites, and viruses. The major pathways of concern are to the groundwater, internal or external contamination of crops, translocation to grazing animals and humans, and off-site transmission via aerosols or runoff. There is no evidence available indicating transmission of parasite diseases from application of wastewater in land treatment systems, because the parasite cysts and eggs will normally settle out during preapplication treatment or in storage ponds. But there are chances of human infection by parasite diseases if the land treatment area is accessible to the public. It has been found that *Ascaris* ova will survive in soil for long periods, and investigations on a site in France proved the presence of *Ascaris* ova in soil.

Concerns with respect to crop contamination focus mainly on surface contamination and then persistence of the pathogens until consumed by humans or animals or the internal infection of the plant via the roots. As shown in Figure 8.13, pathogen survival periods on leaves and fruit crops tend to be shorter than the growth periods of most of these plants (Strauss, 1986). In soil, however, survival of viruses, *Salmonella* bacteria, and *Ascaris* eggs may exceed the growth period of crops (Figure 8.14). Viable pathogens are therefore more likely to be found on root crops and soil than on leaf crops. Human or animal infection can be reduced by taking preventive methods such as prohibiting the consumption of raw vegetables grown on land treatment sites, and allowing a period of 2 weeks or more after application of wastewater before animals are allowed to graze. This will allow sunlight to kill most of the fecal bacteria and viruses. As time and temperatures have significant effects on pathogen die-off (Figure 3.1), prolonged storage of wastewater or sludge to be applied to crops will help minimize the public health

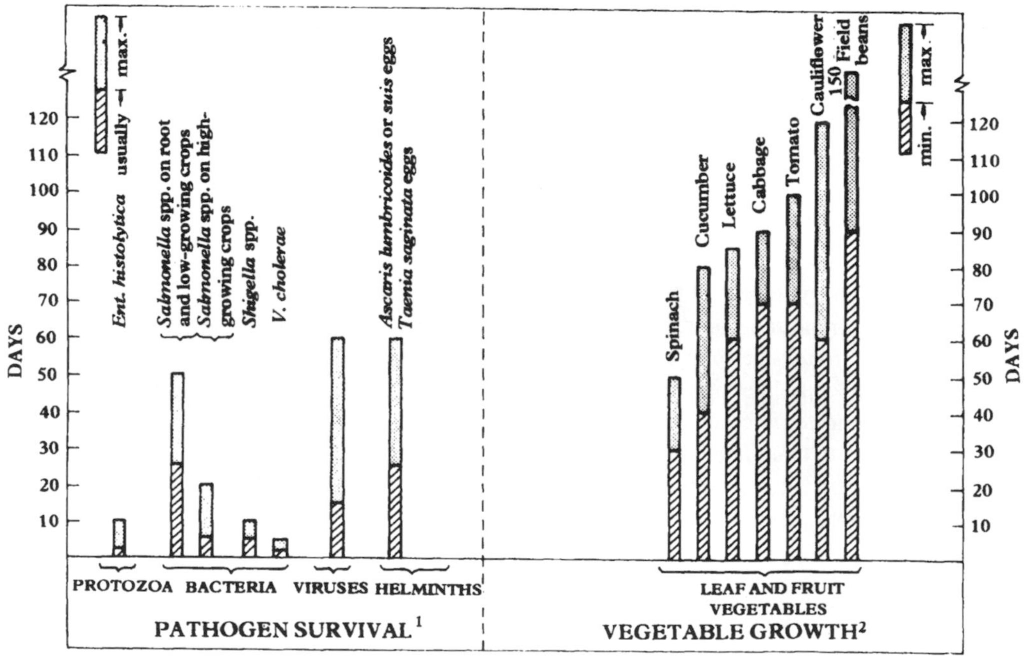

Figure 8.13 Pathogen survival on crops vs vegetable growth periods in warm climates (from Strauss, 1986; reproduced by permission of IRCWD, Switzerland)

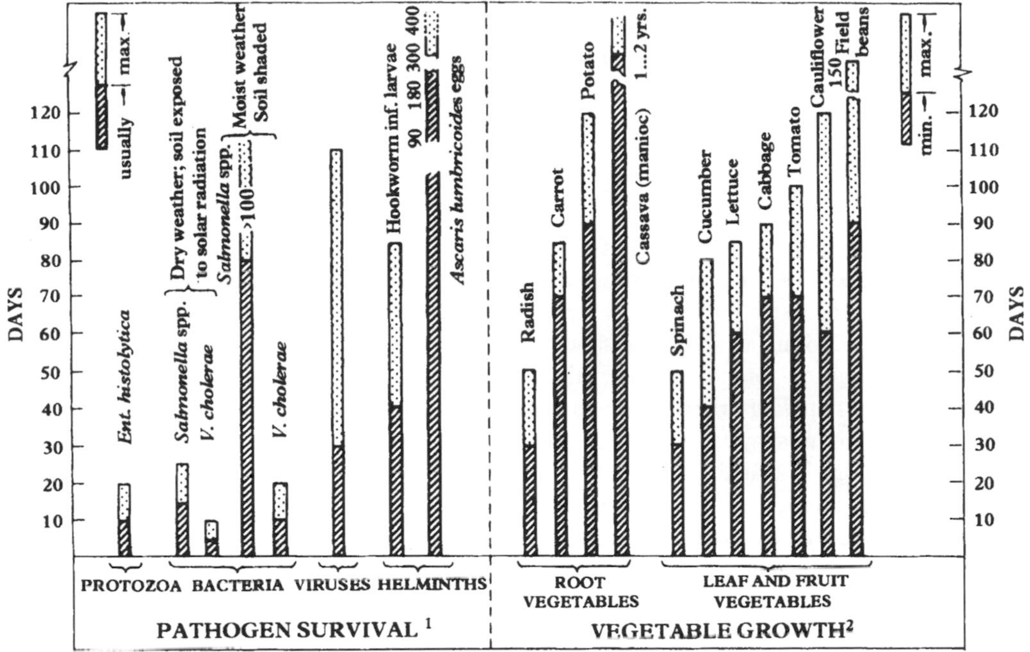

1) Determined under widely varying conditions

2) Maturation period from transplanting or from sowing if not transplanted

Figure 8.14 Pathogen survival in soil vs vegetable growth periods in warm climates (from Strauss, 1986; reproduced by permission of IRCWD, Switzerland)

risks. Strauss (1986) suggested storage periods for pathogen inactivation in excreta, as shown in Table 8.15.

Runoff from a land treatment site might be a potential pathway for pathogen transport; but this is not a problem in an RI system. Groundwater contamination has the highest potential for an RI system due to the high hydraulic loading rate and coarse texture of the receiving soils.

The potential for transport of pathogens by aerosols from land treatment sites has been the most controversial issue in recent years. The main reason for this controversy is an unawareness of the difference between aerosols and sprinkler droplets. To avoid bacterial and virus transport through aerosols it is preferable to

Table 8.15 Tentative recommendations for excreta storage periods in warm climates

Storage period	Hygienic quality achieved
⩾ 2 days	Inactivation of *Clonorchis* and *Opisthorchis* eggs
⩾ 1 month	Complete inactivation of viruses, bacteria, and protozoa (except, possibly, *Salmonella* on moist, shaded soil); inactivation of schistosome eggs
⩾ 4 months	Inactivation of nematode (roundworm) eggs, e.g. hookworm and whipworm (*Trichuris*); survival of a certain percentage (10–30%?) of *Ascaris* eggs
⩾ 12 months	Complete inactivation of *Ascaris* eggs

From Strauss (1986); reproduced by permission of IRCWD

locate the treatment site away and downwind from the residential areas, when sprinklers are used for distribution of wastewater. Depending on wind speed and topography, pathogen-carried aerosols can travel several hundred meters away from the land treatment site.

With respect to epidemiological studies of land treatment of wastewater, it appears from Table 2.26 that the relative health risks due to intestinal nematodes are high, followed by risks from bacterial and viral infections. The microbiological quality guidelines (Table 2.27) suggest that the density of viable eggs of intestinal nematodes (*Ascaris*, *Trichuris*, and hookworms) be equal to or less than 1 per liter of wastewater to be used for both restricted and unrestricted irrigation. With respect to fecal coliforms, their density should be less than 1,000 per 100 mL of wastewater to be used for unrestricted irrigation. A relaxed standard of fecal coliform density is applied to wastewater to be used for restricted irrigation.

Land treatment of wastewater has been in practice in many parts of the world for decades. Where land is available this method of wastewater treatment is generally attractive because it is low-cost, easy to operate and maintain, and can yield some financial return from the sale of marketable crops. Similar to other waste recycling options, the public, depending on the socio-economic conditions, are likely to accept and support this method of wastewater treatment if public health risks are kept at a minimum and the reuse is directed towards non-body contact (see section 1.4). The current effort to improve design and operation criteria for land treatment, together with regulatory standards for effluent discharge, should make wastewater treatment by land a viable method compatible with other conventional methods.

REFERENCES

Ayers, R. S. and Westcot, D. W. (1976). *Water Quality for Agriculture*. FAO irrigation and drainage paper No. 29, Food and Agriculture Organization of the United Nations, Rome.

Booher, L. J. (1974). *Surface Irrigation*. Agricultural development paper No. 95. Food and Agriculture Organization of the United Nations, Rome.

Broadbent, F. E., Pal, D. A., and Aref, K. (1977). Nitrification and denitrification in soil receiving wastewater. In *Wastewater Renovation and Reuse* (ed. F. M. D'Itri), pp. 321–48. Marcel Dekker, New York.

Crites, R. W. and Pound, C. E. (1976). Land treatment of municipal wastewater. *Environ. Sci. Technol.*, **10**, 548–51.

Hinrichs, D. J., Faisst, J. A., Pivetti, D. A., and Schroeder, E. D. (1980). *Assessment of Current Information of Municipal Wastewater*. EPA 430/9–80–002, U.S. Environmental Protection Agency, Washington, D.C.

Lance, J. C. and Gerba, C. P. (1977). Nitrogen, phosphate and virus removal from sewage water during land filtration. *Progr. Water Technol.*, **9**, 157–66.

Loehr, R. C., Jewell, W. J., Novak, J. D., Clarkson, W. W., and Friedman, G. S. (1979). *Land Application of Wastes*, Vol. 1. Van Nostrand Reinhold Co., New York.

Loehr, R. C. and Overcash, M. R. (1985). Land treatment of waste: concepts and general design. *J. Env. Eng.*—ASCE, **111**, 141–60.

Olsen, S. R. and Kemper, W. D. (1968). Movements of nutrients to plant growth. *Adv. Agron.*, **20**, 91–149.

Reed, S. C. and Crites, R. W. (1984). *Handbook of Land Treatment Systems for Industrial and Municipal Wastes*. Noyes, New Jersey.

Smith, R. G. and Schroeder, E. D. (1985). Field studies of the overland flow process for the treatment of raw and primary treated municipal wastewater. *J. Water Pollut. Control Fed.*, **57**, 785–93.

Strauss, M. (1986). *Health Aspects of Nightsoil and Sludge Use in Agriculture and Aquaculture Part II—Pathogen Survival.* International Reference Centre for Waste Disposal (IRCWD), Duebendorf, Switzerland.

Tucker, D. L. and Vivado, N. D. (1980). Design of an overland flow system. *J. Water Pollut. Control Fed.*, **52**, 559–67.

U.S. EPA (1973). *Wastewater Treatment and Reuse by Land Application*, Vols I and II. EPA–660/2–73–006a&b. U.S. Environmental Protection Agency, Washington, D.C.

U.S. EPA (1976). *Land Treatment of Municipal Wastewater Effluents. Design Factors-1.* EPA–625/4–76–010. U.S. Environmental Protection Agency, Washington, D.C.

U.S. EPA (1981). *Process Design Manual for Land Treatment of Municipal Wastewater.* EPA–625/1–81–013. U.S. Environmental Protection Agency, Cincinnati, Ohio.

Witherow, J. L. and Bledsoe, B. E. (1986). Design model for the overland flow process. *J. Water Pollut. Control Fed.*, **58**, 381–6.

EXERCISES

8.1 Discuss the functions of vegetation in the land treatment process. Give the criteria to be used in selecting a type of vegetation (or crop) for a land treatment process.

8.2 **a** You are to find out the indigenous types of soil and crops in your city. Based on this information, suggest a suitable process of land treatment of wastewater you would select for the city. In practice, what is the other information needed for the selection of the type of land treatment process?

b If the city agrees to employ the land treatment process recommended in (a), you are to determine the following:

- land area requirement for wastewater treatment
- method of wastewater application
- wastewater application rate (hydraulic and organic loading rates), and the periods of wastewater application and resting
- expected characteristics of the treated wastewater
- harvesting frequency of the crop
- lay-out (or schematic diagram) of the land treatment system, including the pretreatment units.

8.3 Among the three processes of land treatment, i.e. slow-rate, rapid infiltration and overland flow, discuss which process is most effective in treating wastewaters containing the following constituents:

a N (in organic and ammonium forms)

b P

c heavy metals

d fecal coliform bacteria

e helminthic ova

f toxic organic compounds (e.g. trichloroethylene)

8.4 A wastewater is generated at a flow rate of 1,000 m^3/day. An irrigation (slow-rate) system is to be designed to treat this wastewater with an application rate of 5 cm/week. What is the required land area if the system is operated all year round? What is the required land area if the system is designed for 40 weeks/year?

A sprinkler system is chosen for application of the wastewater. The sprinklers are spaced in a square grid pattern of 20 m × 20 m, and each sprinkler discharges at a rate of 1.6 L/s. How many sprinklers are needed for the year-round operation? What is the application rate in cm/h? How many hours should the sprinkler system be operated each week to achieve the application rate of 5 cm/week?

8.5 A rapid infiltration system is designed for an application rate of 30 m/year. The system is run throughout the year by applying the wastewater every Monday of a week followed by 6 days of resting. The wastewater contains 50 mg/L of BOD_5. What are the average BOD_5 loading rates annually, weekly, and on Monday? Over the 7-day cycle, what is the average BOD_5 loading rate in kg/ha-day? What is the BOD_5 loading on Monday in kg/ha-day? If the wastewater flow is 2,500 m^3/day, with respect to the above application cycle, calculate the land area required.

8.6 A design equation for the overland flow process is given in eq. 8.13:

a Discuss the merits and drawbacks of this equation in predicting the wastewater treatment efficiency.

b Discuss whether eq. 8.13 can be applied to predict fecal coliform removal in the overland flow process.

c Discuss whether eq. 8.13 can be applied to predict wastewater treatment efficiency for the slow-rate and rapid infiltration processes.

9

Land treatment of sludge

Land treatment of sludge has become a method which is being increasingly considered by many municipalities throughout the world. It offers the advantage of recycling nutrients back to the land at low cost, and of reclaiming lands being spoiled by strip mining, deforestation, and over-application of inorganic fertilizers. Sludge is normally stabilized by anaerobic digestion or other suitable means before application on land. Such stabilization eliminates obnoxious odors and fly problems. Yields of grain and forage crops are increased by the nutrients and water supplied by irrigating with digested sludge. Organic matter in digested sludge accumulates in, and imparts favorable characteristics to, soils because of its normally high humus content.

Sludge production in the U.S. in 1982 was estimated to be almost 7 million dry tons, and this quantity is expected to increase in the future (U.S. EPA, 1983). The total production of sewage sludge per year in the European Economic Community is estimated to be 6 million tons of dry matter or 230 million m^3 of raw sludge (Kofoed, 1984).

Table 9.1 shows the current sludge disposal practices in the U.S.A. and the European Community (EC) countries. Beneficial land application generally refers to the spreading of sludge on crop lands, while surface disposal applies to sludge disposal at lagoons, mounds or impoundments. The increased use of sludge for the beneficial land application in the U.S.A. has resulted from the Sewage Sludge Use and Disposal Regulations (Part 503 Standards) which took 15 years to develop and has been implemented by the U.S. EPA since 1993. The term 'biosolids' is also introduced to describe sludge solids that are suitable for beneficial land application. According to Walsh (1995), with sea disposal being phased out under EC legislation and stringent standards being applied to landfills, more sludge would be disposed of through beneficial land application in the EC countries in the next 5–10 years.

Data on sludge production in developing countries are not readily available because most cities do not have adequate sewage systems for wastewater transportation to central treatment plants. However, it can be estimated that each

Table 9.1 Sludge disposal practices in the U.S. and European Community (adapted from Walsh, 1995)

Disposal practice	Percent of total sludge	
	U.S.A.	European Community
Beneficial land application	36	32
Landfills	38	48
Incineration	16	13
Surface disposal	10	2
Sea disposal	—	5

person generates 25–40 kg dry matter of sludge per year or about 800 kg wet sludge (95 percent water content) per year.

9.1 OBJECTIVES, BENEFITS, AND LIMITATIONS

There are four major categories of sludge land application systems:

1. Agricultural utilization: sludge is used as a source of fertilizer nutrients and/or as a soil amendment.
2. Forest utilization: sludge is used to enhance forest productivity.
3. Land reclamation: sludge is used to reclaim disturbed land, such as strip-mined areas, for the purpose of revegetation, or to reclaim seashore land.
4. Land disposal: sludge is applied to soils, with or without vegetation, for the primary purpose of sludge disposal. Crop production or improvement of soil characteristics is of secondary importance.

Although the emphasis of this chapter is on category 1 (agricultural utilization or beneficial land application), the information presented herein should be applicable to the other three categories.

The physical characteristics of various types of sludges are given in Table 9.2. Well-digested sludge is inoffensive and quite suitable for land application. Other types of sludge are mostly not well-stabilized and have offensive odors. Typical chemical compositions of raw and digested sludge are presented in Table 9.3. Digested sludge usually contains smaller percentages of volatile solids and protein than does raw sludge. However, the former type of sludge has high contents of nutrients (N, P, K). Table 9.4 gives the nutrient contents of several sludge types in the U.S.A., and indicates the suitability of using these sludges in agricultural production and land reclamation. As many municipalities can offer their sludge at practically no cost to the farmer, a significant cost-saving is effected both for the farmer and the municipality which produces the sludge. However, the costs of sludge transportation from treatment plant to a land application site and proper application of sludge to land (see section 9.2) can be expensive in some cases.

Table 9.2 Physical characteristics of various types of sludge (Loehr *et al.*, 1979)

Sludge	Color	Other physical properties	Odor	Digestibility (amenability to further biological stabilization)
Primary sedimentation	Gray	Slimy	Extremely offensive	Readily digested
Chemical precipitation (primary)	Black, red surface if high in iron	Slimy, gelatinous, gives off considerable gas	Offensive	Slower rate than primary sedimentation sludge
Activated sludge	Brown, dark if nearly septic	Flocculent	Inoffensive, earthy when fresh; putrefies rapidly	Readily digested
Trickling filter humus	Brownish	Flocculent	Relatively inoffensive, decomposes slowly	Readily digested
Digested sludge	Dark brown to black	Contains very large quantity of gas	Inoffensive if thoroughly digested; like tar or loamy soil	Well stabilized
Septic tank sludge	Black	—	Offensive (H_2S) unless very long storage time	Mostly stabilized

Table 9.3 Chemical composition of raw and digested sludge (Loehr *et al.*, 1979)

Item	Raw primary sludge[a]		Digested sludge[b]	
	Range	Typical	Range	Typical
Total dry solids (TS), %	2.0–7.0	4.0	6.0–12.0	10.0
Volatile solids (% of TS)	60–80	65	30–60	40.0
Grease and fats (ether-soluble, % of TS)	6.0–30.0	—	5.0–20.0	—
Protein (% of TS)	20–30	25	15–20	18
Nitrogen (N, % of TS)	1.5–4.0	2.5	1.6–6.0	3.0
Phosphorus (P_2O_5, % of TS)	0.8–2.8	1.6	1.5–4.0	2.5
Potash (K_2O, % of TS)	0–1.0	0.4	0.0–3.0	1.0
Cellulose (% of TS)	8.0–15.0	10.0	8.0–15.0	10.0
Silica (SiO_2, % of TS)	15.0–20.0	—	10.0–20.0	—
pH	5.0–8.0	6.0	6.5–7.5	7.0
Alkalinity (mg/L as $CaCO_3$)	500–1,500	600	2,500–3,500	3,000

[a] Refers to sludge settled in primary sedimentation tanks
[b] Mostly refers to anaerobically digested sludge

Sewage sludge when applied to soils provides a source of plant nutrients, and is an effective soil amendment. Sludges applied to land provide major plant nutrients such as N, P, K; micro-plant nutrients such as Cu, Fe, and Zn; and organic matter for improving the soil structure (e.g. better aeration and water-holding capacity). The effectiveness of sludge as a soil-improving agent depends upon the composition of the sludge, the characteristics of the soil to which it is applied, and the plant species to be grown.

Table 9.4 Concentrations[a] of organic carbon (C), total nitrogen (N), phosphorus (P), ammonia (NH_4^+–N), nitrate (NO_3^-–N), sulfur (S), and potassium (K) in sewage sludge (U.S. EPA, 1983)

Component	Sludge type[b]	Number of samples	Range	Median[c]	Mean
Organic C, %	Anaerobic	31	18–39	26.8	27.6
	Aerobic	10	27–37	29.5	31.7
	Other	60	6.5–48	32.5	32.6
	All samples	101	6.5–48	30.4	31.0
Total N, %	Anaerobic	85	0.5–17.6	4.2	5.0
	Aerobic	38	0.5–7.6	4.8	4.9
	Other	68	< 0.1–10.0	1.8	1.9
	All samples	191	< 0.1–17.6	3.3	3.9
NH_4^+–N, mg/kg	Anaerobic	67	120–67,600	1,600	9,400
	Aerobic	33	30–11,300	400	950
	Other	3	5–12,500	80	4,200
	All samples	103	5–67,600	920	6,540
NO_3^-–N, mg/kg	Anaerobic	35	2–4,900	79	520
	Aerobic	8	7–830	180	300
	Other	3	—	—	780
	All samples	46	2–4,900	140	490
Total P, %	Anaerobic	86	0.5–14.3	3.0	3.3
	Aerobic	38	1.1–5.5	2.7	2.9
	Other	65	< 0.1–3.3	1.0	1.3
	All samples	189	< 0.1–14.3	2.3	2.5
Total S, %	Anaerobic	19	0.8–1.5	1.1	1.2
	Aerobic	9	0.6–1.1	0.8	0.8
	Other	—	—	—	—
	All samples	28	0.6–1.5	1.1	1.1
K, %	Anaerobic	86	0.02–2.64	0.30	0.52
	Aerobic	37	0.08–1.10	0.39	0.46
	Other	69	0.02–0.87	0.17	0.20
	All samples	192	0.02–2.64	0.30	0.40

[a] Concentrations and percentage composition are on a dried solids basis
[b] 'Other' includes lagooned, primary, tertiary, and unspecified sludges
[c] Median concentrations are reported to be better measures of 'typical' concentration than the arithmetic mean
—, Not reported

Some of the limitations in using sludge as fertilizer include the fluctuation of nutrient content. The nutrient contents of municipal sludges vary considerably, and N, P, and K levels are about one-fifth of those found in typical chemical fertilizers. Much of the N and P in sludge is in organic combination, which should be mineralized before becoming available to plants. The rate of mineralization for N and P in soil is dependent upon local conditions such as soil type, temperature, soil pH, soil water, and other soil chemical and physical characteristics. Certain sludges can be quite inert and create problems in the consideration of ultimate disposal techniques. In addition, sludge application on land may create odor problems and potential health hazards due to the presence of some toxic compounds and pathogens contained in the sludge itself.

9.2 SLUDGE TRANSPORT AND APPLICATION PROCEDURES

The important elements of system design of land treatment of sludge are: modes of sludge transport, sludge application procedures, and sludge application rates. The first two elements will be described in this section, and section 9.3 will discuss the third element.

9.2.1 Mode of sludge transport

Table 9.5 shows the various methods of handling and transporting sludge from a source to a land application or disposal site. Transportation may be accomplished by a pipeline (gravity flow or pressurized), tank truck, barge, or conveyor rail. Sludge characteristics (e.g. solid contents), sludge volume, elevation differences, transport distance, and land availability are important factors in selecting a method of sludge transportation. As shown in Table 9.6, liquid sludge (1–10 percent solid contents) is generally suitable for any mode of sludge transport, while semi-solid or solid sludge, having high solid contents (8–80 percent), should be transported only by trucks or rail hopper cars.

Tank trucks are currently widely used to transport and apply sludge on land because they afford flexibility in the selection of land application sites. Usually, a storage facility for sludge is provided at the land application site.

9.2.2 Sludge application procedures

Similar to the mode of sludge transport, the selection of sludge application systems and equipment depends on: type of sludge (liquid, semi-solid, or solid), quantity, areal application rate, frequency of application, topography of the applied area, and the time of year. Table 9.7 gives the various sludge application methods

Table 9.5 Sludge solids content and handling characteristics (Knezek and Miller, 1976)

Type	Solids content (%)	Handling methods
Liquid	1–10	Gravity flow, pump, tank transport
Semi-solid ('wet' solids)	8–30	Conveyor, auger, truck transport (watertight box)
Solid ('dry' solids)	25–80	Conveyor, bucket, truck transport (box)

Table 9.6 Transport modes for sludge (Knezek and Miller, 1976)

Type	Characteristics
Liquid sludge	
Rail tank car	100 wet tons (24,000 gal.[a]) capacity; suspended solids will settle while in transit
Barge	Capacity determined by waterway; Chicago has used 1,200 wet tons (290,000 gal.) barges
Pipeline	Need minimum velocity of 1 f.p.s. (30 cm/s) to keep solids in suspension; friction decreases as pipe diameter increases (to the fifth power); buried pipeline suitable for year-round use
Vehicles	
Tank truck	Capacity: up to maximum load allowed on road. Can have gravity or pressurized discharge. Field trafficability can be improved by using flotation tires
Farm tank wagon and tractor	Capacity: 800–3,000 gal. Principal use would be for field application
Semi-solid or solid sludge	
Rail hopper car	Need special unloading sites and equipment for field application
Truck	Commercial equipment available to unload and spread on ground; need to level sludge piles if dump truck is used

[a] 1 gallon = 3.785 L

and the equipment required for the application. The irrigation methods such as sprinkling, ridge-and-furrow (Figure 8.2) and overland flow (Figure 8.4), are similar to those described in Chapter 8, 'Land treatment of wastewater'. Sprinkling systems, applicable only to liquid sludge, require less preparation, and can be used with a wide variety of crops. Ridge-and-furrow and overland flow are suitable to level land and sloped land, respectively; they also minimize aerosol-related pathogens and aesthetic problems.

Tank trucks, farm tank wagons, and tractors are also effective in applying sludge on crop lands in which sludge is pumped from a roadway on to the field. Commercial tank trucks with plow furrow cover are available where sludge is discharged in furrow ahead of a plow mounted on the rear of the trucks.

9.3 SYSTEM DESIGN AND SLUDGE APPLICATION RATES

The design of land treatment of wastewater and sludge is usually based on the limiting design parameter (LDP). LDP is the design value in which other constraints such as nutrient requirements or public health problems will be overcome. For example, the annual sludge application rate on an agricultural land may be determined by N or P requirement, and the useful life of the site may be limited by cadmium (Cd) loading. Sludge should be applied to agricultural land at a rate equal to the N uptake rate of the crop, unless lower application rates are required because of Cd limitations. The basis for N loading criteria is to minimize

Table 9.7 Application methods and equipment for liquid and some semi-solid sludges

Method	Characteristics	Topographical and seasonal suitability
Irrigation		
Spray (sprinkler)	Large orifice required on nozzle; large power and lower labor requirement; wide selection of commercial equipment available; sludge must be flushed from pipes when irrigation completed	Can be used on sloping land; can be used year-round if the pipe is drained in winter; not suitable for application to some crops during growing season; odor (aerosol) nuisance may occur
Ridge-and-furrow	Land preparation needed; lower power requirements than spray	Between 0.5 and 1.5% slope depending on percentage solids; can be used between rows of crops
Overland flow	Used on sloping ground with vegetation with no runoff permitted; suitable for emergency operation; difficult to get uniform areal application	Can be applied from ridge roads
Tank truck	Capacity 500 to more than 2,000 gal.; larger-volume trucks will require flotation tires; can use with temporary irrigation set-up; with pump discharge can spray from roadway on to field	Tillable land; not usable with row crops or on soft ground
Farm tank wagon and tractor	Capacity 500 to 3,000 gal.; larger volume will require flotation tires; can use with temporary irrigation set-up; with pump discharge can spray from roadway on to field	Tillable land; not usable with row crops or on soft ground
Flexible irrigation hose with plow furrow or disc cover	Use with pipeline or tank truck with pressure discharge; hose connected to manifold discharge on plow or disc	Tillable land; not usable on wet or frozen ground
Tank truck with plow furrow cover	500-gal. commercial equipment available; sludge discharged in furrow ahead of plow mounted on rear of four-wheel-drive truck	Tillable land; not usable on wet or frozen ground
Farm tank wagon and tractor		
Plow furrow cover	Sludge discharged into furrow ahead of plow mounted on tank trailer—application of 170–225 wet tons/acre; or sludge spread in narrow band on ground surface and immediately plowed under—application of 5–125 wet tons/acre	Tillable land; not usable on wet or frozen ground
Subsurface injection	Sludge discharged into channel opened by a tillable tool mounted on tank trailer; application rate 25–50 wet tons/acre; vehicles should not traverse injected area for several days	Tillable land; not usable on wet or frozen ground

1 acre = 0.405 ha; 1 ton/acre = 2,470 kg/ha; 1 gallon = 3.785 L
From Knezek and Miller (1978)

nitrate leaching to groundwater. The reason for the annual limit on Cd loading is to minimize crop uptake and the potential long-term, subclinical adverse effects on human health. The lifetime limit of a land application site is established on the basis of maximum cumulative loadings of heavy metals such as lead (Pb), zinc (Zn), copper (Cu), nickel (Ni), and Cd. These limits are designed to allow for growth and use of food-chain crops at future dates.

As the liquid content of sludge is typically low, the hydraulic capacity of the site is seldom the LDP for the design. An exception might be toxic and hazardous waste sites where very permeable soils would permit too rapid percolation movement.

A preliminary estimate of the land area required for screening purposes can be determined for municipal sludges with Table 9.8. Estimates of soil treatment area for industrial sludges should be based on the critical LDP such as metal concentration, etc. Slope limitation for the land application of the sludge is given in Table 9.9.

For the agricultural utilization of sludge, the design approach is based on the utilization of sludge as a supplement or replacement for commercial fertilizers. As a result the annual application is based on the N or P needs of the crop in a particular soil. In addition, both the annual and cumulative sludge loadings must be consistent with regulatory limits on pathogens and metals. Table 9.10 summarizes the U.S. EPA limits on cumulative heavy metal loadings for agricultural land as a function of soil cation exchange capacity (CEC). CEC has been employed as an indicator of soil properties because it has been shown experimentally that most metals in sludge-amended soils are not present as an exchangeable cation, or these metals are not exchangeable with a neutral salt. Therefore soil with a higher CEC value would minimize the plant availability of sludge-borne metals more than soil with a lower CEC value (U.S. EPA, 1983). In general the CEC ranges of less than 5, 5–15, and more than 15 meq/100 g correspond to sands, sandy loams, and silt loams, respectively.

Based on scientific risk-assessment evaluation aimed at limiting human and ecological exposure to sludge contaminants, in relation to the current sewage sludge use and disposal regulations, the U.S. EPA has implemented the new land application limits for the 10 heavy metals, as shown in Table 9.11. These limits do not mention about soil CEC, but the cumulative limits for Pb and Ni are lower than those reported in Table 9.10 (at CEC greater than 15 meq/100 g) and Table 9.12

Table 9.8 Estimated land area for municipal sludge applications

Option	Application period	Reported range (dry tons/ha)	Typical rate (dry tons/ha)
Agricultural	Annual	0.4–12	2
Forest	One application or at 3–5 year intervals	1.6–40.5	8
Site reclamation	One application	1.2–81	20

From Reed and Crites (1984); reproduced by permission of Noyes Publications

Table 9.9 Recommended slopes for sludge application sites (U.S. EPA, 1983)

Slope (%)	Comment
0–3	Ideal
3–6	Acceptable for surface application or injection
6–12	Applicable for injected liquid sludge
12–15	Immediate incorporation of all sludges and effective runoff control are necessary

Table 9.10 Recommended cumulative limits for metals of major concern applied to agricultural cropland (U.S. EPA, 1983)

	Soil cation exchange capacity (meq/100 g)[a,b]					
	< 5		5–15		> 15	
Metal	kg/ha	(lb/acre)	kg/ha	(lb/acre)	kg/ha	(lb/acre)
Pb	560	(500)	1,120	(1,000)	2,240	(2,000)
Zn	280	(250)	560	(500)	1,120	(1,000)
Cu	140	(125)	280	(250)	560	(500)
Ni	140	(125)	280	(250)	560	(500)
Cd	5	(4.4)	10	(8.9)	20	(17.8)

[a] Interpolation should be used to obtain values in the CEC range 5–15
[b] Soil must be maintained at pH 6.5 or above

(for Pb), while the cumulative limits for other heavy metals (Cd, Cu, and Zn) are higher. Table 9.11 classifies sludge quality as 'ceiling concentration limits' and 'high-quality pollutant concentration limits'. Sludge quality classifications and application restrictions are defined as follows:

1. If a sludge meets the 'high quality' metal concentration limits, it can be land applied provided that the application rate does not exceed annual pollutant loading rates as shown in Table 9.11.
2. If a sludge does not meet the 'high quality' limits but does meet the ceiling concentration limits, the sludge can be land applied provided that the cumulative pollutant loading rates as defined in Table 9.11 are not exceeded (in addition to annual pollutant loading rates).
3. If a sludge does not meet ceiling concentration limits, it cannot be land applied.

Table 9.13 shows the concentration limits of heavy metals in sewage sludge used for land application in several European countries. These limits put stringent standards on Cd, Hg, and As, similar to those of Table 9.11.

Other guidelines for sludge application for fruit and vegetable production are given in Table 9.12. It is required that soils at sludge application sites be maintained at soil pH of 6.5 or above. Information on fertilizer recommendations for a particular crop in a specific location can be obtained from agricultural

Table 9.11 U.S. land application pollutant limits (Walsh, 1995)[a]. Reproduced by permission of IAWQ

Pollutant	Ceiling concentration limits (mg/kg)	Cumulative pollutant loading rates (kg/ha)	'High quality' pollutant concentration limits (mg/kg)	Annual pollutant loading rates (kg/ha-year)
Arsenic	75	41	41	2.0
Cadmium	85	39	39	1.9
Chromium[b]	3,000	3,000	1,200	150
Copper	4,300	1,500	1,500	75
Lead	840	300	300	15
Mercury	57	17	17	0.85
Molybdenum	75	18[c]	18[c]	0.90
Nickel	420	420	420	21
Selenium	100	100	100[b]	5.0
Zinc	7,500	2,800	2,800	140

[a] All weights are on a dry weight basis and ceiling concentration limits represent absolute values. 'High quality' pollutant loading rates represent monthly averages

[b] In October 1995, the U.S. EPA withdrew all pollutant limits for chromium, and increased the selenium concentration in 'high quality' pollutant concentration limit from 36 to 100 mg/kg

[c] In February 1994, the U.S. EPA withdrew the molybdenum value of 18 mg/kg pending further review of scientific information supporting a higher concentration

experiment extension stations located in the province or country (see Table 8.7). The design is then based on meeting either the N or P needs. Optimum yields and crop production may then require supplementary fertilization for the other nutrients (N, P, and K) (Reed and Crites, 1984).

In addition to Table 9.11, the U.S. EPA also classifies sludge quality based on indicator bacteria. The Class A indicator standard is less than 1000 fecal coliforms per gram of dry sludge solids and the Class B indicator standard is less than 3 million fecal coliforms per gram of dry sludge solids (Walsh, 1995). If a sludge contains about 1 percent solids, the Class A standard is apparently similar to the WHO standard for unrestricted irrigation (Table 2.27), and there are no use restrictions placed on sludges that meet the Class A requirements. Because the Class B sludge poses greater health risks, the U.S. EPA imposes more use restrictions than those outlined in Table 9.12, e.g.

1. food crops that receive a sludge application cannot be harvested for periods ranging from 14 to 38 months afterwards, depending upon the type of crop grown and the method of application;
2. pasture lands that receive sludge cannot be grazed for 30 days;
3. turflands are not allowed to be harvested within 12 months of application; and
4. public lands that receive sludge application will have access restricted for 30 days in low-exposure areas and up to 1 year for high-exposure areas.

Table 9.12 Summary of joint EPA/FDA/USDA guidelines for sludge application for fruit and vegetable production (U.S. EPA, 1983)

Annual and cumulative Cd rates
Annual rate should not exceed 0.5 kg/ha. Cumulative Cd loadings should not exceed 5, 10, or 20 kg/ha, depending on soil pH and CEC values of <5, 5–15, and >15 meq/100 g, respectively.

Soil pH
Soil pH (plow zone—top 15 cm) should be 6.5 or greater at time of each sludge application

PCBs
Sludges with PCB concentrations greater than 10 mg/kg should be incorporated into the soil.

Pathogen reduction
Sludge should be treated by a pathogen reduction process before soil application. A waiting period of 12–18 months before a crop is grown may be required, depending on prior sludge processing and disinfection.

Use of high-quality sludge
High-quality sludge should not contain more than 25 mg/kg Cd, 1,000 mg/kg Pb, and 10 mg/kg PCB (dry-weight basis).

Cumulative Pb application rate
Cumulative Pb loading should not exceed 800 kg/ha.

Pathogenic organisms
A minimum requirement is that crops to be eaten raw should not be planted in sludge-amended fields within 12–18 months after the last sludge application. Further assurance of safe and wholesome food products can be achieved by increasing the time interval to 36 months.

Physical contamination and filth
Sludge should be applied directly to soil and not directly to any human food crop. Crops grown for human consumption on sludge-amended fields should be processed with good food industry practices, especially for root crops and low-growing fresh fruits and vegetables.

Soil monitoring
Soil monitoring should be performed on a regular basis, at least annually for pH. Every few years a soil test should be run for Cd and Pb.

Choice of crop type
The growing of plants which do not accumulate heavy metals is encouraged.

FDA = U.S. Food and Drug Administration
USDA = U.S. Department of Agriculture

9.3.1 Sludge application rates

The application rates of sludges to cropland vary according to the nutrient requirements of the crop, existing soil characteristics (e.g. drainage, nutrient level, heavy metal content), climate, and the characteristics of the sludge. In all cases the application rate should be such that:

Table 9.13 Concentration limits of heavy metals in sewage sludge used for land application in some European countries (cited in Rulkens *et al.*, 1989). Reproduced by permission of Chapman & Hall

	Concentration limit (mg/kg dry solids)							
	Zn	Cu	Pb	Cr	Ni	Cd	Hg	As
Netherlands	2,000	600	500	500	100	5	5	10
France	3,000	1,000	800	1,000	200	40	10	—
Germany	3,000	1,200	1,200	1,200	200	20	25	—
Norway	3,000	1,500	300	200	100	10	7	—
Sweden	10,000	3,000	300	1,000	500	15	8	—
Finland	5,000	3,000	1,200	1,000	500	30	25	—
Scotland	10,000	1,000	800	800	250	20	7.5	150

1. crop production and quality are not decreased;
2. excessive organic material or heavy metals are not built up in the soil;
3. nutrients and excessive salts do not leak into surface or subsurface water supplies;
4. crops are not contaminated with pathogens that will become health risks to the farm operators and consumers.

If the heavy metal and fecal coliform limits, listed above, are met, the sludge application rates are based mainly on N loading. As much of the sludge N is in organic form, as shown in Tables 9.3 and 9.4, it is not all readily available to the plants. Other forms of N in sludge are ammonia N and very little nitrate. When applied on land, part of the organic N will be mineralized or biologically converted into ammonia and/or nitrate to become available to the plants. Table 9.14 contains suggested mineralization rates for organic N present in a few types of sludge. The organic N not mineralized in the first year will be mineralized in subsequent years according to the mineralization rates, shown in Table 9.14.

Table 9.14 Mineralization rates for organic N in wastewater sludges (U.S. EPA, 1983)

	Mineralization rate (%)		
Year after sludge application	Raw sludge	Anaerobically digested sludge	Composted sludge
First	40	20	10
Second	20	10	5
Third	10	5	3
Fourth	5	3	3
Fifth	3	3	3
Sixth	3	3	3
Seventh	3	3	3
Eighth	3	3	3
Ninth	3	3	3
Tenth	3	3	3

Under a high pH condition some ammonia N will be lost through volatilization (eq. 6.1 and Figure 6.8). Ammonia N is readily absorbed by soil, but nitrate N is not. Nitrates that are not taken up by the plants, and not biologically denitrified (or converted into gaseous N), may be leached into groundwater.

The following example shows a method of calculating sludge application rate in the first year based on N uptake rate of a crop, and considering mass balance of N in soil.

Example 9.1 If corn is to be grown in a sandy loam soil, estimate the application rate of a digested sludge which has the following N analysis: organic N = 20,000 mg/L, ammonia N = 1,500 mg/L, nitrite N = 5 mg/L, nitrate N = 50 mg/L. This sludge contains 21,555 mg/L total N or 2.15 percent N by weight. For simplicity of calculation, assume that the sludge's bulk density is 1 kg/L.

The annual N uptake rate for corn is about 224 kg/ha. Assume the N mineralized during the first year is 20 percent of the organic N, and that the volatilized fraction is 50 percent of the ammonia N. Also assume that no N flows into groundwater, no denitrification occurs, and initial available N content of the soil is negligible. The mass balance is:

$$\text{N applied} = \text{N volatilized} + \text{N leached to groundwater} + \text{N used by plants} \\ + \text{N lost through denitrification} \tag{9.1}$$

The amount of N applied is calculated as kg/ha. N is volatilized in the form of ammonia N. As 20 percent of the organic N is mineralized (to ammonia), the total available ammonia N is 1,500 + 0.2 (20,000) = 5,500 mg/L, and the volatilized ammonia is 0.5 (5,500) = 2,750 mg/L (or about 0.275 percent N is lost through volatilization). If the annual application rate of sludge is x kg/ha dry weight, the total N volatilized is 2.75×10^{-3} (x) kg/ha.

The amounts of N loss to groundwater and due to denitrification are zero. The amount of N used by the corn is 224 kg/(ha-year).

The total N applied is the product of dry solids application rate, x (ka/ha) times the fraction of total N or 21.555×10^{-3} (x) kg/ha. From the mass balance equation (eq. 9.1)

$$21.555 \times 10^{-3}\ (x) = 2.75 \times 10^{-3}\ (x) + 0 + 224 + 0$$
$$\text{and } x = 11{,}912\ \text{kg/ha}$$

The amount of sludge to be applied to corn crop in the first year is 11,912 kg/ha (dry weight).

9.3.2 Sludge loading determination

The calculation for sludge loading on an N basis is a three-step procedure:

1. Determine the plant-available N (N_p) in the sludge during the application year:

$$N_p = (1{,}000)[NO_3 + K_v(NH_4) + f_1(ON)] \quad (9.2)$$

where N_p = plant-available N in sludge during application year, kg/dry ton of sludge;
NO_3 = percentage nitrate in sludge, as a decimal (e.g. 1,000 mg/L (or ppm) = 0.1 percent = 0.001);
K_v = volatilization factor: 0.5 for surface-applied liquid sludge, 1.0 for incorporated (or injected) liquid sludge and dewatered digested sludge applied in any manner;
NH_4 = percentage ammonia N in sludge, as a decimal;
f_1 = mineralization factor for first year, as a decimal (Table 9.14);
ON = percentage organic N in sludge, as a decimal.

Equation 9.2 will determine the plant-available N in sludge in kg/ton. To convert the concentrations of NO_3, NH_4 and ON into the same unit, the value of each of these N constituents has to be in decimal terms. For example, an NH_4 concentration of 1,000 mg/L (or ppm) is equivalent to 0.1 percent, or sludge 100 kg contains NH_4 0.1 kg. Therefore, sludge 1,000 kg (or 1 ton) will contain NH_4 = 1,000 times (0.1/100). The term in parentheses is the decimal value.

2. Determine the plant-available N (N_{PR}) from mineralization of the residual sludge in subsequent years:

$$N_{PR1} = 1{,}000[f_2(ON)_2 + f_3(ON)_3 + \cdots] \quad (9.3)$$

where N_{PR1} = plant-available N from mineralization of the first-year sludge in subsequent years, kg/dry ton of sludge;
f = mineralization rate (Table 9.14), as a decimal—subscripts refer to the relevant year;
ON = percentage organic N remaining in the sludge in a particular year, as a decimal—subscripts refer to the relevant year.

A system with continuous annual applications will have to solve eq. 9.3 for each of the subsequent years, i.e. N_{PR2}, N_{PR3}, N_{PR4}, etc. The calculations will converge on a relatively constant value after 5–6 years if the sludge composition remains the same.

3. The annual sludge loading (S_{Nt}) based on N is determined with:

$$S_{Nt} = \frac{U_N}{(N_p + N_{PR1} \cdots)} \quad (9.4)$$

where S_{Nt} = annual sludge loading in year (t) of concern, dry ton/ha
U_N = crop uptake of N, kg/(ha-year)
N_p, N_{PR1} = from eqs 9.2 and 9.3.

In addition to the above three-step procedure, attention should be given to possible loss of sludge or nutrients due to surface runoff. This sludge loss can be significant in areas having high rainfall intensity and soil erosion. According to

Batelle Memorial Institute (1987), the extent of sludge loss is dependent upon rainfall intensity, soil erosivity, soil slope, type of crops being grown, and erosion control practice implemented at the site. Under conditions where high rainfall and steep slope prevail, or when sludge application is conducted during flooding and raining, the magnitude of sludge loss due to surface runoff should be taken into consideration.

Example 9.2 Determine the land area for application of anaerobically digested municipal sludge with the following conditions:

(a) Sludge production: 22 dry tons/day.
(b) Sludge characteristics: Pb = 500 ppm, Cd = 50 ppm, Ni = 100 ppm, Cu = 500 ppm, Zn = 2,000 ppm, total N = 2.5 percent, NH_4 = 1.0 percent, $NO_3^- = 0$.
(c) Sludge will be incorporated, so $K_v = 1$, CEC of 10 meq/100 g for site soils. Corn is the intended crop, $U_N = 180$ kg/(ha-yr).
(d) The design can be based on N fertilization rates or heavy metal loading rates.

Mineralization rates (Table 9.14) for digested sludge are:

Year	*f*
1	0.20
2	0.10
3	0.05
4–10	0.03

Solution

1. Organic N (ON) = Total N − NH_4
   = (2.5 − 1.0) percent = 0.015
   (a) From eq. 9.2 available N in the first year:
   $N_p = (1{,}000)[0 + 1(0.01) + 0.20(0.015)]$
   $= 13$ kg/ton of dry sludge
   (b) From the first-year sludge, the organic N remaining in second year:
   $ON_2 = (0.015) - (0.20)(0.015)$
   $= 0.012$
   From eq. 9.3 the amount of first-year sludge mineralized in second year:
   $N_{PR2} = (1{,}000)(f_2)(ON_2) = (1{,}000)(0.1)(0.012)$
   $= 1.2$ kg/ton of dry sludge
   (c) From the first-year sludge, organic N remaining in third year:
   $ON_3 = (0.012) - (0.1)(0.012) = 0.0108$
   Mineralization of the first-year sludge in third year:
   $N_{PR3} = (1{,}000)(f_3)(ON_3) = (0.05)(0.0108)(1{,}000)$
   $= 0.54$ kg/ton of dry sludge

(d) In the fourth year:

$$ON_4 = (0.0108) - (0.05)(0.0108) = 0.0103$$

$$N_{PR4} = (1{,}000)(f_4)(ON_4) = (1{,}000)(0.03)(0.0103) = 0.309 \text{ kg/ton of dry sludge}$$

(e) The N_{PR} for year 5 and beyond converges on a value of about 0.3 kg/ton of dry sludge.

2. Repeat the calculations for sludge applied in years 2 through 10 and tabulate results.

(a) For all applications of sludge in each year.

Year after application	N_{PR}, kg/ton
2	1.2
3	0.54
4	0.31
5–10	0.3

(b) The total available N for each year in kg/ton will therefore increase yearly as shown in column 2 of Table 9.15. As the crop uptake of N is fixed at 180 kg/(ha-year), the amount of sludge applied has to be decreased yearly according to eq. 9.4, as shown in column 3 in Table 9.15.

3. Based on Table 9.12, Cd application rate to crop should not exceed 0.5 kg/(ha-year), for this sludge containing 50 ppm Cd, the recommended loading is:

$$\frac{0.5 \text{ kg}}{\text{ha-year}} \cdot \frac{1}{0.00005} \cdot \frac{\text{ton}}{(1{,}000 \text{ kg})} = 10 \text{ tons/(ha-year)}$$

This Cd application is lower than the N application rate. Thus the sludge loading of 10 tons/(ha-year) is the LDP to be used in determining the land requirement of sludge application. To maximize corn growth, chemical fertilizer needs to be applied to the crop to supplement for the N requirement.

4. In a case where Cd is LDP, the land area requirement for the sludge production rate of 22 tons/day is:

$$\frac{(22 \text{ tons/day})(365 \text{ days/year})}{10 \text{ tons/(ha-year)}} = 800 \text{ ha.}$$

5. At the sludge application rate of 10 tons/(ha-year), Cd content in sludge = 50 ppm, and soil CEC of 10 meq/100 g, the life period of sludge application that will not allow Cd to exceed the cumulative limit as given in Table 9.10 is:

$$\frac{10 \text{ kg/ha}}{10 \text{ tons/(ha-year)}} \cdot \frac{\text{ton}}{1{,}000 \text{ kg}} \cdot \frac{1}{(0.00005)} = 20 \text{ years}$$

The design period of this land application site based on cumulative limit of Cd loading is 20 years. The design period based on cumulative limits of other heavy metals will be several times higher than that of Cd for this type of sludge.

Table 9.15 Total available N and sludge application rates for Example 9.2

Year	Total available N, kg/dry ton sludge	Annual sludge loading, (eq. 9.4), S_{Nt}, dry ton sludge/(ha-yr)
1	$13 = 13$	13.85
2	$13 + 1.2 = 14.2$	12.68
3	$13 + 1.2 + 0.54 = 14.74$	12.21
4	$13 + 1.2 + 0.54 + 0.31 = 15.05$	11.96
5	$13 + 1.2 + 0.54 + 0.31 + 0.3 = 15.35$	11.73
6	$13 + 1.2 + 0.54 + 0.31 + 0.3 + 0.3 = 15.65$	11.50
7	$13 + 1.2 + 0.54 + 0.31 + 0.3 + 0.3 + 0.3 = 15.95$	11.28
8	$13 + 1.2 + 0.54 + 0.31 + 0.3 + 0.3 + 0.3 + 0.3 = 16.25$	11.08
9	$13 + 1.2 + 0.54 + 0.31 + 0.3 + 0.3 + 0.3 + 0.3 + 0.3 = 16.55$	10.88
10	$13 + 1.2 + 0.54 + 0.31 + 0.3 + 0.3 + 0.3 + 0.3 + 0.3 + 0.3 = 16.85$	10.68

Note. If the Cd application rate of 1.9 kg/(ha-year) as listed in Table 9.11 is used, the sludge loading rate would be 38 tons/(ha-year). However, to avoid excessive N loading to corn crops, the land area requirements should be calculated from the sludge loading rates given in Table 9.15.

9.3.3 Monitoring program

As the design is normally based on limiting both N and heavy metals, monitoring for these two parameters should not be necessary. Routine soil testing every 2–4 years for plant-available P and K, and determination of lime requirements for pH maintenance at 6.5, are all that should be necessary.

The schedule for sludge applications will depend on the type of crop and on the climate for the area. Sludge is not usually applied when the ground is flooded or frozen, to reduce risk of runoff contamination. Sludge can be applied to the fields for row crops prior to planting and after harvest. Sludge application to forage grasses is possible in all months of the year when the ground is not flooded.

Periodic sludge analysis is needed to provide nutrient and heavy metal concentrations so that rates of application can be determined to meet crop nutrient needs, and total heavy metal loadings can be recorded from year to year.

9.3.4 Case studies

The following are two case studies of agricultural application of sludge in the U.S.A. and Canada.

CASE STUDY I: U.S.A.

The city of Salem, Oregon, generates 121,120 m^3/year of anaerobically digested sludge containing 2.7 percent solids, and 90–95 percent of this sludge is recycled to local farmland (U.S. EPA, 1984). Typical sludge characteristics after digestion are shown in Table 9.16. The sludge N levels are raised by the addition of ammonia during the treatment process because the raw sewage contains a high percentage of food processing wastes, which are deficient in nutrients. The sludge is applied to about 1,200 ha of local agricultural land. At virtually all application sites, the sludge is applied only once per year. Sludge application rates are based on the N needs of the crop and the nutrient content of the sludge. They vary from 2.2 to 6.3 dry tons/ha, averaging approximately 3.4 dry tons/ha.

The sludge is applied primarily to fields used to produce grains, grasses, pasture, and silage corn. Sludge-amended sites are also used to produce seed crops, Christmas tree farms, commercial nurseries, and filbert orchards. No sludge is applied to fruit and vegetable crops. For poorly drained soils, sludge is only applied from April to October. For well-drained soils, sludge can be applied anytime except during or immediately after rainstorms. Schedules for application of sludge to soils with intermediate drainage capacity fall between these two extremes. Cation exchange capacity is used to limit cumulative metal loadings added by sludge application. However, if soil pH is less than 6.5, as it is in most of the Salem area, then cumulative Cd addition is limited to 4 kg/ha, regardless of soil cation exchange capacity. As the sludge generated by Salem is very low in metals, application sites generally have a life well over 25 years.

Table 9.16 Typical characteristics of digested sludge at Salem, Oregon (U.S. EPA, 1984)[a]

Constituent	Concentration
pH	7.3
Total solids (%)	2.5
Total nitrogen (%)	10.3
Ammonia nitrogen (%)	5.9
Phosphorus (%)	2.0
Potassium (%)	0.96
Zinc (mg/kg)	980
Copper (mg/kg)	470
Nickel (mg/kg)	43
Cadmium (mg/kg)	7
Lead (mg/kg)	230
Chromium (mg/kg)	60
Magnesium (mg/kg)	200
Calcium (mg/kg)	12,200
Sodium (mg/kg)	3,000

[a] All constituents except pH are reported on a dry-weight basis

Each sludge application site is investigated by the Oregon Department of Environmental Quality, which makes recommendations on a case-by-case basis. General guidelines for sludge application sites are as follows:

1. Minimum distance to domestic wells = 61 m.
2. Minimum distance to surface water = 15 m.
3. Minimum rooting depth (effective depth of soil) = 0.61 m.
4. Minimum depth to groundwater at time that sludge is applied = 1.22 m.
5. Minimum distance of sludge application to public access areas varies with the method of sludge application:

   If sludge is incorporated into soil = 0

   If sludge is not incorporated into soil = 30.5 m

   If sludge is pressure-sprayed ('big gun'-type sprayer) over the soil = 91–152 m
6. Sludge application is not approved close to residential developments, schools, parks, and similar areas.
7. Minimum slope is largely left to the investigator's discretion. Where no surface waters are endangered, slopes as high as 30 percent have been approved. Generally, however, the maximum allowable slope is 12 percent and, in cases where sensitive surface waters are nearby, maximum slopes may be held at 7 percent or less.

In general, pasture and grass land receive sludge applications during the winter months; agricultural land growing seasonal crops receives sludge during the warmer months, before planting or after harvesting. When weather prevents sludge application, the sludge is stored in lagoons at the treatment plant.

Sludge is usually applied by haul trucks themselves. If the application soil is too wet or otherwise unsuitable for direct truck access, then the sludge is sprayed on to the application site.

Analyses of the sludge-amended soil showed virtually no change in soil chemical and physical characteristics. Groundwater from wells on or within 150 m of sludge application sites was sampled and analyzed before and after application. Results showed no significant changes in groundwater quality over a period of 3 years and now only selected wells are sampled approximately every 3 years. Crop tissue sampling and analysis during the initial years of the sludge application showed no significant difference between crops grown on sludge amended soils and control crops.

CASE STUDY II: CANADA

The Halton Region in Ontario Province operates seven municipal wastewater treatment plants treating over 220,000 m^3/day of wastewater. The primarily treated sludge is stabilized in anaerobic digesters and the stabilized sludge, rich in N, P, and organic matter, is produced about 220,000 m^3/year.

The sludge is applied on farmland (growing field crops, e.g. wheat) by tractors equipped with injectors for direct incorporation of sludge into the ground. Another mode of sludge application by pressure spread on the field surface is also employed. At the Halton Region, sludge application rate is based on the NH_3 and NO_3^-–N loading which should not exceed 135 kg/ha during the 5 year period; this practice also results in about 220 kg/year of P applied on the farm land in 5 years, a cost-saving for N and P for the farmers.

According to the Ontario Ministry of Agriculture and Food (1993), the application of this sludge on the farmland follows the provincial guidelines for sewage sludge utilization on agricultural lands, and has proven to be safe. Sludge is routinely analyzed for N, P, and heavy metal contents, and strict control is maintained to prevent heavy metal overloading to the soils.

During wet weather which could cause soil compaction and run-off problems, the sludge is taken to a temporary storage site with a sludge storage capacity of 63,000 m^3.

9.4 TOXIC COMPOUNDS VS CROP GROWTH

Toxic substances in sludge can be categorized as heavy metals and toxic organic compounds. The U.S. EPA had collected sludge samples from more than 200 representative wastewater treatment facilities and analyzed for more than 400 pollutants. They found that many toxic compounds traditionally subject to regulation were either not present or were present at levels that would not pose a health or environmental risk (Walsh, 1995). Hence, only the heavy metal limits (see Table 9.11) and fecal coliform standards, reported in section 9.3, are being emphasized.

Some of the factors affecting plant uptake of heavy metals are as follows (Bitton *et al.*, 1980):

1. Levels of toxic elements in the wastewater or sludge and their characteristics;
2. species of plants grown, their age, conditions, and rooting depth—some plants are known to be metal accumulators;
3. background concentration of toxic elements in the soil and their distribution;
4. ability of chemical constituents in soil to convert toxic elements to non-available chemical compounds—this ability is in turn affected by the nature of toxic element and the type and characteristics of soil, for example:
   (a) pH of the soil solution,
   (b) organic and clay content and type of soil,
   (c) phosphate level in the soil,
   (d) cation exchange capacity (CEC) of soil,
   (e) adsorption and precipitation.

The above factors determine the deleterious effect of heavy metals and other substances on plants and soils. In soils with high pH, most heavy metals occur in precipitated, unavailable forms. Similarly, the P uptake by plants is also retarded at high pH. Thus, calcareous soils are generally less amenable to manifestations of heavy metal toxicity than are the neutral or acidic soils.

Based on Tables 9.11–9.13, serious consideration should be given to the fate of mercury, molybdenum, Cd, selenium, and arsenic in the food chain. The other heavy metals do not seem to accumulate in edible portions of crops, and are essentially phytotoxins or causing toxicity to plants (some substances in trace amounts may, in fact, be advantageous for crop growth). Zinc is more readily absorbed than most other heavy metals. The presence of Cu somewhat inhibits Zn transport through the plant.

There is some uncertainty at the present time concerning the safe levels of many elements accumulating in plants. Little is known concerning the long-range effects of toxic elements applied to agricultural lands through the continuous use of wastewater and sludges. The guidelines given in Tables 9.11 and 9.13 have been proposed to minimize the health impact resulting from the heavy metals present in sludge. These guidelines should be followed wherever possible in the design and operation of land treatment of sludge.

At present, polychlorinated biphenyls (PCBs) are the only group of toxic organic compounds addressed in the joint U.S. EPA/FDA/U.S. DA guidelines for sludge application for fruit and vegetable production (Table 9.12). Sludge containing more than 10 mg PCB/kg must be soil incorporated, and sludges containing more than 50 mg PCB/kg may not be land applied at any rate, nor mixed with less contaminated sludges to lower the mixture's PCB content (O'Conner *et al.*, 1990).

The principal problem arising from PCBs is direct ingestion by animals grazing on forages treated with surface-applied sludge. Dairy cattle are most susceptible to PCB contamination of forages, because PCBs do not adsorb on to the surface of root crops such as carrots, hence they are not a major concern for this kind of crop (U.S. EPA, 1983).

9.5 MICROBIAL ASPECTS OF SLUDGE APPLICATION ON LAND

With the practice of land disposal of sludge becoming more popular and widespread, the survival of pathogenic organisms in the soil, water, and on crops grown in sludge assumes increasing importance. Pathogens in soil are destroyed by the natural environmental conditions which favor native soil organisms. Some pathogens are entrapped and adsorbed at the soil surface and undergo rapid die-off in the soil matrix.

In general, microorganisms may survive in the soil for a period varying from a few hours to several months, depending on factors such as: (i) type of organism; (ii) temperature: lower temperature increases survival time; (iii) moisture: longevity is greater in moistened soils than in dry soils; (iv) type of soil: neutral pH, high-moisture-holding soils favor survival; (v) organic matter: the type and amount of organic matter present may serve as food or energy to sustain the microorganisms; and (vi) the presence of other microorganisms can have an antagonistic effect on the pathogens. Pathogens do not generally survive as long on vegetation as they do in soil, because they are exposed to adverse environmental conditions such as UV radiation, desiccation, and temperature extremes (see Figures 8.13 and 8.14).

There are various forces acting to retain or facilitate the movement of microorganisms through soil. Filtration by the soil at the soil–water interface is the primary means of retaining bacteria in the soil, or in some cases in an additional biological mat formed in the top 0.5 cm of soil. Other mechanisms that retain bacteria in the top few centimeters of fine soil are intergrain contacts, sedimentation, and adsorption by soil particles. The soils containing clay remove most microorganisms through adsorption, while soils containing sand remove them through filtration at the soil–water interface. The movement of microorganisms through soil relates directly to the hydraulic infiltration rate, and inversely to the size of soil particles and the concentration of cations in the solute. Microorganisms will travel quickly through fissured zones, such as limestone, to the groundwater.

The potential for pathogen transmission exists and can cause a public health problem if land application is done improperly. Transmission can occur through groundwater, surface runoff, aerosols formed during application, and direct contact with the sludge or raw crops from the application site.

Because bacteria, viruses, and parasites do not enter plant tissue, transmission of pathogens via crops grown on the land application site results from contamination of the plant surface. If contaminated crops are consumed raw, disease transmission is possible; however, disease transmission due to application of sludge on to farmland is rare. Reported outbreaks of disease have generally been the result of the application of inadequately treated sludges to gardens or other crops which were eaten raw.

Pathogens in land-applied sludge will usually die rapidly, depending on temperature, moisture, and exposure to sunlight. In general, pathogen survival is shorter on plant surfaces than in the soil. To prevent disease transmission, sludge should not be applied to land during a year when edible crops are to be grown. Where humans have little physical contact with the sludge and the crops, the presence of pathogens may be of less concern. The soil can filter and inactivate bacteria and viruses. Sludge application methods and rates should be designed based on the soil characteristics to reduce public health concerns.

The microorganisms present in human and animal wastes and in sludge, as described in Chapter 2, can pose potential health risks when these wastes are applied to land and agricultural crops. From a review of epidemiological effects of the use of sludge on land, the transmissions of hookworm, *Ascaris*, *Trichuris*, and *Schistosoma japonicum* infections should be of major concern (Blum and Feachem, 1985). The above helminths are present in ova form in sludge, hence their long survival time in soil or on crops. They also become high risks to human health (see Table 2.26). In general, the microbiological guidelines as given in Table 2.27 and in section 9.3 for the class B sludge should be followed to minimize this public health impact.

REFERENCES

Battelle Memorial Institute (1987). *Manual for Sewage Application to Crop Lands and Orchards*. Report submitted to U.S. Environmental Protection Agency. Contract No. 68-01-6986. Battelle's Environmental Program Office, Washington, D.C.
Bitton, G., Damron, B. L., Edds, G. T., and Davidson, J. M. (eds.) (1980). *Sludge—Health Risks of Land Application*. Ann Arbor Science, Michigan.
Blum, D. and Feachem, R. G. (1985). *Health Aspects of Nightsoil and Sludge Use in Agriculture and Aquaculture: Part III. An Epidemiological Perspective*. International Reference Centre for Waste Disposal (IRCWD), Duebendorf, Switzerland.
Crites, R. W. and Richard, D. (1987). Land application of sludge in San Diego. *J. Water Pollut. Control Fed.*, **59**, 774–80.
Knezek, B. D. and Miller, R. H. (1978). *Application of Sludges and Wastewater on Agricultural Land: A Planning and Education Guide*. MCD-35, U.S. Environmental Protection Agency, Washington, D.C.
Kofoed, A. D. (1984). Optimum use of sludge in agriculture. In *Utilization of Sewage Sludge on Land: Rates of Application and Long-Term Effects of Metals* (eds. S. Berglund, R. D. Davis, and P. L'Hermite), pp. 2–21. D. Reidel, Dordrecht.
Loehr, R. C., Jewell, W. J., Novak, J. D., Clarkson, W. W., and Friedman, G. S. (1979). *Land Application of Wastes*, Vol. 1. Van Nostrand-Reinhold, New York.
O'Conner, G. A., Kiehl, D., Eiceman, G. A., and Ryan, J. A. (1990). Plant uptake of sludge-borne PCBs. *J. Environ. Qual.*, **19**, 113–8.
Ontario Ministry of Agriculture and Food (1993). *Halton Sludge Utilization Program*. A brochure, Halton, Ontario.
Reed, S. C. and Crites, R. W. (1984). *Handbook of Land Treatment Systems for Industrial and Municipal Wastes*. Noyes Publications, New Jersey.
Rulkens, W. H., Voorneburg, F. V., and Joziasse, J. (1989). Removal of heavy metals from sewage sludges. In *Sewage Sludge Treatment and Use—New Developments, Technological Aspects and Environmental Effects* (eds. A. H. Dirkzwager and P. L'Hermite). Elsevier Applied Science, London.
U.S. EPA (1983). *Process Design Manual for Land Application of Municipal Sludge*. EPA-625/1-83-016, United States Environmental Protection Agency, Cincinnati, Ohio.
U.S. EPA (1984). *Use and Disposal of Municipal Wastewater Sludge*. EPA 625/10-84-003, United States Environmental Protection Agency, Washington, D.C.
Walsh, M. J. (1995). Sludge handling and disposal—An American perspective. *Water Qual. Int.*, **2**, 20–3.

EXERCISES

9.1 Discuss the advantages and limitations of agricultural utilization of sludge, and compare them with aquacultural utilization (or fish production) of sludge.

9.2 Why is it important to maintain the soil pH at 6.5 or greater in sludge application on land? What should be done if the land for sludge application has a pH lower than 6.5?

9.3 Objectionable odors of sludge could result in unfavorable public relations and reduced acceptance of land application options. Therefore, all sludge management systems must consider objectionable odor as a potential problem. What measures should be taken to reduce or eliminate the odor problems during sludge application on land?

9.4 A piece of land received anaerobically digested sludge containing 3 percent organic N at a rate of 15 dry tons/ha in 1995. In 1996, a composted sludge containing 2 percent organic N was applied to the same land at a rate of 15 dry tons/ha. Determine the available N in the year 1997 from the sludge applications in the year 1995 and 1996.

9.5 An analysis of a composted sludge reveals the following information:

production	100 tons/day
dry solids	15%
total N	3.5%
NO_3–N	0.1%
NH_4–N	1.6%
P	1.2%
K	0.3%
Zn	2,000 ppm
Cu	1,000 ppm
Ni	100 ppm
Cd	20 ppm
Pb	500 ppm

Determine the land area needed for application of this sludge for a period of 10 years and the amount of additional chemical fertilizers required. The sludge is to be incorporated into cotton fields with a CEC of 8 meq/100 g soil. Nutrient uptake rates of cotton are given in Table 8.7.

10

Planning, institutional development and regulatory aspects

10.1 PLANNING FOR WASTE RECYCLING PROGRAMS

Waste recycling programs have been practiced in many countries for centuries, but proper planning procedures and an institutional set-up are yet to be developed. The overall success of a waste recycling program depends greatly on its planning and implementation.

Planning for a waste recycling program requires the following aspects for consideration.

1. *Raw materials*: wastes from households, industries, and public institutions are the ingredients for a waste recycling program, and the outcome is the useful products as well as other benefits, such as public health improvement and pollution abatement. The program has to envisage the limitation of the waste quality and quantity over the planning period.
2. *Manpower*: various levels of skill are needed for the execution of the program. Present and future manpower requirements need to be estimated, and future development of manpower planned.
3. *Capital*: a waste recycling program or project has to be financially viable. It may be necessary initially to fund the programs by local authorities or by international agencies. Later on, collection of fees for waste collection and treatment, and sale of the recycled products, have to be implemented. This should be accepted by the public.
4. *Technology*: collection, handling, processing, and final disposal or reuse of the waste and by-products need efficient methods so as to maximize the benefits and minimize the environmental pollution.
5. *Market study*: all the resources and effort devoted to recover and recycle the wastes will not be useful if there is no market for the recycled products. Waste recycling technology and facilities must be able to consistently meet

the user's quality, quantity, and reliable delivery period, when required.

6. *Political will*: this is an important aspect of a waste recycling program. Education of the people at all levels to promote public awareness, and to realize the importance of waste recycling, has to be done continuously. Uncertainties and myths have to be clarified, such as the aspects of public health and environmental impacts due to waste recycling programs.

The above-mentioned aspects interrelate and influence each other with respect to a decision to proceed with a waste recycling program. In general, steps that should be considered in the planning of a waste recycling program (Figure 10.1) are:

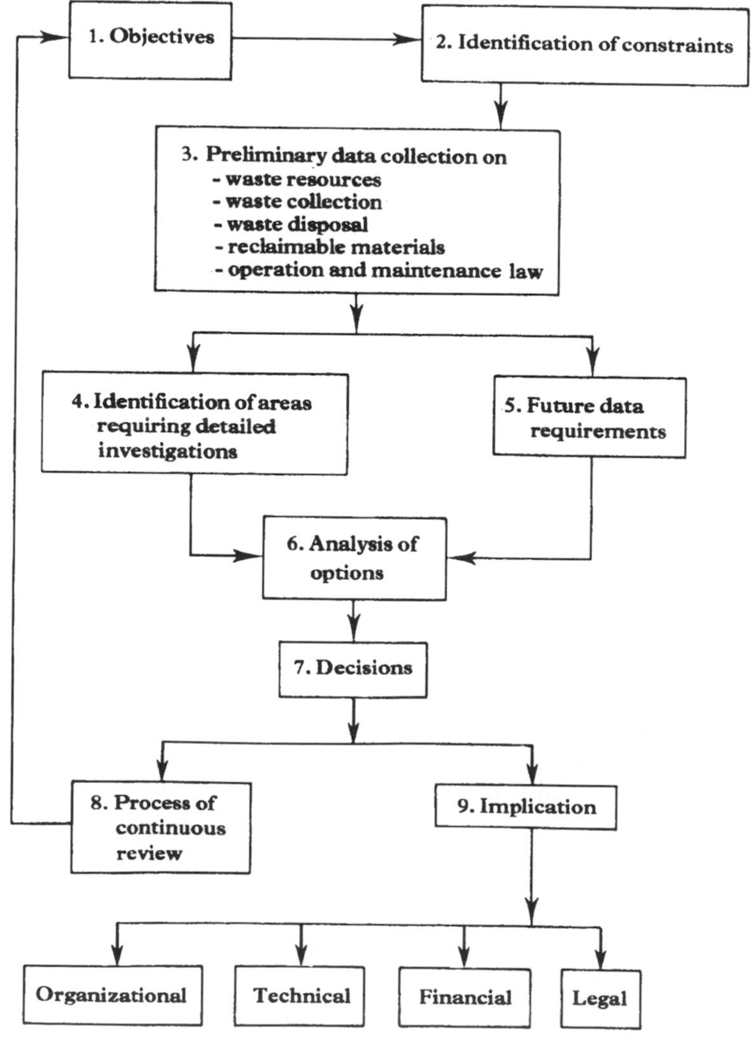

Figure 10.1 Principal steps in the development of a waste recycling program (Thanh *et al.*, 1979)

1. identification of objectives
2. constraints
3. data collection
4. analysis of principal options
5. the decisions and process of review

10.1.1 Identification of objectives

The objectives should identify the purpose of the waste recycling program, that is to be able to reclaim, reuse, and at the same time to treat the waste to a degree so that it does not cause health hazards or environmental pollution. In considering the cost-effectiveness of a waste recycling scheme the benefits to be gained from pollution control and public health improvement should also be taken into account.

The objective should, in particular, include what the plan is to encompass in terms of:

1. geographical area, where waste collection and treatment/recycling can be planned on a regional basis to minimize expenditures and to yield effective management of the system;
2. types of waste, so that an appropriate waste recycling program can be selected;
3. time period of the plan, which depends on local circumstances and needs such as the waste quantity and characteristics and long-term needs of the recycled products.

10.1.2 Constraints

There are a number of constraints that will tend to limit the available waste recycling alternatives. By recognizing these at the outset a great deal of unnecessary work can be avoided; their recognition will also be of assistance in the eventual selection of an appropriate recycling scheme. These constraints are:

1. *Financial constraints*, which may limit the use of capital-intensive high technology for recycling the waste, unless the advantage, or market, for the recycled products is promising. For example, it may not be appropriate to employ a DANO composting unit (Figure 3.18) to produce compost fertilizer where farmers still prefer to use chemical fertilizers available cheaply. In addition, the production of virgin materials often enjoy a wide range of subsidies and incentives which make them cheaper than the recycled products.
2. *Manpower constraints:* the manpower constraints in the planning of a waste recycling program should be taken into account, such as the need for skilled technicians to operate some biogas digesters and high-rate algal ponds.
3. *Land use constraints:* some waste recycling systems are not suitable for

urban and suburban areas due to the shortage of land. Existing land uses will provide the best guidelines that are likely to prove more or less acceptable.
4. *Environmental constraints:* hydrogeological characteristics of the waste recycling area must be considered, so that the environment of the site and nearby is not affected by this practice.
5. *Public acceptance constraints:* the recycled products from the waste recycling processes should be acceptable to the public, otherwise the recycling process is of no use. In general, people do not want to have waste recycling sites located close to residential areas.

10.1.3 Data collection

PRELIMINARY INFORMATION REQUIREMENT

The following information will be required for the planning of a waste recycling program:

1. population, housing, industries, or agricultural activities;
2. waste quantity, composition, and source;
3. waste storage, collection, and transportation;
4. treatment and disposal/reuse methods, existing methods, future capacities, plant's life, local hydrogeology, land availability;
5. existing waste recycling system and benefits achieved, if any;
6. public acceptance of reclaimed products;
7. management and law, organization, regulations, and their enforcements.

DETAILED INVESTIGATION

In particular there is often a shortage of information on the quantity and composition of waste, and the amount of reclaimed products. Hence, detailed surveying, and estimation of the quantity and composition of waste, should be carried out before planning a waste recycling program.

Both waste quantity and composition are subject to daily and seasonal fluctuation. Waste recycling facilities must be provided to cope with peak waste load rather than the average, and seasonal variations in the quantities of available materials may have impact upon the extent of resource recovery. In addition, data on national average characteristics of the wastes can be significantly different from local characteristics of the wastes of concern.

For the treatment and recovery of products (e.g. composting—material recovery; anaerobic digestion—energy and materials; pond system—food production), the waste composition is an important factor.

Where specific problems have been identified a more detailed investigation on waste composition and quantity will be required to provide sufficient data for the options or decisions to be made. The detailed investigations should also take into account the amount and value of reclaimed products, as well as their public acceptance. Long-term outlets for these reclaimed products should also be considered.

10.1.4 Analysis and decisions

Where a number of options for a waste recycling scheme are available it is necessary to evaluate each option for capital and operating costs, the benefits, and associated health and environmental impacts.

Alternative technological options and waste recycling plans are assessed against a set of criteria which will normally encompass economic, technical, environmental, and political objectives. Possible subdivisions of each category are suggested in Table 10.1. The selection and definition of criteria are dynamic, with feedback occurring from later stages of the planning process. For example, the criteria may be revised or the emphasis given to a particular criterion may be altered when practical implications rather than initial abstractions are considered.

The recycling program should not be selected only on the benefits and impacts in monetary terms, but more so on the benefits gained from pollution control and public health improvement.

With regard to cost and benefits, all elements in the systems should be costed for comparisons on the options and for evaluation of improvements in performance. Costs and benefits should be expressed in terms of cost and benefit per ton of waste recycled in the system. The costs include capital and operating costs in terms of its present value, whereas the benefits should be expressed not only in terms of financial benefits but also considering pollution control and public health impacts.

10.2 GUIDELINES FOR TECHNOLOGY SELECTION

10.2.1 Definition of technology and its choice

Technology is often identified as knowledge of machines and processes; extending to skill, knowledge, and procedures for making, using, and doing useful things. It includes the nature and specification of what is produced, as well as how it is produced. A complete description of technology used in a country must include the organization of productive units in terms of scale and ownership.

10.2.2 Limiting factors in the choice of technology

In general, people would normally select a technology that is suitable to the environment, and economical to operate. However, technology selection is not an easy job. There are some factors that limit the choice of technology. The first is incomplete knowledge of techniques and methods. Technology is generally developed in industrialized countries. Although in some instances a certain

Table 10.1 Criteria for the assessment of waste recycling plans (Wilson, 1981). Reproduced by permission of Oxford University Press

Economic
Capital costs
Land costs
Operating costs
Revenues:
  Extent of market commitment
  Stability of markets
Net recycling cost per ton of wastes
Net present cost of the recycled products
Sensitivity of costs to market or other fluctuations
Uncertainty in cost estimates, i.e. financial risk
Financing arrangements

Technical
Adequacy of the technology:
  Feasibility
  Operating experience
  Adaptability to local conditions
  Reliability
  Interdependence of components (can the system operate if one component fails?)
  Safety
Potential for future development
Flexibility to cope with changes in:
  Waste quantities
  Waste composition
  Source separation of materials
Dependence on outside systems:
  e.g. vulnerability of strikes

Environmental
Public health
Water pollution
Air pollution:
  Dust
  Noxious gases
  Odors
Quality and quantity of residual wastes
  Noise
  Traffic
  Aesthetics

Political
Equity between communities or interest groups
Flexibility in location of facilities
Public acceptance
Number of jobs created
Employee acceptance

Use and conservation of resources
Products covered:
  Market potential
Net effect on primary energy supply:
  Energy requirements
Net effect on supply of materials:
  Raw materials usage
Land usage:
  Volume reduction
  Land reclamation
Water requirements

technology can be adapted to developing country environment, no country is willing to try it because of the lack of knowledge of that technology. The absence of individuals or groups involved in using the technology in the community also hinders technology selection. This is true when the individual or the community lacks the entrepreneurial spirit to innovate or try something new. Unless technologies from other countries are tried and tested, the choice of technology will be very limited.

Guidelines for technology selection are proposed as a basis for selecting and/or adopting technology from other countries. Although some techniques, such as the use of excreta for composting, are acceptable in China or Vietnam, they may not be socially acceptable in other Asian countries. The following guidelines are recommended for technology selection in waste recycling and recovery, but should also be applicable to other categories of technology:

1. The technology should be suited to the local environment economically and culturally.
2. The operation and maintenance should be easily undertaken using local manpower.
3. The technology should use, whenever possible, local materials and energy sources. Where imported materials are recommended, they must be affordable and easily obtainable.
4. The technology should be simple and easily understood by the local people, and have a certain flexibility for possible changes.
5. The technology should be innovative in order to improve the human and material conditions of the local people through the use of new organizational types and new technological devices.
6. The technology, whenever possible, could be sited in existing high-density areas.
7. The technology should not directly or indirectly contribute to the pollution and destruction of the existing ecology.
8. The technology should enhance health and sanitation aspects, and upgrade the economic well-being of the community.

10.3 MONITORING OF FACILITY PERFORMANCE

One of the essential aspects of a waste recycling program is monitoring the facility performance. This is required for analysis and evaluation of the chosen technology in achieving its objective. One of the many reasons why some of the objectives of a planned program are not attained is a failure of the concerned persons or agencies to monitor the project implementation and evaluate operational problems which are generally dealt with only at the critical state. To avoid such failures and to prevent the problem from becoming a crisis, it is obligatory for all projects to formulate a monitoring program. A monitoring set-up for facility performance depends on:

1. objectives defined for monitoring;
2. data evaluation, analysis, and documentation;
3. equipment monitoring efficiency; and
4. organizational infrastructure.

10.3.1 Objectives defined for monitoring

In view of the present socio-economic situations of most societies presently involved with finding economical ways of waste recovery and recycling, the pragmatic objectives of such a monitoring system can be stated as:

1. efficient control and regulation of the present system;
2. emission analysis and evaluation;
3. to cater for future needs; and
4. to analyze the public response to system interaction.

EFFICIENT CONTROL AND REGULATION OF THE SYSTEM

Monitoring is a present-day scientific tool for efficient management of a system. As in most systems, efficiency is evaluated on an economical scale, and is continuously monitored based on cost–benefit analysis to make sure that the system is working on the positive side. Based on the technology selected, the monitoring process is regulated according to the inflow of suitable wastes and demand for the recycled product. Examples of this are the slow-rate (irrigation) system where certain crops are grown from wastewater application (see Chapter 8) and fish production in waste-fed ponds (see Chapter 6).

EMISSION ANALYSIS AND EVALUATION

The purpose of monitoring in this context is to check the addition of pollutants to the environment from waste recycling programs. In practice it can be better explained as the analysis and evaluation of recycled products such as pathogen contents in the wastewater-irrigated crops or fish raised in waste-fed ponds and nutrient balances in the composted products or digester slurry.

TO CATER FOR FUTURE NEEDS

Modern technologies change rapidly with time. Any system has to be flexible to cope with future changes, and this is so in the case of waste recycling and recovery. One of the solutions, therefore, is to foresee the changing trends, and to restructure the waste recycling system to cope with future demands. The increasing number of methods of reusing and recycling organic wastes may be cited as examples of this. Organic wastes have been used in the past as manure-cum-land conditioner and are now used in biogas generation and in aquaculture. The present trend can be foreseen; the emphasis lies more on refined methods of nutrient recovery in the form of protein.

TO ANALYZE THE PUBLIC RESPONSE TO SYSTEM INTERACTION

A system can exist and grow only when it is patronized by the people for whom it has been established. In the context of the poor developing countries, the waste recycling system adopted should conform with social attitudes, and must be acceptable to the people. Any reluctance can seriously affect the system. As regards development, a change has to come in the traditional outlook, and the best that can be done is to mold the public temperament gradually. A typical example of this is the gradually diminishing public reluctance in Asian countries to consume fish grown on human or animal wastes.

An efficient monitoring system, then, must have its feedback from end-users, and should consider this with due concern.

10.3.2 Data evaluation, analysis, and documentation

Monitoring relies heavily on information based on data and statistics. Statistical information is generated from the continuous study of the various parameters of the waste recycling system and their thorough analysis. Some common parameters used in a waste recycling system are BOD_5, nutrients, and pathogen contents. These parameters give information on the quality and characteristics of the incoming waste in the process and the output of recycled products. These data help to identify trends in performance and phenomena of the process and, in many cases, to establish empirical formulae.

Documentation of data is equally important to keep relevant and reliable data for future reference and decision-making. In all cases easy-to-handle and systematic storage is essential, as the long-term evaluation of the process is mostly based on these data.

10.3.3 Equipment monitoring

A major part of waste recycling technology is the various items of equipment used in the process. These can vary widely in nature according to use. Monitoring of these assets is equally important as they can influence the system's efficiency to a great extent. A simple example can be given: a leak in the gas-holder tank of a digester can significantly reduce its efficiency.

10.3.4 Organizational infrastructure

The organization for monitoring in the case of waste recycling and recovery depends largely on the extent and intensity of the program. In most cases of waste recovery and recycling the program is not for a centrally based system but for large-scale participation by the people. In the case of a centrally based recovery and recycling unit, reasonably well-qualified technicians, but less in number, are required. Where the program is more public-intensive, and emphasis is placed on making it a social habit, a large number of people who may not necessarily have higher qualifications but having more dedication and commitment, are required. In such cases large-scale training and apprenticeships are more beneficial.

The financial aspects of these establishments depend on their size and manpower involvement. But in all cases the central organization should be adequately furnished with the basic requirements such as laboratory and training units so that the feedback from the field, and the present knowledge, can be better blended to raise the living standard of the poor masses.

10.4 INSTITUTIONAL ARRANGEMENTS

A critical dimension in the success of waste recovery and recycling programs concerns an institutional arrangement between the government agencies, municipalities, private corporations, and non-government organizations. All these institutions must join together and coordinate all activities in order to avoid conflicts and duplication of work and duties. The institutional arrangements may be considered for financing, administering, and operating recovery and recycling projects. In developing countries, without government guidance, it seems unlikely that other institutions can be expected to handle waste materials efficiently and optimally. Some government control, guidance, and incentive are necessary to secure success in pursuing the program. A strategy or approach has to be developed, either regional or grassroots, to lay the foundations of work. The private corporations and non-government organizations, including international agencies and foundations, should be approached for financial or technical support. The municipality must cooperate with the government for the promotion and implementation of planning programs. Cooperatives or districts should be established to be directly involved in construction, operation, and maintenance of recycling facilities, if there are any.

The development and application of science and technology should be conducive to waste recycling. Furthermore, a policy that takes into account the natural, economic, and social characteristics of each region or locality, if taken widely, would secure more cooperation in creating recycling policy.

To make steady progress in waste recycling policy and to produce a more satisfactory result in this direction, all parties interested in recycling issues (such as government agencies, municipalities, non-profit corporations, entrepreneurs, and inhabitants) should be aware of their respective roles and make continuous efforts befitting their positions. Government agencies, among others, have a leading role in making the program successful.

Close cooperation among interested parties (including government agencies, municipalities, non-profit agencies, entrepreneurs, and inhabitants) is indispensable to enforce a recycling policy that takes into account either local or regional characteristics. Each of the institutions has to have a role. The following institutional arrangements are suggested for the success of a recycling and reuse program.

10.4.1 Government agency

Its role is to:

1. formulate policy, prepare guidelines, and undertake planning programs on waste recovery and recycling;
2. extend technical, financial, and management support to all participants of the program;
3. promote and coordinate all activities related to waste recovery and recycling to assure public acceptance of the program and to avoid duplication of work;
4. monitor success and failures of the program, formulate alternate programs to reduce losses (in terms of investment);
5. undertake manpower development and training programs for people involved in the program.

10.4.2 Municipality

Its role is to:

1. assist the government agencies in the selection of the most appropriate recovery and recycling programs to establish in the community;
2. act as a link between government agencies and the community for the continuous flow of information necessary to ensure the success of a recovery and recycling program;
3. initiate acceptance and participation of the people on the government program through dissemination of information;
4. assume a leading role in the formation of cooperatives or a local district which would have a direct responsibility for the implementation of a recovery and recycling program in the community;
5. assist the cooperative and district or individual in identifying problem areas—recommend remedial solutions to prevent failure of the program.

10.4.3 Private corporation

Its role is to:

1. assist the government agency and municipality in the waste recovery and recycling program by extending financial, and possibly technical, support;
2. embark on projects that would transform their wastes into useful products (this is possible for private corporations dealing with agro-based industry where large quantities of agro-industrial wastes are produced, see Chapter 2);
3. undertake some research for the development of sound technology in a recovery and recycling program;
4. participate in government programs by providing loans at low interest rates to a community or individual who needs capital for construction of equipment, and facilities needed in recovery and recycling;
5. assist in the promotion of the programs.

10.4.4 Non-government organization

Its role is to:

1. assist government agencies and municipalities in waste recycling and recovery programs by extending financial and technical assistance;
2. finance projects that may have low profitability but would ensure the abatement of pollution and degradation of the environment;
3. assist in the promotion of pioneer or innovative technology in waste recovery and recycling;
4. assist the government in manpower development and training, as well as in technology transfer;
5. undertake research for the improvement of recycling technology.

10.4.5 Biogas promotion in Nepal

The following is an example of how a biogas program was promoted in Nepal over a period of 4 years in the 1970s (Figure 10.2). It may seem to be a slow and demanding program, but has proven to be successful. Not only are new biogas digesters in demand, but most existing biogas digesters have been in operation satisfactorily (U.N., 1980).

YEAR ONE

1. A study was made of existing biogas digester designs. The best one was selected to suit local needs and was modified as necessary.

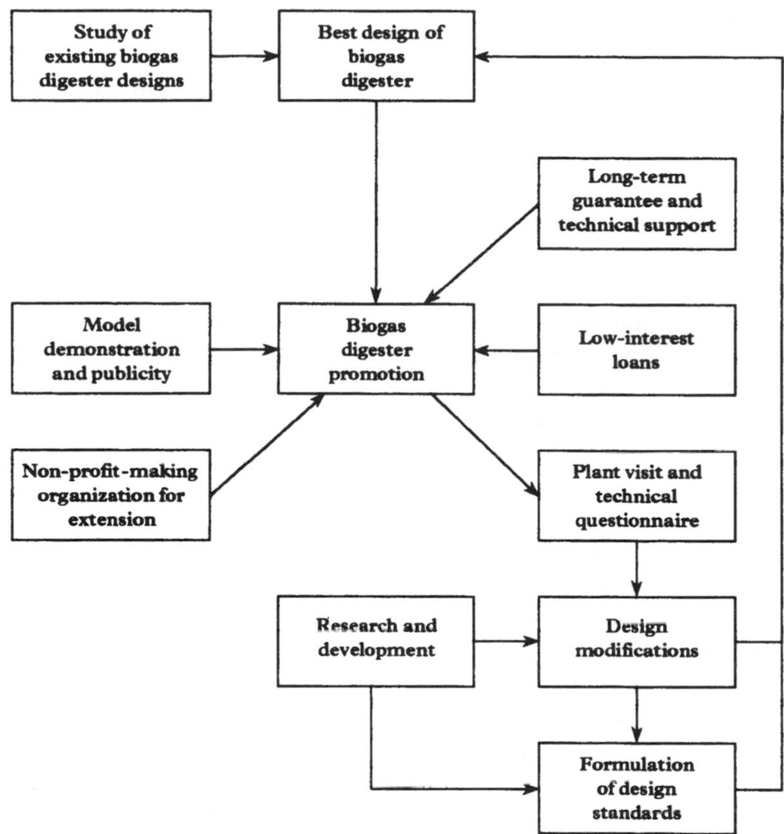

Figure 10.2 Biogas promotion in Nepal

2. A small portable demonstration biogas digester was shown at fairs, etc.
3. A few biogas digesters were built at a fixed price on the basis that if it did not work there would be no charge. A 1-year guarantee was given together with regular, good technical support.

YEAR TWO

4. Low-interest loans were made available from the national agricultural development bank. Publicity and information were given over the radio.
5. Twenty to 50 biogas digesters were built in a restricted area (for ease of control). A guarantee and technical support were also given.
6. The biogas digesters were revisited at intervals to ensure that they were operated properly and that there were no serious technical problems. The guarantee was honored. A technical questionnaire was filled in by the biogas digester operators every 6 months.

YEAR THREE

7. In the light of information gained in (6) above, designs and specifications for materials were modified as necessary and the building and servicing of the biogas digesters continued.

YEAR FOUR

8. With the confidence gained from the work already done, a private non-profit-making company was formed. The shareholders were the national agricultural development bank, an appropriate government agency and a building organization. This company was to work in the whole country (or in one state or province, as appropriate).
9. Good quality biogas digesters were built. A 1-year guarantee was given on the biogas digesters. The digester and gas holder were guaranteed for a period not less than the payback period for the loans (probably 5–10 years). These two items accounted for about 70 percent of the total cost of the biogas digester. To do this the development organization had to collect sufficient money to pay for its own staff to visit the biogas digester and paint the gas holder at regular intervals throughout the guarantee period. All biogas digesters were visited three times in the first year and a technical report was made. After the first year, an annual visit and report on every biogas digester were made for a period of 20 years. At the time of visit, free technical advice was given and an on-the-spot estimate for any repairs or modifications needed.
10. Financial help was given in the form of reduced interest or interest-free loans. No direct subsidies were given at all.
11. Research and development is continuing. Any new ideas are thoroughly tested before general use. Standards are established for biogas digesters, gas holder designs, protective coatings, fittings, and appliances.

10.5 REGULATORY ASPECTS

Appropriate legislation establishes binding policies and standards, provides the base for substantive and procedural regulations, and creates institutions to implement policies and enforce the rules. Legislation is a measure of the public acceptance of the need for waste recycling and enhancement through institutions; it is also a measure of how adequately waste recycling institutions are entrusted with political clout and legal justifiability such that environmental regulations are better enforced. Waste recycling legislation and waste recycling administration reinforce each other, and in the process become practically more effective.

10.5.1 Legislation in developed countries

It is well known that developed countries, such as the U.S.A., Japan, and Germany, have paid much attention to environmental quality control for decades. Environmental legislations in these countries are very stringent and have been progressively developed for several years. Application of waste recycling technology in pollution control has been widely practiced in these countries. For example, Japan has achieved a waste recycling rate of 55.4 percent (Environment Agency, 1993) and the U.S.A. is expected to achieve a waste recycling rate of 50 percent by the year 2000 (Grogan, 1993). Table 10.2 lists the water quality standards set by the U.S. EPA, WHO, and Japan, which, together with Table 2.4, can be used as guidelines for setting waste recycling objectives. Table 10.3 gives some examples of measures that could be used to encourage reclamation. Table 10.4 shows some measures in European Community countries for promoting reclamation. Solid waste legislations including waste recycling goals in some states of the U.S.A. are presented in Table 10.5.

10.5.2 Legislation in developing countries

One of the most difficult problems facing governments of developing countries is how to set up appropriate legislation on environmental quality control. Examples of such legislation include water quality standards, treated effluent standards for industrial and domestic wastes, ambient air quality standards, waste reuse and recycling regulation. Because the setting of this legislation influence the cost of development projects (both capital and operation and maintenance costs), the development of appropriate legislation requires knowledge of the receiving environment, available technology, and affordability.

UNEP (1991) reported that deaths due to infectious and parasitic diseases are six times more frequent in developing countries than in developed countries. One possible cause of the diseases is the severe degradation of water quality, which is due to less stringent water quality standards and ineffective implementation. Experiences in implementing the water quality standards in Japan reveal that actual effectiveness of a water pollution control system depends not on values of the regulated parameters, but on the supporting measures to implement the system itself. Table 10.6 lists water quality standards being in use in three Asian developing countries, namely, Thailand, China, and India.

In general, developed countries have rich experiences in waste recycling legislation, the most cited example being the Resource Conservation and Recovery Act (RCRA) of the U.S.A. On the contrary, most of the developing countries do not have any laws concerning waste recycling. Recently, however, waste recycling has been included in the national policies of some developing countries, either in the promotion stage and or in the drafting of the legislation. Therefore, investigation and evaluation of legislative instruments used in the developed countries should be made prior to formulating the legislation.

Table 10.2 Water quality standards of U.S. EPA, WHO, and Japan (Lohani and Evans, 1993)

Parameter (Average maxima, except [ ] = minima) (ND = non-detectable)	U.S. EPA					WHO	Japan				
	CWS	RWS[a]	FS	RC	AG	RWS	CWS	RWS	FS	RC	EF
1. Physical parameters and solids											
1.1 Temperature (°C)											
1.2 Color (units)						300					
1.3 Turbidity (units)	1	75									
1.4 Suspended solids (mg/L)								25	25–50		200
1.5 Total dissolved solids (mg/L)	500					1,500					
1.6 Dissolved oxygen (mg/L)			5.5					7.5	5–7.5		
1.7 Transparency (Secchi depth, m)											
1.8 Settleable solids (mg/L)											
2. Mineral water quality parameters											
2.1 Buffer capacity/pH/hardness											
2.1.1 Alkalinity ($CaCO_3$) (mg/L)			[20]								
2.1.2 pH	5–9	6.5–9.0						6.5–8.5	6.5–8.5	6.5–8.5	5.8–8.6
2.1.3 Hardness ($CaCO_3$) (mg/L)	150[b]										
2.2 Major cations (mg/L)											
2.2.1 Calcium (Ca)											
2.2.2 Magnesium (Mg)											
2.2.3 Sodium (Na)	250										
2.2.4 Potassium (K)											
2.3 Minor anions (mg/L)											
2.3.1 Bicarbonate (HCO_3)											
2.3.2 Sulfate (SO_4)	250										
2.3.3 Chloride (Cl)	250										
2.3.4 Nitrate (NO_3)	45		225			45					
2.4 Iron and manganese (mg/L)											
2.4.1 Iron (Fe, soluble)	0.3		1			50					
2.4.2 Manganese (Mn, soluble)	0.05					5					

3. *Nutrient/eutrophication parameters*										
3.1 Nitrogen (mg/L, as N)										
3.1.1 Nitrate (NO_3)			80							
3.1.2 Nitrite (NO_2)			5							
3.1.3 Total nitrogen										
3.1.4 Organic N										
3.1.5 Ammonia (NH_3)										
3.2 Phosphate (mg/L, as P)										
3.2.1 Ortho-phosphate										
3.2.2 Total phosphate				0.025–0.05[c]						
3.3 Chlorophyll *a*										
4. *Agricultural parameters*										
4.1 Boron (mg/L)					0.75					
4.2 Sodium adsorption ratio (SAR)					4–8					
4.3 Total dissolved solids (mg/L)					500					
5. *Aquatic ecology*										
5.1 Dissolved oxygen (mg/L)										
5.2 Transparency (Secchi depth, m)										
6. *Recreation/swimming*										
6.1 Coliforms, total (no./100 mL)							50	1,000–5,000	300,000	
6.2 *E. coli* (no./100 mL)										
7. Pollution parameters										
7.1 Coliforms, total (no./100 mL)	1			1,000		5,000				
7.2 Coliforms, *E. coli* (no./100 mL)				126						
7.3 BOD (mg/L)						6	1	2.5	2	12
7.4 COD (mg/L)						10				
7.5 Oil/grease (mg/L)		ND		ND		1				
7.6 Floatables (mg/L)		ND		ND		0				
7.7 MBAS	0.5					0.5				

(*continued overleaf*)

Table 10.2 Water quality standards of U.S. EPA, WHO, and Japan (Lohani and Evans, 1993) (*continued*)

Parameter (Average maxima, except [ ] = minima) (ND = non-detectable)	U.S. EPA					WHO	Japan				
	CWS	RWS[a]	FS	RC	AG	RWS	CWS	RWS	FS	RC	EF
8. *Heavy metals*											
8.1 Barium (mg/L)	1										
8.2 Cadmium (mg/L)	10	10	0.7–2[d]			10	10				100
8.3 Chromium (VI) (mg/L)	0.05		0.11[d]				0.05				0.5
8.4 Chromium (total) (mg/L)	0.1	0.17				0.05					2
8.5 Copper (mg/L)	1	1	0.006–0.02[d]			1.5					3
8.6 Lead (μg/L)	50	50	1.8[d]			50	100				100
8.7 Mercury (μg/L)	0.14		1.012[d]				0.5				0.5
8.8 Nickel (mg/L)	0.013	0.13	0.06–0.16[d]								
8.9 Selenium (mg/L)	0.01		0.035			0.01					
8.10 Silver (μg/L)	50		1–1.3								
8.11 Zinc (mg/L)	50		0.2–0.6[d]								5
9. *Other toxics*											
9.1 Ammonia (NH_3) (mg/L)											
9.2 Arsenic (mg/L)	0.05		0.19[d]			0.05	0.05				0.5
9.3 Chlorine, free (mg/L)	0.1					0.05					
9.4 Cyanide (mg/L)	0.2	0.2	0.005[d]			0.2	ND				1
9.5 Fluoride (mg/L)	1.1–2.4					1.5					15
9.6 Phenol (mg/L)	0.3		2.6			0.002					5
9.7 Sulfide (H_2S) (mg/L)											

[a] Raw water supply suitable for treatment by conventional rapid sand filtration to produce finished CWS
[b] Moderate hardness
[c] 0.025 for reservoirs, and 0.05 for streams
[d] 4-day average, once every 3 years
CWS = community water supply; RWS = raw water supply; RC = recreation (fresh/marine); AG = agriculture; FS = fishery; EF = effluents; MBAS = methylene-blue-active substances

Table 10.3 Examples of measures for encouraging reclamation (Environmental Resources Ltd, 1978)

Stage of material handling initially affected	Fiscal measures		Regulations	
	Taxes	Subsidies[a]	By legislation	By official procurement
Waste discharge	Tax on waste discharged in uncontrolled manner	Publicity campaign on the need for reclamation. Payments for waste received	Prohibition of uncontrolled discharge	
Initial recovery	Tax on waste not recovered	Finance for data bank for information on recovery methods. Financial assistance for research and development or pilot plants, new operating plants, etc. Payments per unit of waste handled for recovery in an acceptable manner. Finance for stockpiling of reclaimed material. Special assistance for cooperative projects	Requirement that waste be recovered and/or particular recovery methods be used. Requirement that authorities and other organizations cooperate in various ways. Requirement that a certain level of stocks of reclaimed material be held	
Further treatment	Tax on material not subjected to further treatment		Requirement that recovered materials be treated in certain ways. Requirement for a necessary degree of cooperation between waste handlers/reclaimers and material users. Requirement that a certain level of stocks of reclaimed material be held	
Production and sale	Production charges (tax) on items contributing to waste or difficult to recover. Deposit refundable on return for reclamation. Tax on virgin materials	Publicity campaign and data bank for information on nature of reclaimed materials and products made from them. Payments in relation to use of reclaimed material (e.g. direct grants, assistance with new plants, etc.). Finance for stockpiling of reclaimed material	Requirement that products meet certain standards of recoverability and/or incorporate a certain proportion of reclaimed material. Prohibition of certain products that create non-recoverable wastes. Requirement that a certain level of stocks of reclaimed material be held	Setting of standards of recoverability and/or reclaimed material content for products purchased

[a] Including reliefs from existing taxes

Table 10.4 General measures in selected European countries for promoting reclamation (Environmental Resources Ltd, 1978)

Type of measure	Country	Action
1. Procurement policies	France	Government intends to specify minimum proportions of secondary materials in products procured for public bodies
	Germany	Government intends to specify proportions of secondary materials in products procured for public bodies
	Netherlands	Government examining possibilities for specifying proportions of secondary materials in products procured for public bodies
2. Specifications to encourage demand for reclaimed materials	France	For categories of products defined by the Government, it is forbidden to discriminate against the presence of reclaimed materials in products, which satisfy regulations and standards. Law enables Government to fix minimum proportions of secondary materials contained in products
	Germany	Quality control and standardization of secondary materials under investigation
3. Financial assistance to reclamation projects	Belgium	Investment grants up to 60% of construction costs available to municipalities for waste treatment and recovery plants
	Denmark	Limited *ad hoc* research and development grants for private and public sector projects
	France	National Agency can make available research grants
	Germany	Grants available for innovative investments and research and development into reclamation. Grants available for plants to reclaim energy from waste
4. Other measures to assist with separation and use of reclaimed material	Belgium	New company planned to handle reclamation from refuse
	Germany	Provision under tax law for accelerated depreciation in investments of immediate and exclusive benefit to the environment. Government is considering regulations to make recovery from certain types of waste obligatory
	Netherlands	Under proposed law, Government may order the separate collection of domestic wastes
	Italy	Law of March 1941 requires communities with over 50,000 population to segregate refuse (not in use)
5. Public education programs	France	Public education program planned
	Germany	Public education program on waste and recovery planned

Table 10.5 Solid waste legislation by some states of the U.S.A. (Grogan, 1993)

Date passed	WR/R goal	Completion date	Product and disposal bans	Funding	Other unique characteristics
				California	
1989	25% 50%	1995 2000	None	Statewide disposal surcharge. Advance disposal free on tires	Tax credit on purchasing of recycling equipment. Newspaper publishers to use news with 25% recycled content by 1991 and 50% by 2000. Commission to study market opportunities. Plastics coding
				Florida	
1988	30%	1994	Product ban on detachable pull-rings, some nonbiodegradable plastic packaging. Disposal ban on tires, and lead-acid batteries	Advance disposal fees, sales tax, business registration. Oil overcharge monies. Solid waste management trust fund	Additional advance disposal fees if recycling targets not met. Plastics coding. State agencies to use compost and recovered construction materials
				Indiana	
1990	35% 50%	1996 2001	Disposal ban on designated recyclables will be implemented if deemed necessary	Statewide disposal surcharge	Counties required to form solid waste management districts. Low-interest loans for recycling organizations. Educational programs are being developed for students, consumers and business

(*continued overleaf*)

Table 10.5 Solid waste legislation by some states of the U.S.A. (Grogan, 1993) (*continued*)

Date passed	WR/R goal	Completion date	Product and disposal bans	Funding	Other unique characteristics
				Maine	
1989	25% 50%	1992 1994	Product ban on multilayer juice containers, plastic cans, 6-pack yokes	Disposal surcharge. Advance disposal fee on special wastes	30% tax credit on recycling equipment purchases. Creates Waste Management Agency. State agencies to purchase recycled paper. Commercial sector to recycle
				Washington	
1989	50%	1995	Disposal ban on lead-acid batteries	Advanced disposal fee on tires. Solid waste collection tax. County disposal surcharge	County solid waste management plans to include WR/R education programs and market strategies. Market development committee established
				Pennsylvania	
1988	25%	1997	None	Disposal surcharge	Municipal recycling programs required. Businesses, government offices, hospitals and schools required to source separate. Funding for market research and development. Low interest loans available to recycling companies. Development of recycling curriculum

WR/R = waste reuse and recycling

Table 10.6 Water quality standards of Thailand, China, India (Lohani and Evans, 1993)

Parameter (Average maxima, except [ ] = minima) (ND = non-detectable)	Thailand					P. R. China			India				
	CWS	RWS[a]	FS	AG	EF	CWS	RWS	FS	CWS	RWS	FS	AG	EF
1. Physical parameters and solids													
1.1 Temperature (°C)													
1.2 Color (units)	20					15	10–15		10	25			
1.3 Turbidity (units)	5					5			10				
1.4 Suspended solids (mg/L)								10					100
1.5 Total dissolved solids (mg/L)	500				5% increase								
1.6 Dissolved oxygen (mg/L)		[4–6]	[6]	[4]			4.0–6.0	[5]	[6.0]	[4.0]	[4.0]		
1.7 Transparency (Secchi depth, m)													
1.8 Settleable solids (mg/L)					1								
2. Mineral water quality parameters													
2.1 Buffer capacity/pH/hardness													
2.1.1 Alkalinity ($CaCO_3$) (mg/L)													
2.1.2 pH	6.5–8.5	5–9	5–9	5–9		6.5–8.5	6.5–8.5	6.5–8.3					5.5–9
2.1.3 Hardness ($CaCO_3$) (mg/L)	100					250							
2.2 Major cations (mg/L)													
2.2.1 Calcium (Ca)													
2.2.2 Magnesium (Mg)													
2.2.3 Sodium (Na)													
2.2.4 Potassium (K)													
2.3 Minor anions (mg/L)													
2.3.1 Bicarbonate (HCO_3)													
2.3.2 Sulfate (SO_4)													
2.3.3 Chloride (Cl)	250											250	
2.3.4 Nitrate (NO_3)	18	22.5	22.5	22.5									
2.4 Iron and manganese (mg/L)													
2.4.1 Iron (Fe, soluble)	0.5				15	0.3							
2.4.2 Manganese (Mn, soluble)	0.05	1	1	1	1	0.1							

(continued overleaf)

Table 10.6 Water quality standards of Thailand, China, India (Lohani and Evans, 1993) (*continued*)

Parameter (Average maxima, except [ ] = minima) (ND = non-detectable)	Thailand					P. R. China			India				
	CWS	RWS[a]	FS	AG	EF	CWS	RWS	FS	CWS	RWS	FS	AG	EF
3. *Nutrient/eutrophication parameters*													
3.1 Nitrogen (mg/L, as N)													
3.1.1 Nitrate (NO_3)													
3.1.2 Nitrite (NO_2)		0.5	0.5	0.5									
3.1.3 Total nitrogen							1						
3.1.4 Organic N													
3.1.5 Ammonia (NH_3)													
3.2 Phosphate (mg/L, as P)													
3.2.1 Orthophosphate							0.1						
3.2.2 Total phosphate													
3.3 Chlorophyll *a*													
4. *Agricultural parameters*													
4.1 Boron (mg/L)													
4.2 Sodium adsorption ratio (SAR)					5							2	
4.3 Total dissolved solids (mg/L)												1,000	
5. *Aquatic ecology*													
5.1 Dissolved oxygen (mg/L)													
5.2 Transparency (Secchi depth, m)													
6. *Recreation/swimming*													
6.1 Coliforms, total (no./100 mL) (80%)													
6.2 *E. coli* (no./100 mL) (80%)													
7. *Pollution parameters*													
7.1 Coliforms, total (no./100 mL) (8)	2.2	5,000	5,000	20,000		300	500–50,000		50	5,000	5,000		
7.2 Coliforms, *E. coli* (no./100 mL)	0	1,000	1,000	4,000									
7.3 BOD (mg/L)		1.5–2	1.5	2			1.0–5.0	5	2	3	6		30
7.4 COD (mg/L)							2.0–6.0						
7.5 Oil/grease (mg/L)					10–20			0.05					10
7.6 Floatables (mg/L)													
7.7 MBAS						0.3			ND				

8. *Heavy metals*									
8.1 Barium (mg/L)	1				2				
8.2 Cadmium (mg/L)		5–50	5–50	5–50	100	10	1–10	5	2
8.3 Chromium (VI) (mg/L)		0.05	0.05	0.05	0.1	0.05	0.01–0.05		0.1
8.4 Chromium (total) (mg/L)	0.05				2			1	
8.5 Copper (mg/L)	1	0.1	0.1	0.1		1	0.005–0.03	0.01	3
8.6 Lead (μg/L)		50	50	50	500	100	10–100	100	100
8.7 Mercury (μg/L)	2	2	2	2	10		0.1	0.5	10
8.8 Nickel (mg/L)		0.1	0.1	0.1	2,000	1		0.1	3
8.9 Selenium (mg/L)	0.01					0.01			0.05
8.10 Silver (μg/L)	50				200				
8.11 Zinc (mg/L)	5	1	1	1				0.1	5
9. *Other toxics*									
9.1 Ammonia (NH_3) (mg/L)		0.5	0.5	0.5					50
9.2 Arsenic (mg/L)	0.05	0.01	0.01	0.01		0.04	0.01–0.08	0.1	0.2
9.3 Chlorine, free (mg/L)						0.05			1
9.4 Cyanide (mg/L)		0.005	0.005	0.005		0.05	0.01–0.1	0.02	0.2
9.5 Fluoride (mg/L)	1.5				10	10		1	2
9.6 Phenol (mg/L)	0.001	0.005	0.005	0.005			0.001–0.01	1	1
9.7 Sulfide (H_2S) (mg/L)						10		0.2	2

[a] Raw water supply suitable for treatment by conventional rapid sand filtration to produce finished CWS
CWS = community water supply; RWS = raw water supply; EF = effluents; AG = agriculture; FS = fishery; MBAS = methylene-blue-active substances

Porteous (1977) proposed the following fiscal/regulatory measures, which may be used either singly or more likely, jointly, to be implemented for recycling programs:

1. resource depletion levy, which is a negative incentive applied by the producers to prevent the rapid run-down of virgin materials;
2. virgin materials tax, e.g. an import levy on groundwood pulp to encourage the use of a higher recycled paper content in newsprint;
3. recycled materials utilization incentives; these can be a straightforward subsidy or tax allowance for the consumers of recycled as opposed to virgin materials;
4. accelerated depreciation allowances; these would be made available to the recycling industries on plant and buildings used wholly or partly for reclamation and recycling;
5. price support, i.e. excess stocks of recycled materials are not allowed to drive down the price, thus allowing recycling efforts to continue at predetermined levels.

Other legislative instruments for regulating and promoting waste recycling programs, as listed in Tables 10.3–10.5, which could be applicable to developing countries are:

1. mandatory source separation
2. mandatory deposit-refund systems
3. payments per unit of waste handled
4. advance disposal fees
5. disposal bans
6. tax credit
7. subsidies
8. market development initiatives.

Each instrument has some substantial advantages and constraints in implementations, depending on areas, economic, people, waste characteristics, public participation, and public awareness of each country. Some developed countries like U.S.A. and Japan, whose gross national products are considerably high, would have capacity of implementing monetary instruments, such as advance disposal fees, payments per unit of waste handled, subsidies. Furthermore, people in these countries are willing to incorporate waste recycling program because the public environmental awareness is somewhat high.

To utilize effectively these legislative instruments in developing countries, the legislators should realize the realistic status of the country as well as crucial roles of the institutes/agencies involved, as previously described in section 10.4.

REFERENCES

The materials presented in this Chapter were obtained partly from the following references.

Congdon, R. J. (ed.) (1977). *Introduction to Appropriate Technology: Towards a Simple Life Style*. Rodale Press, Emmaus.

Environment Agency (1993). *Quality of the Environment in Japan, 1992*. Environmental Agency of Government of Japan, Tokyo.
Environmental Resources Ltd. (1978). *The Economics of Recycling*. Graham & Trotman Ltd., London.
Feachem, R., McGarry, M., and Mara, D. (eds.) (1977). *Water, Waste and Health in Hot Climates*. John Wiley & Sons, Chichester.
Grogan, P. L. (1993). Legislative evaluations. In *The McGraw-Hill Recycling Handbook* (ed. H. F. Lund). McGraw-Hill, New York.
Lohani, B. N. and Evans, J. W. (1993). *Appropriate Environmental Standards for Developing Countries*. Environmental Systems Reviews No. 35. Environmental Sanitation Information Center, Asian Institute of Technology, Bangkok.
Polprasert, C. and Edwards, P. (1981). Low-cost waste recycling in the tropics. *Biocycle, J. Waste Recycling*, **22**, 30–5.
Porteous, A. (1977). *Recycling Resources Refuse*. Longman, London.
Stewart, F. (1978). *Technology and Under-development*, 2nd edition. Macmillan, London.
Thanh, N. C., Lohani, B. N., and Tharun, G. (eds.) (1979). *Waste Disposal and Resources Recovery*. Proceeding of the Seminar on Solid Waste Management, Bangkok, Thailand, 25–30 September, 1978. Asian Institute of Technology, Bangkok.
U.N. (1980). *Guidebook on Biogas Development*. Energy Resource Development Series No. 21. United Nations, New York.
UNEP (1991). *Environmental Data Report*. 3rd edition. United Nations Environment Program. Basil Blackwell, Oxford.
U.S. EPA (1971). *Guidelines for Local Governments on Solid Waste Management*. U.S. Government Printing Office, U.S. Public Health Service Publication No. 2084, Washington, D.C.
White, L. P. and Plaskett, L. G. (1981). *Biomass as Fuel*. Academic Press, London.
Wilson, D. C. (1981). *Waste Management*. Oxford University Press, New York.

EXERCISES

10.1 Find out the existing regulations and measures dealing with waste recycling in your country and compare them with those in Tables 10.3–10.5.

10.2 You are to develop a program of wastewater land treatment in your country. Draw a strategic plan for implementation (e.g. similar to that shown in Figure 10.2) so that this waste recycling program will be sustainable.

10.3 Find out the basis for development of regulatory standards for effluent discharges in your country. How can these standards be applied for waste recycling activities?

10.4 In your opinion, should organic waste recycling programs be carried out in rural or urban areas of your country. Give reasons.

10.5 What are the present institutional set-ups for waste recycling programs in your country? Give recommendations for improvement of these set-ups.

Index